Papers on Topology

Analysis Situs and Its Five Supplements

History of Mathematics • Volume 37
SOURCES

Papers on Topology

Analysis Situs and Its Five Supplements

Henri Poincaré

Translated by

John Stillwell

AMS
American Mathematical Society
Providence, Rhode Island

The London Mathematical Society
London, England

2000 *Mathematics Subject Classification.* Primary 01–XX, 55–XX, 57–XX.

Front cover photograph courtesy of Laboratoire d'Histoire des Sciences et de Philosophie—Archives Henri Poincaré (CNRS—Nancy-Université), http://poincare.univ-nancy2.fr.

For additional information and updates on this book, visit
www.ams.org/bookpages/hmath-37

Library of Congress Cataloging-in-Publication Data

Poincaré, Henri, 1854–1912.
Papers on topology : analysis situs and its five supplements / Henri Poincaré ; translated by John Stillwell.
p. cm. — (History of mathematics ; v. 37)
Includes bibliographical references and index.
ISBN 978-0-8218-5234-7 (alk. paper)
1. Algebraic topology. I. Title.
QA612.P65 2010
514′.2—22

2010014958

10 9 8 7 6 5 4 3 2 1 15, 14, 13, 12, 11, 10

Contents

Translator's Introduction vii
  Topology before Poincaré vii
  Poincaré before topology viii
  The *Analysis situs* paper x
  The five supplements xii
  The Poincaré conjecture xiv
  Comments on terminology and notation xv
  Acknowledgements xvi

Translator's Bibliography xix

On Analysis Situs 1

Analysis Situs 5
  Introduction 5
  §1. First definition of manifold 7
  §2. Homeomorphism 9
  §3. Second definition of manifold 10
  §4. Oppositely oriented manifolds 13
  §5. Homologies 15
  §6. Betti numbers 15
  §7. The use of integrals 16
  §8. Orientable and nonorientable manifolds 19
  §9. Intersection of two manifolds 23
  §10. Geometric representation 31
  §11. Representation by a discontinuous group 36
  §12. The fundamental group 38
  §13. Fundamental equivalences 41
  §14. Conditions for homeomorphism 44
  §15. Other modes of generation 53
  §16. The theorem of Euler 61
  §17. The case where p is odd 68
  §18. Second proof 70

Supplement to Analysis Situs 75
  §I. Introduction 75
  §II. Schema of a polyhedron 78
  §III. Reduced Betti numbers 81
  §IV. Subdivision of polyhedra 84
  §V. Influence of subdivision on reduced Betti numbers 85

§VI. Return to the proofs of paragraph III 89
§VII. Reciprocal polyhedra 92
§VIII. Proof of the fundamental theorem 96
§IX. Various remarks 100
§X. Arithmetic proof of a theorem of paragraph VII 103
§XI. The possibility of subdivision 105

Second Supplement to Analysis Situs 111
Introduction. 111
§1. Review of the principal definitions 111
§2. Reduction of tables 114
§3. Comparison of the tables T_q and T'_q 117
§4. Application to some examples 120
§5. Generalization of a theorem in the first supplement 125
§6. Internal torsion of manifolds 131

On Certain Algebraic Surfaces;
Third Supplement to Analysis Situs 135

Cycles on Algebraic Surfaces;
Fourth Supplement to Analysis Situs 151
§1. Introduction 151
§2. Three-dimensional cycles 155
§3. Two-dimensional cycles 162
§4. One-dimensional cycles 169
§5. Various remarks 172

Fifth Supplement to Analysis Situs 179
§1 179
§2 179
§3 187
§4 200
§5 207
§6 215

Index 225

Translator's Introduction

Topology before Poincaré

Without much exaggeration, it can be said that only *one* important topological concept came to light before Poincaré. This was the Euler characteristic of surfaces, whose name stems from the paper of Euler (1752) on what we now call the *Euler polyhedron formula.* When writing in English, one usually expresses the formula as

$$V - E + F = 2,$$

where

$$\begin{aligned} V &= \text{number of vertices,} \\ E &= \text{number of edges,} \\ F &= \text{number of faces,} \end{aligned}$$

of a convex polyhedron or, more generally, of a subdivided surface homeomorphic to the two-dimensional sphere $\mathbb{S}^2$. In Poincaré (for example, in §16 of his *Analysis situs* paper) one finds the French version

$$S - A + F = 2,$$

where S stands *sommets* and A for *arêtes.*

The formula is usually proved by showing that the quantity $V - E + F$ remains invariant under all possible changes from one subdivision to another. It follows that $V - E + F$ is an invariant of any surface, not necessarily homeomorphic to $\mathbb{S}^2$. This invariant is now called the *Euler characteristic.* Thus $\mathbb{S}^2$ has Euler characteristic 2, whereas the torus has Euler characteristic 0.

Between the 1820s and 1880s, several different lines of research were found to converge to the Euler characteristic.

(1) The classification of polyhedra, following Euler (and even before him, Descartes). Here the terms *edges* and *faces* have their traditional meaning in Euclidean geometry.

(2) The classification of surfaces of constant curvature, where the "edges" are now geodesic segments. Here one finds that surfaces of positive Euler characteristic have positive curvature, those of zero Euler characteristic have zero curvature, and those of negative Euler characteristic have negative curvature.

(3) More generally, the *average* curvature of a smooth surface is positive for positive Euler characteristic, zero for zero Euler characteristic, and negative for negative Euler characteristic. This follows from the *Gauss–Bonnet theorem* of Gauss (1827) and Bonnet (1848).

(4) The study of algebraic curves, revolutionized by Riemann (1851) when he modelled each complex algebraic curve by a surface—its *Riemann surface.*

Under this interpretation, a number that Abel (1841) called the *genus* of an algebraic curve turns out to depend on the Euler characteristic of its Riemann surface. Genus g is related to Euler characteristic χ by

$$\chi = 2 - 2g.$$

Moreover, the genus g has a simple geometric interpretation as the number of *holes* in the surface. Thus $\mathbb{S}^2$ has genus 0 and the torus has genus 1.

(5) The topological classification of surfaces, by Möbius (1863). Möbius studied closed surfaces in $\mathbb{R}^3$ by slicing them into simple pieces by parallel planes. By this method he found that every such surface is homeomorphic to a standard surface with g holes. Thus closed surfaces in $\mathbb{R}^3$—that is, all orientable surfaces—are classified by their genus, and hence by their Euler characteristic. (Despite his discovery of the nonorientable surface that bears his name, Möbius did not classify nonorientable surfaces. This was done by Dyck (1888).)

(6) The study of *pits, peaks, and passes* on surfaces in $\mathbb{R}^3$ by Cayley (1859) and Maxwell (1870). A family of parallel planes in $\mathbb{R}^3$ intersects a surface S in curves we may view as curves of *constant height* (contour lines) on S. If the planes are taken to be in general position, and the surface is smooth, then S has only finitely many pits, peaks, and passes relative to the height function. It turns out that

$$\text{number of peaks} - \text{number of passes} + \text{number of pits}$$

is precisely the Euler characteristic of S.

All of these ideas admit generalizations to higher dimensions, but the only substantial step towards topology in arbitrary dimensions before Poincaré was that of Betti (1871). Betti was inspired by Riemann's concept of *connectivity of surfaces* to define connectivity numbers, now known as *Betti numbers* $P_1, P_2, \ldots$, in all dimensions. The connectivity number of a surface S may be defined as the maximum number of disjoint closed curves that can be drawn on S without separating it. This number P_1 is equal to the genus of S, hence it is just the Euler characteristic in disguise.

For a three-dimensional manifold M one can also consider the maximum number P_2 of disjoint closed surfaces in M that fail to separate M as the *two-dimensional connectivity number* of M. However, the idea of separation fails to explain the *one-dimensional connectivity* of M, since no finite set of curves can separate M. Instead, one takes the maximum number of curves that can lie in M without forming the *boundary of a surface* in M. (For a surface M, this maximum is the same as Riemann's connectivity number.) Betti defined P_m similarly, in a manifold M of arbitrary dimension, as the maximum number of m-dimensional pieces of M that do not form the boundary of a connected $(m + 1)$-dimensional piece of M. Thus Betti brought the concept of *boundary* into topology in order to generalize Riemann's concept of connectivity.

This was Poincaré's starting point, but he went much further, as we will see.

Poincaré before topology

In the introduction to his first major topology paper, the *Analysis situs*, Poincaré (1895) announced his goal of creating of creating an n-dimensional geometry. As he memorably put it:

> ... geometry is the art of reasoning well from badly drawn figures; however, these figures, if they are not to deceive us, must satisfy certain conditions; the proportions may be grossly altered, but the relative positions of the different parts must not be upset.

Because "positions must not be upset," Poincaré sought what Leibniz called *Analysis situs*, a geometry of position, or what we now call topology. He cited as precedents the work of Riemann and Betti, and his own experience with differential equations, celestial mechanics, and discontinuous groups. Of these, I believe the most influential was the last, which stems from his work (and the related work of Klein) on Fuchsian functions in early 1880s.

Poincaré's major papers on Fuchsian functions may be found translated into English in Poincaré (1985). The ideas relevant to topology may be summarized as follows.

One considers a group Γ of translations of the plane which is *fixed-point-free* (that is, nonidentity group elements move every point) and *discontinuous* (that is, there is a nonzero lower bound to the distance that each point is moved by the nonidentity elements). A special case is where the plane is $\mathbb{C}$ and Γ is generated by two Euclidean translations in different directions. Generally the *plane* is the *hyperbolic plane* $\mathbb{H}^2$, which may be modeled by either the upper half plane of $\mathbb{C}$ or the open unit disk $\{z : |z| < 1\}$.

In either case, Γ has a *fundamental domain* $\mathcal{D}$ which is a polygon, and the plane is filled without overlapping by its translates $\gamma\mathcal{D}$ for $\gamma \in \Gamma$. In the special case where the plane is $\mathbb{C}$, the fundamental domain can be taken to be a parallelogram, the translations of which in the two directions fill $\mathbb{C}$. In the hyperbolic case, $\mathcal{D}$ is a polygon with $4g$ sides for some $g \geq 2$, and Γ is generated by $2g$ elements, each of which translates $\mathcal{D}$ to a polygon with just one side in common with $\mathcal{D}$.

It follows that the quotient $\mathbb{C}/\Gamma$ in the special case is a torus (obtained by identifying opposite sides of the fundamental parallelogram), while in the hyperbolic case the quotient $\mathbb{H}^2/\Gamma$ is a surface of genus $g \geq 2$, obtained by identifying sides of the fundamental $4g$-gon in certain pairs.[1]

Each pair of identified sides come together on the quotient surface as a closed curve. For example, the identified sides of a fundamental parallelogram become two closed curves a and b on the torus $\mathbb{C}/\Gamma$, as shown in Figure 1. The curve a

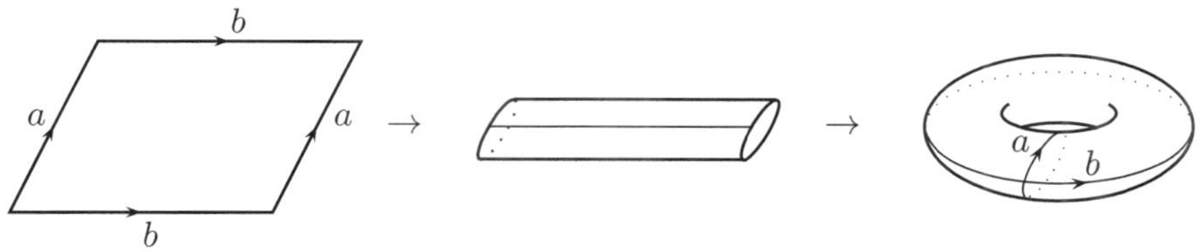

FIGURE 1. Constructing a torus from a fundamental parallelogram.

corresponds to the translation of $\mathbb{C}$ with direction and length of the sides marked a of the parallelogram, and curve b similarly corresponds to the translation with

[1]The connection with Fuchsian functions, often mentioned by Poincaré but not very relevant to topology, is that there are functions f that are *periodic* with respect to substitutions from the group Γ; that is, $f(\gamma(z)) = f(z)$ for all $\gamma \in \Gamma$. In the special case of the torus, these functions are the famous elliptic functions.

direction and length of the sides marked b. Thus one is led to think of a *group of curves* on the torus, isomorphic to the group of translations of $\mathbb{C}$ generated by the translations a and b.

This group is what Poincaré later called the *fundamental group*, and we can see why he viewed it as a group of *substitutions*—in this case, translations of $\mathbb{C}$—rather than as a group of (homotopy classes of) closed curves with fixed origin on the torus, as we now do. Indeed, Poincaré in the 1880s much preferred to work with fundamental polygons in the plane, and it was Klein (1882) who realized that insight could be gained by looking at the quotient surface instead. In particular, Klein used the Möbius classification of surfaces into canonical forms to find canonical *defining relations* for Fuchsian groups.

Here is how we find the defining relation in the case of the torus (in which case one relation suffices). Clearly, if we perform the translations a, b, a^{-1}, b^{-1} of $\mathbb{C}$ in succession, where a^{-1} and b^{-1} denote the inverses of a and b, respectively, the whole plane arrives back at its starting position. We write this relation symbolically as

$$aba^{-1}b^{-1} = 1,$$

where 1 denotes the identity translation. On the torus surface, $aba^{-1}b^{-1}$ denotes a closed curve that bounds a parallelogram (reversing the process shown in Figure 1), and hence this curve is contractible to a point. This is the *topological* interpretation of the relation

$$aba^{-1}b^{-1} = 1.$$

With either interpretation it is quite easy to show that *all* relations between a and b follow from the single relation $aba^{-1}b^{-1} = 1$. This is why we call $aba^{-1}b^{-1} = 1$ the *defining* relation of the torus group.

In a similar way, we find the defining relation of any surface group by cutting the surface along closed curves so as to produce a polygon. Equating the sequence of edges in the boundary of this polygon to 1 then gives a valid relation, and again it is not hard to show that the relation thus obtained is a defining relation. The curves most commonly used for the surface S_g of genus g are called $a_1, b_1, a_2, b_2, \ldots, a_g, b_g$, and cutting S_g along them produces a polygon with boundary

$$a_1b_1a_1^{-1}b_1^{-1}\cdots a_gb_ga_g^{-1}b_g^{-1}.$$

Consequently, the group of S_g may be generated by elements $a_1, b_1, \ldots, a_g, b_g$ and it has defining relation

$$a_1b_1a_1^{-1}b_1^{-1}\cdots a_gb_ga_g^{-1}b_g^{-1} = 1.$$

Figure 2 shows the curves a_1, b_1, a_2, b_2 on the surface S_2, and the resulting polygon.

The *Analysis situs* paper

Poincaré set the agenda for his 1895 *Analysis situs* paper with a short announcement, Poincaré (1892), a translation of which is also included in this volume. In it he raises the question whether the Betti numbers suffice to determine the topological type of a manifold, and introduces the fundamental group to further illuminate this question. He gives a family of three-dimensional manifolds, obtained as quotients of $\mathbb{R}^3$ by certain groups with a cube as fundamental region, and shows that certain of these manifolds have the same Betti numbers but *different* fundamental

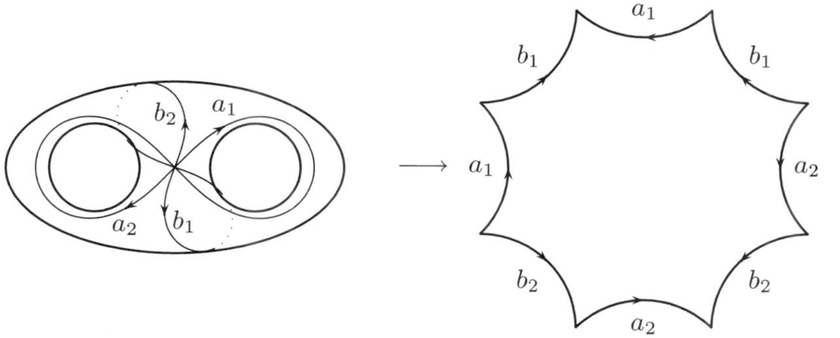

FIGURE 2. Genus 2 surface and its fundamental polygon

groups. It follows, assuming that the fundamental group is a topological invariant, that the Betti numbers do not suffice to distinguish three-dimensional manifolds.

In *Analysis situs*, Poincaré develops these ideas in several directions.

(1) He attempts to provide a new foundation for the Betti numbers in a rudimentary *homology theory*, which introduces the idea of computing with topological objects (in particular, adding, subtracting, testing for linear independence). As Scholz (1980), p. 300, puts it:

> The first phase of algebraic topology, inaugurated by Poincaré, is characterized by the fact that its algebraic relations and operations always deal with *topological objects* (submanifolds).

(2) Using his homology theory, he discovers a *duality theorem* for the Betti numbers of an n-dimensional manifold:

$$P_m = P_{n-m} \quad \text{for} \quad m = 1, 2, \ldots, n-1.$$

In words: "the Betti numbers equidistant from the ends are equal." He later called this the *fundamental theorem* for Betti numbers (p. 96).[2]

(3) He generalizes the Euler polyhedron formula to arbitrary dimensions and situates it in his homology theory.

(4) He constructs several three-dimensional manifolds by identifying faces of polyhedra, observing that this leads to natural presentations of their fundamental groups by generators and relations.

(5) Recognizing that the fundamental group first becomes important for three-dimensional manifolds, Poincaré asks whether it suffices to distinguish between them. He is not able to answer this question.

Analysis situs is rightly regarded as the origin of algebraic topology, because of Poincaré's construction of homology theory and the fundamental group. The fundamental group is the more striking of the two, because it is a blatantly abstract structure and generally noncommutative, yet surprisingly easy to grasp via generators and relations. Homology theory reveals an algebraic structure behind the bare Betti numbers and Euler characteristics of Poincaré's predecessors, but it is not easy to say what this structure really is. Indeed, Poincaré did not realize that the Betti numbers are only part of the story, and he had to write Supplements 1

[2] All page numbers referenced in Poincaré's topology papers are the page numbers in *this* volume, not the page numbers in the original papers.

and 2 to *Analysis situs* before the so-called torsion coefficients came to light. And it was only in 1925 that Emmy Noether discovered the homology *groups*, which we now view as the proper home of the Betti and torsion numbers. She announced this discovery in Noether (1926).[3]

Thus, along with great breakthroughs, there is also confusion in *Analysis situs.* The confusion extends to the very subject matter of algebraic topology, the manifolds (or "varieties" as Poincaré calls them). His definitions suggest that he is generally thinking of differentiable manifolds, but most of his three-dimensional examples are defined combinatorially, by identifying faces of polyhedra, without checking their differentiability. His definition of Betti numbers needs revision, as he discovers in Supplements 1 and 2, and imprecise arguments are frequently used.

Another source of confusion concerns *simply connected manifolds*, and it ultimately led to the famous Poincaré conjecture in Supplement 5. In *Analysis situs*, §14, he defines a manifold to be *simply connected* if its fundamental group is trivial. It follows easily that a sphere of any dimension is simply connected, but Poincaré sometimes forgets that the converse is not obvious. On occasion, he assumes that any simply connected manifold is homeomorphic to a sphere (for example, on p.111), and on other occasions he even assumes (wrongly) that a region with trivial *homology* is simply connected (for example, on p. 39). These errors have been flagged by footnotes, some by the original editors of Volume VI of Poincaré's *Œuvres*, René Garnier and Jean Leray (the actual author of these footnotes is not identified), and some by myself.

Actually, it is not surprising that Poincaré made mistakes, given the novelty and subtle nature of the subject, and his style of work. Darboux (1952), p. lvi, describes Poincaré's working method as follows:

> Whenever asked to resolve a difficulty, his response came with the speed of an arrow. When he wrote a memoir, he drafted it all in one go, with only a few erasures, and did not return to what he had written.

It is perhaps wise to read Poincaré's memoirs in the same style: try to take in their general sweep without lingering too long over gaps and errors.

The five supplements

Poincaré (1899) wrote a *Complément a l'analysis situs* in response to the criticism of Heegaard (1898). Heegaard had become interested in three-dimensional manifolds, and he found an example where Poincaré's definition of Betti numbers comes into conflict with his duality theorem. To save the theorem, Poincaré revised his homology theory in the *Complément*, moving towards a more combinatorial theory in which manifolds are assumed to have a polyhedral structure and computing Betti numbers from the incidence matrices of this structure. He also arrived at a clearer explanation of the duality theorem in terms of the dual ("reciprocal") subdivision of a polyhedron, in which cells of dimension m in the original polyhedron correspond to cells of dimension $n-m$ in its dual. He concluded the *Complement* with a (rather unconvincing) attempt to prove that every differentiable manifold

[3]To be fair to Poincaré, he came close to discovering the first homology group H_1 as the abelianization of the fundamental group π_1 in *Analysis situs*, §13. There he considered the homologies obtained by allowing the generators of π_1 to commute and observed that "knowledge of these homologies immediately yields the Betti number P_1."

has a polyhedral subdivision. This theorem was first proved rigorously by Cairns (1934).

Poincaré may have thought that the *Complément* would complete his *Analysis situs* paper, but four more "complements" were to follow, as further gaps and loose ends came to light. For this reason, I have chosen to use the word "supplement" rather than "complement" (as Poincaré himself did on occasion).

In the second supplement, Poincaré (1900) dug more deeply into the problems of his original homology theory, uncovering the existence of *torsion*, and expanding his technique for computing Betti numbers to one that also computes torsion coefficients. He motivated his choice of the word "torsion" by showing that torsion occurs only in manifolds, such as the Möbius band, that are nonorientable and hence "twisted onto themselves" in some way (p. 134). When Emmy Noether built the Betti numbers and torsion numbers into the homology groups in 1926, the word "torsion" took up residence in algebra, much to the mystification of group theory students who were not informed of its origin in topology.

Having now attained some mastery of homology theory, Poincaré was emboldened to conjecture (p. 134) that *the three-dimensional sphere is the only closed three-dimensional manifold with trivial Betti and torsion numbers.* This was his first (and incorrect) version of the Poincaré conjecture.

The third and fourth supplements (Poincaré (1902a), Poincaré (1902b)) hark back to the first major application of Betti numbers to classical mathematics, the work of Picard (1889) on the connectivity of algebraic surfaces. An algebraic surface (or "algebraic function of two variables" as Picard called it) is taken to have complex values of the variables, hence it has four real dimensions. By a mixture of analytic and topological arguments, Picard succeeded in finding the first Betti number P_1 of algebraic surfaces, but he had less success in finding P_2. Invoking his new homology theory, Poincaré pushed on to P_2 in his fourth supplement (as far he needed to go, since $P_3 = P_1$, by Poincaré duality). Like Picard, Poincaré also appealed to results from analysis, in his case referring to his work on Fuchsian functions and non-Euclidean geometry from the early 1880s. An exposition of Poincaré's argument (in German) may be found in Scholz (1980), pp. 365–371.

The return to non-Euclidean geometry paid off unexpectedly in the fifth supplement, Poincaré (1904), with an interesting geometric algorithm (p. 200) to decide whether a curve on a surface is homotopic to a simple curve. Poincaré's result is that, in the case of genus greater than 1 where the surface can be given a non-Euclidean metric, a homotopy class contains a simple curve if and only if the *geodesic representative* is simple. Informally speaking, one can decide whether a curve κ is homotopic to simple curve by "stretching κ tight" on the surface and observing whether the stretched form of κ is simple. It seems likely that Poincaré's application of non-Euclidean geometry to surface topology inspired the later work of Dehn and Nielsen between 1910 and the 1940s, and the work of Thurston in the 1970s.

In the fifth supplement, the result on simple curves is just part of a rather meandering investigation of curves on surfaces, and their role in the construction of three-dimensional manifolds (*Heegaard diagrams*). In the final pages of the paper this investigation leads to a spectacular discovery: the *Poincaré homology sphere.* By pasting together two handlebodies of genus 2, H_1 and H_2 say, so that certain carefully chosen curves on H_1 become identified with canonical disk-spanning curves on H_2, Poincaré obtains a three-dimensional manifold V whose fundamental group

$\pi_1(V)$ he can write down in terms of generators and relations. To the reader's astonishment, the presentation of $\pi_1(V)$ implies relations that hold in the icosahedral group, so $\pi_1(V)$ *is nontrivial.* On the other hand, by allowing the generators of $\pi_1(V)$ to commute one finds (in our language) that $H_1(V) = 0$, so that V *is a closed three-dimensional manifold with trivial homology but nontrivial fundamental group* (hence V is *not* simply connected.)

The Poincaré homology sphere therefore refutes the conjecture made at the end of the second supplement, and it prompts the revised Poincaré conjecture (now known to be correct): *the three-sphere is the only closed three-dimensional manifold with trivial fundamental group.*

Poincaré prudently concludes the fifth supplement by remarking that investigation of the revised conjecture "would carry us too far away."

The Poincaré conjecture

In the *Analysis situs* and its five supplements, Poincaré opened up a vast new area of mathematics. It is not surprising that he left it incompletely explored. Among the most important gaps in his coverage were:

(1) The topological invariance of dimension, first proved by Brouwer (1911).
(2) The topological invariance of the Betti and torsion numbers, first proved by Alexander (1915).
(3) The existence of nonhomeomorphic three-dimensional manifolds with the same fundamental group, first proved by Alexander (1919).
(4) The existence of a polyhedral structure on every differentiable manifold, first proved by Cairns (1934).
(5) The existence of topological manifolds *without* a polyhedral structure, first proved by Kirby and Siebenmann around 1970, and published in Kirby and Siebenmann (1977).

But the deepest of the unsolved problems left by Poincaré was one he first thought was trivial—the Poincaré conjecture.

In the beginning, there was no conjecture, because Poincaré thought it obvious that a simply connected closed manifold was homeomorphic to a sphere. In the second supplement he came up with a sharper claim that was less obvious, hence in his view worth conjecturing: a closed manifold with trivial *homology* is homeomorphic to a sphere. But on occasions thereafter he forgot that there is a difference between trivial homology and trivial fundamental group. Finally, the discovery of Poincaré homology sphere in Supplement 5 opened his eyes to the real problem, and the Poincaré conjecture as we know it today was born: *a closed* 3*-manifold with trivial fundamental group is homeomorphic to the* 3*-sphere.*

The existence of homology spheres shows that three dimensions are more complicated than two, but just *how much* more complicated they are was not immediately clear. Further results on three-dimensional manifolds came with glacial slowness, and they often revealed new complications. Dehn (1910) found infinitely many homology spheres, and Whitehead (1935) found an *open* three-dimensional manifold that is simply connected but not homeomorphic to $\mathbb{R}^3$.

In the 1950s and 1960s there was at last some good news about three-dimensional manifolds: for example, they all have a polyhedral structure (Moise (1952)). The news did not include a proof of the Poincaré conjecture, however. Instead, progress on the conjecture came in higher dimensions, with a proof by Smale (1961) of the analogous conjecture for the n-dimensional sphere $\mathbb{S}^n$ for $n \geq 5$. Unfortunately, while three dimensions are harder than two, five are *easier* than three in some respects. So Smale's proof did not throw much light on the classical Poincaré conjecture, or on the analogous conjecture for $\mathbb{S}^4$ either.

The analogue of the Poincaré conjecture for four-dimensional manifolds was finally proved by Freedman (1982). Freedman's proof was a *tour de force* that simultaneously solved several longstanding problems about four-dimensional manifolds. That his approach worked at all was a surprise to many of his colleagues, and finding a similar approach to the classical Poincaré conjecture seemed out of the question.

Indeed, an entirely new approach to the Poincaré conjecture had already been taking shape in the hands of William Thurston in the late 1970s. Thurston, like Poincaré and Dehn, was interested in *geometric* realizations of manifolds, exemplified by the surfaces of constant curvature that realize all the topological forms of closed surfaces. He conjectured that all 3-manifolds may be realized in a similar, though more complicated, way. Instead of the three 2-dimensional geometries of constant curvature, one has eight "homogeneous" 3-dimensional geometries. (The eight geometries were discovered by Bianchi (1898), and rediscovered by Thurston.) And instead of a single geometry for each 3-manifold M one has a *decomposition* of M into finitely many pieces, each carrying one of the eight geometries.

Thurston's *geometrization conjecture* states that each closed and connected 3-manifold is homeomorphic to one with such a decomposition. The Poincaré conjecture follows from a special case of the geometrization conjecture for manifolds of positive curvature. For more details on the evolution of the Poincaré conjecture up to this point, see Milnor (2003).

Thurston was able to prove many cases of his geometrization conjecture, but geometrization seemed to run out of steam in the early 1980s. This was not entirely disappointing to some topologists, who still hoped for a proof of the Poincaré conjecture by purely topological methods. However, *more* geometry was to come, not less, and *differential* geometry at that. It was not enough to consider manifolds with "homogeneous" geometry; one had to consider manifolds with *arbitrary* smooth geometry, and to let the geometry "flow" towards homogeneity.

The idea of "flowing towards homogeneity" was initiated by Hamilton (1982), using what is called the *Ricci curvature flow.* Hamilton was able to show that the Ricci curvature flow works in many cases, but he was stymied by the formation of singularities in the general case. The difficulties were brilliantly overcome by Grigory Perelman in 2003. Perelman published his proof only in outline, in three papers posted on the Internet in 2002 and 2003, but experts later found that these papers contained all the ideas necessary to construct a complete proof of the geometrisation conjecture. Perelman himself, apparently sure that he would be vindicated, published nothing further and seems to have gone into seclusion.

For a very thorough and detailed account of Perelman's proof of the Poincaré conjecture, see Morgan and Tian (2007).

Comments on terminology and notation

Poincaré's topology papers pose an unusual problem for the translator, inasmuch as they contain numerous errors, both large and small, and misleading notation. My policy (which is probably not entirely consistent) has been to make only small changes where they help the modern reader—such as correcting obvious typographical errors—but to leave serious errors untouched except for footnotes pointing them out.

The most serious errors must be retained because they were a key stimulus to the development of Poincaré's thought in topology. As mentioned above, some of the five supplements exist only because of mistakes in the *Analysis situs* paper. It is more debatable whether one should retain annoying notation, such as the $+$ sign Poincaré uses to denote the (generally noncommutative) group operation in the fundamental group, or the $\equiv$ sign and the word "congruence" he uses for the (asymmetric) boundary relation. I have opted to retain these, partly to assist readers who wish to compare the translation with the original papers, and also because they may be a clue to what Poincaré was thinking when he first applied algebra to topology.

I have also retained the word "conjugate" that Poincaré uses for the paired sides of a polyhedron, or the paired sides of a polygon, that are to be identified to form a manifold. One can replace "conjugate" by "identified" in many cases, but sometimes "paired" is better, so I thought it safest not to meddle. Fortunately, Poincaré does not use the word "conjugate" in the group-theoretic sense, even though the *concept* of conjugacy in group theory briefly arises.

On the other hand, I consistently use the word "manifold" where Poincaré uses "variety," and I call manifolds "orientable" where he calls them "two-sided" and "nonorientable" where he calls them "one-sided." The word "variety" always suggests algebraic geometry today, whereas Poincaré is really thinking about the topology of manifolds (even though many of them are in fact algebraic varieties), so "manifold" is the right word for the modern reader. The words "two-sided" and "one-sided" are less misleading than "variety," but Poincaré also uses them in a second sense, to describe separating and nonseparating curves on a surface, which have "two sides" and "one side," respectively. Calling manifolds "orientable" and "nonorientable" therefore removes a possible source of confusion.

Acknowledgements

I have drawn on the work of Sarkaria (1999), which gives a detailed summary, with some comments and corrections, of Poincaré's papers in topology. The commentary of Dieudonné (1989) on Poincaré's *Analysis situs* and its first two supplements is also useful, as is Chapter VII of Scholz (1980).

I translated *Analysis situs* and the the first, second, and fifth supplements in the 1970s, when I was first learning topology. At the time, I did not think there would be an opportunity to publish these papers, so I did not bother to translate the remaining two supplements, which were further from my interests at the time. Thirty years later, I was pleasantly surprised to be contacted by Andrew Ranicki and Cameron Gordon about their classic papers project. With their encouragement, I translated the two missing supplements and edited the whole sequence into roughly the form you see today. I am delighted that Poincaré's topological work is finally appearing in English, and I thank Andrew and Cameron for making this possible.

The publication of this book in the AMS/LMS History of Mathematics series is a great bonus, given that I did not originally expect to publish these translations at all. I am grateful to Rob Kirby for suggesting to me that the AMS might be interested, and to Ina Mette for enthusiastically taking up the suggestion. Finally, I thank Jennifer Wright Sharp for her extraordinary editing, which made an enormous number of improvements to the text.

John Stillwell
Monash University and University of San Francisco
July 2010

Translator's Bibliography

Abel, N. H. (1826/1841). Mémoire sur une propriété générale d'une classe trés-étendue de fonctions transcendantes. *Mém. Acad. des Sciences de Paris*, 176–264. Received in 1826, published in 1841.

Alexander, J. W. (1915). A proof of the invariance of certain constants of analysis situs. *Trans. Amer. Math. Soc. 16*, 148–154.

Alexander, J. W. (1919). Note on two three-dimensional manifolds with the same group. *Trans. Amer. Math. Soc. 20*, 339–342.

Betti, E. (1871). Sopra gli spazi di un numero qualunque di dimensioni. *Annali di Matematica pura ed applicata 4*, 140–158.

Bianchi, L. (1898). Sugli spazi a tre dimensioni che ammettono un gruppo continuo di movimenti. *Memorie di Matematica e di Fisica della Societa Italiana delle Scienze 11*, 267–352. English translation by Robert Jantzen in *General Relativity and Gravitation 33*, (2001), pp. 2171–2253.

Biggs, N. L., E. K. Lloyd, and R. J. Wilson (1986). *Graph Theory. 1736–1936* (Second ed.). New York: The Clarendon Press Oxford University Press.

Bonnet, O. (1848). Mémoire sur la théorie générale des surfaces. *J. Éc. Polytech. 19*, 1–146.

Brouwer, L. E. J. (1911). Beweis der Invarianz der Dimensionzahl. *Math. Ann. 70*, 161–165.

Cairns, S. S. (1934). On the triangulation of regular loci. *Ann. of Math. 35*, 579–587.

Cayley, A. (1859). On contour and slope lines. *Phil. Mag. 18*, 264–268.

Darboux, G. (1913/1952). Éloge Historique d'Henri Poincaré. In *Œuvres de Henri Poincaré, Tome II*, pp. vii–lxxi. Paris: Gauthier-Villars.

Dehn, M. (1910). Über die Topologie des dreidimensional Raumes. *Math. Ann. 69*, 137–168. English translation in Dehn (1987).

Dehn, M. (1987). *Papers on Group Theory and Topology.* New York: Springer-Verlag. Translated from the German and with introductions and an appendix by John Stillwell, and with an appendix by Otto Schreier.

Dieudonné, J. (1989). *A History of Algebraic and Differential Topology. 1900–1960.* Boston, MA: Birkhäuser Boston Inc.

Dombrowski, P. (1979). *150 Years after Gauss' "Disquisitiones generales circa superficies curvas".* Paris: Société Mathématique de France. With the original text of Gauss.

Dyck, W. (1888). Beiträge zur Analysis situs I. *Math. Ann. 32*, 457–512.

Euler, L. (1752). Elementa doctrinae solidorum. *Novi Comm. Acad. Sci. Petrop. 4*, 109–140. In his *Opera Omnia*, ser. 1, 26: 71–93. English translation in Biggs et al. (1986).

Freedman, M. H. (1982). The topology of four-dimensional manifolds. *J. Differential Geom. 17*, 357–453.

Gauss, C. F. (1827). *Disquisitiones generales circa superficies curvas.* Göttingen: König. Ges. Wiss. Göttingen. English translation in Dombrowski (1979).

Hamilton, R. S. (1982). Three-manifolds with positive Ricci curvature. *J. Differential Geom. 17*, 255–306.

Heegaard, P. (1898). *Forstudier til en topologisk Teorie for de algebraiske Fladers Sammenhæng.* Copenhagen: University of Copenhagen Dissertation. French translation: *Sur l'Analysis situs* in *Bull. Math. Soc. France 44* (1916), pp. 161–242.

Kirby, R. C. and L. C. Siebenmann (1977). *Foundational Essays on Topological Manifolds, Smoothings, and Triangulations.* Princeton, NJ: Princeton University Press.

Klein, F. (1882). Neue Beiträge zur Riemanschen Funktionentheorie. *Math. Ann. 21*, 141–218. In his *Gesammellte Mathematische Abhandlungen* 3: 630–710.

Maxwell, J. C. (1870). On hills and dales. *Phil. Mag. 40*, 421–427.

Milnor, J. (2003). Towards the Poincaré conjecture and the classification of 3-manifolds. *Notices Amer. Math. Soc. 50*, 1226–1233.

Möbius, A. F. (1863). Theorie der Elementaren Verwandtschaft. *Werke* 2: 433-471.

Moise, E. E. (1952). Affine structures in 3-manifolds. V. The triangulation theorem and Hauptvermutung. *Ann. of Math. (2) 56*, 96–114.

Morgan, J. and G. Tian (2007). *Ricci Flow and the Poincaré Conjecture*, Volume 3 of *Clay Mathematics Monographs.* Providence, RI: American Mathematical Society.

Noether, E. (1926). Ableitung der Elementarteilertheorie aus der Gruppentheorie. *Jber. Deutsch. Math. Verein. 36*, 104.

Picard, E. (1889). Mémoire sur la théorie des fonctions algébriques de deux variables. *J. de Math. 5*, 135–319.

Poincaré, H. (1892). Sur l'analysis situs. *Comptes rendus de l'Academie des Sciences 115*, 633–636.

Poincaré, H. (1895). Analysis situs. *J. Éc. Polytech., ser. 2 1*, 1–123.

Poincaré, H. (1899). Complément à l'analysis situs. *Rendiconti del Circulo Matematico di Palermo 13*, 285–343.

Poincaré, H. (1900). Second complément à l'analysis situs. *Proc. London Math. Soc. 32*, 277–308.

Poincaré, H. (1902a). Sur certaines surfaces algébriques; troisième complément à l'analysis situs. *Bull. de la Soc. math. de France 30*, 49–70.

Poincaré, H. (1902b). Sur les cycles des surfaces algébriques; quatrième complément à l'analysis situs. *J. de math. pures et appliquées 8*, 169–214.

Poincaré, H. (1904). Cinquième complément à l'analysis situs. *Rendiconti del Circolo matematico di Palermo 18*, 45–110.

Poincaré, H. (1985). *Papers on Fuchsian Functions.* New York: Springer-Verlag. Translated from the French and with an introduction by John Stillwell.

Riemann, G. F. B. (1851). Grundlagen für eine allgemeine Theorie der Functionen einer veränderlichen complexen Grösse. *Werke*, 2nd ed., 3–48.

Sarkaria, K. S. (1999). The topological work of Henri Poincaré. In *History of Topology*, pp. 123–167. Amsterdam: North-Holland.

Scholz, E. (1980). *Geschichte des Mannigfaltigkeitsbegriffs von Riemann bis Poincaré.* Boston, MA: Birkhäuser Boston.

Smale, S. (1961). Generalized Poincaré conjecture in dimensions greater than four. *Ann. of Math. 74*, 391–406.

Whitehead, J. H. C. (1935). A certain open manifold whose group is unity. *Quart. J. Math. 6*, 268–279.

On Analysis Situs

Comptes rendus de l'Académie des Sciences 115 (1892), pp. 633–636.

One knows what is meant by the connectivity of a surface, and the important role this notion plays in the general theory of functions, despite being borrowed from an entirely different branch of mathematics, namely the geometry of situation or *Analysis situs.*

It is because researches of this kind can have applications outside geometry that it is of interest to pursue them to spaces of more than three dimensions. Riemann understood this well; and he wished to extend his beautiful discovery and apply it to the *analysis situs* of general spaces, but unfortunately he left the subject in a very incomplete state. Betti, in volume IV, series 2 of *Annali di Matematica*, recovered and completed Riemann's results. He considered a surface (manifold of dimension n) in the space of $n+1$ dimensions, and defined $n-1$ numbers

$$p_1, \quad p_2, \quad \ldots, \quad p_{n-1}$$

that he called the orders of connection of the surface.

Persons who recoil from geometry of more than three dimensions may believe this result to be useless and view it as a futile game, if they have not been informed of their error by the use made of Betti numbers by our colleague M. Picard in pure analysis and ordinary geometry.

Meanwhile, the field is by no means exhausted. One may ask whether the Betti numbers suffice to determine a closed surface from the viewpoint of *analysis situs*. That is, given two surfaces with the same Betti numbers, we ask whether it is possible to pass from one to the other by a continuous deformation. This is true in the space of three dimensions, and we may be inclined to believe that it is again true in any space. The contrary is true.

In order to explain, I want to approach the question from a new viewpoint. Let $x_1, x_2, \ldots, x_{n+1}$ be the coordinates of a point on the surface. These $n+1$ quantities are connected by the equation of the surface. Now let

$$F_1, \quad F_2, \quad \ldots, \quad F_p$$

be any p functions of the $n+1$ coordinates x (which I always suppose to be connected by the equation of the surface, and which I suppose to take only real values).

I do not assume that the functions F are uniform, but I suppose that if the point $(x_1, x_2, \ldots, x_{n+1})$ describes an *infinitely small* contour on the surface then each of the functions F returns to its initial value. This being so, we suppose that our point now describes a *finite* closed contour on the surface. It may then happen that the p functions do not return to their initial values, but instead become

$$F_1', \quad F_2', \quad \ldots, \quad F_p'.$$

In other words, they undergo the substitution

$$(F_1, F_2, \ldots, F_p; F'_1, F'_2, \ldots, F'_P).$$

All the substitutions corresponding to the different closed contours that we can trace on the surface form a group which is discontinuous (at least as far as its form is concerned).

This group evidently depends on the choice of functions F. We suppose first that these functions are the most one can imagine, other than being subject to the condition imposed above, and let G be the corresponding group. If G' is the group corresponding to another choice of functions, then G' will be isomorphic to G—holoedrically in general but meriedrically in special cases.[1]

The group G can then serve to define the form of the surface, and it is called the group of the surface.[2] It is clear that if two surfaces can each be transformed to the other by a continuous transformation, then their groups are isomorphic. The converse, though less evident, is again true for closed surfaces, so that *what defines a closed surface, from the viewpoint of* analysis situs, *is its group.*[3]

This leads us to pose the following question: *Do two surfaces with the same Betti numbers always have the same group?*

To resolve this question, we make use of a simple mode of representation in ordinary space when we want to define a surface in four-dimensional space. We consider a properly discontinuous group G in ordinary space. The space is thereby decomposed into infinitely many fundamental domains, each the transform of some other by one of the transformations in the group. I suppose that the fundamental domain does not extend to infinity and that each [nonidentity] substitution in the group has no fixed point.

Let

$$X_1, \quad X_2, \quad X_3, \quad X_4$$

be four functions of the coordinates x, y, z of ordinary space that are invariant under substitutions in the group G. If we consider X_1, X_2, X_3, X_4 as the coordinates of a point in four-dimensional space, this point describes a closed surface whose group is isomorphic to G, and holoedrically so if the functions X are the most general possible among those that are invariant under G.

We consider, in particular, the group generated by the three substitutions

$$\begin{gathered}(x, y, z; x+1, y, z),\\ (x, y, z; x, y+1, z),\\ (x, y, z; \alpha x+\beta y, \gamma x+\delta y, z+1),\end{gathered}$$

where $\alpha, \beta, \gamma, \delta$ are integers such that $\alpha\delta - \beta\gamma = 1$. I call this the group $(\alpha, \beta, \gamma, \delta)$ for short.

It has a cube as fundamental domain.[4]

---

[1]Here Poincaré is using the 19th-century terminology, where a "holoedric isomorphism" is what we call an isomorphism, and a "meriedric isomorphism" is a homomorphism. (Translator's note.)

[2]Later called the *fundamental* group by Poincaré in §12 of his *Analysis situs* paper. (Translator's note.)

[3]This is true for surfaces in the traditional, two-dimensional, sense, but not for manifolds of three dimensions. Near the end of §14 of *Analysis situs*, Poincaré raised the question whether two manifolds with the same group are necessarily homeomorphic. (Translator's note.)

[4]Moreover, we see that the vertical sides of the cube are mapped onto each other by unit translations in the x- and y-directions, so that each horizontal cross-section of the cube becomes

We first observe that the two groups $(\alpha, \beta, \gamma, \delta)$ and $(\alpha', \beta', \gamma', \delta')$ cannot be isomorphic unless the two transformations

$$(x, y; \alpha x + \beta y, \gamma x + \delta y), \quad (x, y; \alpha' x + \beta' y, \gamma' x + \delta' y)$$

are transforms[5] of each other by a linear transformation with integer coefficients.

This does not happen in general.

We now seek to determine the Betti numbers for the surface with group $(\alpha, \beta, \gamma, \delta)$. We see that one of the orders of connectivity is always quadruple and the other is

double in the general case;
triple if $\alpha + \delta = 2$;
quadruple if $\alpha = \delta = 1$, $\beta = \gamma = 0$.

It follows that the Betti numbers can be the same for two surfaces without their groups being isomorphic and, consequently, without it being possible to pass from one surface to the other by a continuous deformation.

This remark throws some light on the theory of ordinary algebraic surfaces and makes less strange the discovery of M. Picard, according to which the surfaces have no one-dimensional cycle if they are the most general of their degree.

---

a torus. The top and bottom faces of the cube are also mapped onto each other (by the more complicated third generator of G), so that the manifold obtained is what we would now call a torus bundle over the circle. (Translator's note.)

[5]What we would now call conjugates. (Translator's note.)

Analysis Situs

Journal de l'École Polytechnique 1 (1895), pp. 1–121.

Introduction

Nobody doubts nowadays that the geometry of n dimensions is a real object. Figures in hyperspace are as susceptible to precise definition as those in ordinary space, and even if we cannot represent them, we can still conceive of them and study them. So if the mechanics of more than three dimensions is to be condemned as lacking in object, the same cannot be said of hypergeometry.

Geometry, in fact, has a unique *raison d'être* as the immediate description of the structures which underlie our senses; it is above all the analytic study of a group; consequently, there is nothing to prevent us from proceeding to study other groups which are analogous but more general.

But why, it may be said, not preserve the analytic language and replace the language of geometry, as this will have the advantage that the senses can no longer intervene. It is that the new language is more concise; it is the analogy with ordinary geometry which can create fruitful associations of ideas and suggest useful generalizations.

Perhaps these reasons are not sufficient in themselves? It is not enough, in fact, for a science to be legitimate; its utility must be incontestable. So many objects demand our attention that only the most important have the right to be considered.

Also, there are parts of hypergeometry which do not have a place of great interest; for example, researches on the curvature of hypersurfaces in the space of n dimensions. We are certain in advance of obtaining the same results as in ordinary geometry, and we need not undertake a long voyage to view a spectacle like the one we encounter at home.

But there are problems where the analytic language is entirely unsuitable.

We know how useful geometric figures are in the theory of imaginary functions and integrals evaluated between imaginary limits, and how much we desire their assistance when we want to study, for example, functions of two complex variables.

If we try to account for the nature of this assistance, figures first of all make up for the infirmity of our intellect by calling on the aid of our senses; but not only this. It is worthy repeating that geometry is the art of reasoning well from badly drawn figures; however, these figures, if they are not to deceive us, must satisfy certain conditions; the proportions may be grossly altered, but the relative positions of the different parts must not be upset.

The use of figures is, above all, then, for the purpose of making known certain relations between the objects that we study, and these relations are those which occupy the branch of geometry that we have called *analysis situs*, and which describes the relative situation of points and lines on surfaces, without consideration of their magnitude.

The fact that relations of the same nature hold between the objects of hypersurface, so that there is then an *analysis situs* of more than three dimensions, is due, as we have shown, to Riemann and Betti.

This science enables us to know the nature of these relations, although this knowledge is less intuitive, since it lacks a counterpart in our senses. Indeed, in certain cases it renders us the service that we ordinarily demand of geometrical figures.

I shall confine myself to three examples.

The classification of algebraic curves into types rests, after Riemann, on the classification of real closed surfaces from the point of view of *analysis situs*. An immediate induction shows us that the classification of algebraic surfaces and the theory of their birational transformations are intimately connected with the classification of real closed hypersurfaces in the space of five dimensions from the point of view of *analysis situs*. M. Picard, in a memoir honored by the Academy of Sciences, has already insisted on this point.

Then again, in a series of memoirs in Liouville's journal, entitled *Sur les courbes définies par les équations différentielles*, I have employed the ordinary *analysis situs* of three dimensions in the study of differential equations. The same researches have been pursued by M. Walther Dyck. We can easily see that generalized *analysis situs* will permit us to treat higher order equations in the same way, in particular, the equations of celestial mechanics.

M. Jordan has determined analytically the groups of finite order contained in the linear group of n variables. Before that, M. Klein had resolved the same problem for the linear group of two variables by a geometric method of rare elegance. Could not the method of M. Klein be extended to the group of n variables, or *any continuous group*? I have not been able to succeed so far, but I have thought a great deal about the question, and it seems to me that the solution must depend on a problem of *analysis situs* and that the generalization of the celebrated Euler polyhedron theorem must play a role.

I do not think then that I have engaged in useless work in writing the present memoir; I regret only that it is too long, but when I try to restrict myself I fall into obscurity; I prefer to be considered a little loquacious.

Table of Contents

§1.	First definition of manifold	7
§2.	Homeomorphism	9
§3.	Second definition of manifold	10
§4.	Oppositely oriented manifolds	13
§5.	Homologies	15
§6.	Betti numbers	15
§7.	Use of integrals	16
§8.	Orientable and nonorientable manifolds	19
§9.	Intersection of two manifolds	23
§10.	Geometric representation	31
§11.	Representation by a discontinuous group	36
§12.	The fundamental group	38
§13.	Fundamental equivalences	41
§14.	Conditions for homeomorphism	44
§15.	Other modes of generation	53
§16.	The theorem of Euler	61
§17.	The case where p is odd	68
§18.	Second proof	70

§1. First definition of manifold

Let $x_1, x_2, \ldots, x_n$ be n variables, which can be regarded as the coordinates of a point in n-dimensional space.

For the time being I assume that these n variables are always real.

Any sequence of n variables will be called a *point*. We consider the following system consisting of p equations and q inequalities.

$$(1)\qquad \begin{cases} F_1(x_1, x_2, \ldots, x_n) & = & 0 \\ F_2(x_1, x_2, \ldots, x_n) & = & 0 \\ \ldots \\ F_p(x_1, x_2, \ldots, x_n) & = & 0 \\ \varphi_1(x_1, x_2, \ldots, x_n) & > & 0 \\ \varphi_2(x_1, x_2, \ldots, x_n) & > & 0 \\ \ldots \\ \varphi_q(x_1, x_2, \ldots, x_n) & > & 0 \end{cases}$$

I assume that the functions F and φ are uniform and continuous, and that they have continuous derivatives; in addition I assume that if we form the matrix

$$\begin{matrix} \frac{\partial F_1}{\partial x_1} & \frac{\partial F_1}{\partial x_2} & \cdots & \frac{\partial F_1}{\partial x_n} \\ \frac{\partial F_2}{\partial x_1} & \frac{\partial F_2}{\partial x_2} & \cdots & \frac{\partial F_n}{\partial x_n} \\ \cdots & \cdots & \cdots & \cdots \\ \frac{\partial F_p}{\partial x_1} & \frac{\partial F_p}{\partial x_2} & \cdots & \frac{\partial F_p}{\partial x_n} \end{matrix}$$

and form the determinants obtained by taking any p columns, then these determinants are never all zero simultaneously.

I shall say that the set of points which satisfies the conditions (1) constitutes a *manifold of* $n-p$ *dimensions.* If in particular $p = 0$, so that there are no equations, I have an n-dimensional manifold which is nothing but a portion of the n-dimensional space; ordinarily I shall describe this manifold as a *domain.*

Two cases can occur. Let $\alpha_1, \alpha_2, \ldots, \alpha_n; \beta_1, \beta_2, \ldots, \beta_n$ be two points satisfying the conditions (1). Then either it is possible to vary $x_1, x_2, \ldots, x_n$ in a continuous manner from $\alpha_1, \alpha_2, \ldots, \alpha_n$ to $\beta_1, \beta_2, \ldots, \beta_n$ without violating the conditions (1) for any values $\alpha_1, \alpha_2, \ldots, \alpha_n$ and $\beta_1, \beta_2, \ldots, \beta_n$ which satisfy these conditions, or else this is not always possible.

In the first case we say that the manifold defined by the conditions (1) is *connected.*

In what follows I shall generally consider connected manifolds, and, to the extent that nonconnected manifolds concern us, I confine myself to observing that they can always be decomposed into a finite or infinite number of connected manifolds.

Consider for example the manifold

$$x_2^2 + x_1^4 - 4x_1^2 + 1 = 0.$$

Here $n = 2$ and the point x_1, x_2 is a point of the plane; our manifold is then none other than a 4th degree curve; however, since that curve is composed of two closed branches, our manifold is not connected.

But we can decompose it into two others, namely

$$x_2^2 + x_1^4 - 4x_1^2 + 1 = 0, \qquad x_1 > 0,$$

and

$$x_2^2 + x_1^4 - 4x_1^2 + 1 = 0, \qquad x_1 < 0,$$

and each of these, being a single closed branch of the curve, is connected.

I shall say that a manifold is *finite* if all its points satisfy the condition

$$x_1^2 + x_2^2 + \cdots + x_n^2 < K^2,$$

where K is a given constant.

We now consider the system of relations

$$(2) \qquad \begin{cases} F_\alpha = 0 & (\alpha = 1, 2, \ldots, p) \\ \varphi_\beta = 0 & \\ \varphi_\gamma > 0 & (\gamma \gtrless \beta) \end{cases}$$

consisting of $p + 1$ equations and $q - 1$ inequalities.

It can happen that there is no point satisfying the conditions (2), or there may be, in which case such points constitute a manifold of less than $n - p$ dimensions.

The set of points which satisfy one of the q systems of relations

$$(3) \qquad \begin{cases} F_\alpha = 0, & \varphi_1 = 0, & \varphi_\gamma > 0 & (\gamma \gtrless 1) \\ F_\alpha = 0, & \varphi_2 = 0, & \varphi_\gamma > 0 & (\gamma \gtrless 2) \\ \ldots & \ldots & \ldots & \ldots \\ F_\alpha = 0, & \varphi_q = 0, & \varphi_\gamma > 0 & (\gamma \gtrless q) \end{cases}$$

is called the *boundary* of the manifold defined by the conditions (1). However, we shall sometimes take another point of view and only consider those which have $n - p - 1$ dimensions as true boundaries.

It may happen that there is no manifold of $n-p-1$ dimensions satisfying any of the q systems (3). In that case, the manifold defined by the conditions (1) will be called *unbounded.* In the contrary case it will be called *bounded.*

If a manifold is simultaneously finite, connected and bounded, it will be called *closed.*

To abbreviate our language a little, we shall give the name (*hyper*) *surfaces* to manifolds of $n-1$ dimensions, except for the case $n=2$, in which case we give them the name *curves.*

§2. Homeomorphism

Consider a substitution which changes $x_1, x_2, \dots, x_n$ into $x'_1, x'_2, \dots, x'_n$, subject only to the following conditions.

We have

$$x'_i = \varphi_i(x_1, x_2, \dots, x_n) \quad (i = 1, 2, \dots, n). \tag{4}$$

In a certain domain the functions φ_i are uniform, finite and continuous; they have continuous derivatives and their Jacobian is nonzero.

If we solve the equations (4) for $x_1, x_2, \dots, x_n$ we get

$$x_k = \varphi'_k(x'_1, x'_2, \dots, x'_n) \quad (k = 1, 2, \dots, n),$$

and the functions φ'_k satisfy the same conditions as the functions φ_i.

It is clear that the set of substitutions which satisfy these conditions constitutes a group, and this group is one of the most general that we can imagine. The science whose object is the study of this group and its analogues receives the name *analysis situs.*

It is clear that a substitution of the group transforms a manifold of m dimensions into a manifold of m dimensions, and that the new manifold will be connected, or finite, or unbounded according as this is the case for the original manifold (and conversely).

Consider two manifolds V and V' of the same number of dimensions defined by the respective conditions

$$\begin{cases} F_\alpha = 0 & (\alpha = 1, 2, \dots, p) \\ \varphi_\beta > 0 & (\beta = 1, 2, \dots, q) \end{cases} \tag{1}$$

and

$$\begin{cases} F'_\alpha = 0 & (\alpha = 1, 2, \dots, p) \\ \varphi'_\beta > 0 & (\beta = 1, 2, \dots, q). \end{cases} \tag{1$'$}$$

Suppose that we can make a point $x_1, x_2, \dots, x_n$ of the manifold V correspond to a point $x'_1, x'_2, \dots, x'_n$ of the manifold V', in such a way that we have

$$x'_k = \psi_k(x_1, x_2, \dots, x_n) \quad (k = 1, 2, \dots, n). \tag{5}$$

I consider the domain D defined by the inequalities

$$F_\alpha > -\varepsilon, \quad F_\alpha < \varepsilon, \quad \varphi_\beta > 0.$$

The manifold V is evidently contained entirely in the domain D.

I assume that *in the domain* D the functions ψ_k are finite, continuous and uniform, that they have continuous derivatives and that their Jacobian is nonzero.

Solving the equations (5) we find

$$x_k = \psi'_k(x'_1, x'_2, \dots, x'_n) \quad (k = 1, 2, \dots, n). \tag{6}$$

I consider the domain D' defined by the inequalities

$$F'_\alpha > -\varepsilon, \;\; F'_\alpha < \varepsilon, \;\; \varphi'_\beta > 0,$$

and I assume that in the domain D' the functions ψ'_k are finite, continuous and uniform, that they have continuous derivatives and that their Jacobian is nonzero.

It follows from these hypotheses that each point of V corresponds to exactly one point of V', and conversely; to every manifold W contained in V there corresponds a manifold W', of the same number of dimensions, contained in V'; if W is connected, finite, or unbounded, the same is true of W' and conversely.

If all these conditions are satisfied we say that the two manifolds V and V' are equivalent from the point of view of *analysis situs*, or simply *homeomorphic*, that is to say, of the same form.

I may also say that two more complicated figures, composed of any number of manifolds, are homeomorphic when one passes into the other by a transformation of the form (5).

Thus two polygons with the same number of sides are homeomorphic, two polyhedra with the same arrangement of faces and the same number of sides on each face are homeomorphic, etc.

§3. Second definition of manifold

One can define manifolds in an entirely different manner. Consider n equations

$$\tag{8} \begin{cases} x_1 = \theta_1(y_1, y_2, \dots, y_m) \\ x_2 = \theta_2(y_1, y_2, \dots, y_m) \\ \dots\dots\dots\dots\dots\dots \\ x_n = \theta_n(y_1, y_2, \dots, y_m). \end{cases}$$

It is clear that these equations (if the y are regarded as independent variables) represent a manifold of m dimensions.

We again have a manifold of m dimensions if we adjoin to the equations (8) a certain number of inequalities of the form

$$\psi(y_1, y_2, \dots, y_m) > 0$$

which limit the field of variation of the variables y.

I assume that the functions θ are finite and continuous; but I shall make a further hypothesis which does lessen generality in an essential fashion (moreover, this could have been done likewise with my first definition of manifold).

I shall suppose, namely, that the functions θ are analytic. If, in fact, the functions θ are finite and continuous, we can find functions θ' which are analytic and differ from the θ by as little as we wish.

I assume, in addition, that the Jacobians of the m functions θ with respect to the y are never simultaneously zero.

We obtain a manifold no different from the first by replacing the y in the equations (8) by any m analytic functions of the m variables $z_1, z_2, \dots, z_m$.

The power of this new definition would be limited if it were not for the fact that we could augment by the procedure of *analytic continuation*.

We consider two manifolds V and V' defined by equations analogous to (8). For example, let the equations

$$\tag{8} x_i = \theta_i(y_1, y_2, \dots, y_m)$$

define V, and let the equations

$$x_i = \theta'_i(y_1, y_2, \ldots, y_m) \tag{8'}$$

define V'.

It can happen that the two manifolds have a common part V'' also of m dimensions.

In that case, in the interior of V'', the y will be analytic functions of the y' and conversely.

We then say that the two manifolds V and V' are *analytic continuations* of each other.

In this way we can form a chain of manifolds

$$V_1, \quad V_2, \quad \ldots, \quad V_n$$

such that each is an analytic continuation of its predecessor, and there is a common part between any two consecutive manifolds of the chain. I shall call this a *connected chain.*

It can also happen that the chain is closed, i.e., that V_n is the same as V_1.

We can then have a network of manifolds, that is to say, a set of manifolds in which each is the continuation of certain others, and we can pass from any one to any one by analytic continuation; I shall call this a *connected network.*

We could then consider the set of all manifolds of the same chain or the same network as forming a unique manifold.

This is a broader definition than the first one.

There are, in fact, manifolds (and we shall see examples later) which can be decomposed into a certain number of partial manifolds forming a connected chain or network and such that each of them can be defined by equations of the form (8) (manifolds which, consequently, are covered by our second definition), which nevertheless cannot be defined by relations of the form (1) and hence are not covered by our first definition.

On the other hand, each manifold which satisfies the first definition also satisfies the second.

In fact, by a well-known theorem, if $y_1, y_2, \ldots, y_n$ are defined by n relations of the form

$$\left\{\begin{array}{l} y_1 = F_1(x_1, x_2, \ldots, x_n) \\ y_2 = F_2(x_1, x_2, \ldots, x_n) \\ \ldots\ldots\ldots\ldots\ldots\ldots\ldots \\ y_n = F_n(x_1, x_2, \ldots, x_n) \end{array}\right. \tag{α}$$

and if, in a certain domain, the F are holomorphic functions of the x, with nonzero Jacobian, then we can solve the equations for the

$$x_i = \theta_i(y_1, y_2, \ldots, y_n), \tag{β}$$

where the θ_i are holomorphic functions of the y.

Now consider a manifold V satisfying our first definition, i.e., defined by relations

$$F_\alpha = 0, \quad \varphi_\beta > 0. \tag{1}$$

Referring to the equations (α), we take for $F_1, F_2, \ldots, F_p$ the first members of the p equations (1). As regards the other functions

$$F_{p+1}, \quad F_{p+2}, \quad \ldots, \quad F_n,$$

we take any $n-p$ holomorphic functions of the x. I subject them to only one condition.

Let $x_1^0, x_2^0, \ldots, x_n^0$ be any point M_0 of the manifold V. I arrange the Jacobian of the n functions F in such a way that it does not vanish for

$$x_i = x_i^0.$$

This is evidently possible, since I assumed that the Jacobians of the p functions

$$F_1, \quad F_2, \quad \ldots, \quad F_p$$

with respect to any p of the variables x do not simultaneously vanish.

I can also assume that $F_{p+1}, F_{p+2}, \ldots, F_n$ vanish at the point M_0, i.e., for

$$x_i = x_i^0.$$

Then, by the theorem cited above, we can solve the equations (α) and we find that the x_i's are expressible in series of powers of

$$y_1, \quad y_2, \quad \ldots, \quad y_n,$$

convergent when these quantities satisfy certain inequalities.

Then let

$$(\beta) \qquad x_i = \theta_i$$

be these equations, and let

$$\lambda_k(y_1, y_2, \ldots, y_n) > 0$$

be the conditions the y must satisfy to guarantee convergence of the series.

If we now make

$$y_1 = y_2 = \cdots = y_p = 0,$$

the initial p equations (α) are the same as the p equations (1); and the functions θ_i, not dependent on more than $n - p = m$ variables, are of the form (8).

Then the set of relations

$$x_i = \theta_i, \quad \varphi_\beta > 0, \quad \lambda_k > 0,$$

represents a manifold v defined in the second manner which is the same as the part of V defined by

$$F_\alpha = 0, \quad \varphi_\beta > 0, \quad \lambda_k > 0.$$

The point M_0, which is an *arbitrary* point of V, is part of v. We can then construct, around any point of V, a manifold analogous to v.

The simplest case is where the convergence conditions $\lambda_k > 0$ are consequences of the inequalities $\varphi_\beta > 0$. Then v is the same as V, and to define it we can be content with the equations (8) and the inequalities

$$(9) \qquad \psi_\beta > 0.$$

We remark in passing that the Jacobians of m of the functions θ with respect to the y are not simultaneously zero.

If the conditions $\lambda_\beta > 0$ are not consequences of the inequalities $\varphi_\beta > 0$, we decompose the manifold V into partial manifolds; thus for example when ψ is any function of $x_1, x_2, \ldots, x_n$ the manifold V can evidently be decomposed into two partial manifolds

$$F_\alpha = 0, \quad \varphi_\beta > 0, \quad \psi > 0,$$

and

$$F_\alpha = 0, \quad \varphi_\beta > 0, \quad \psi < 0.$$

Given that we can always decompose V into partial manifolds, or better, construct a number of partial manifolds on V which overlap each other to an arbitrarily small extent, we find for each of them a system of auxiliary variables which permit that partial manifold to be represented by equations and inequalities of the form (8) and (9) satisfying all the conditions enunciated above. Each point M_0 of V belongs to one of these partial manifolds and the set of these manifolds forms a connected net. Thus the first definition is restored to the second.

Nevertheless, it remains to remark that it can happen that two different systems of values of $y_1, y_2, \ldots, y_n$ correspond to the same system of values of the functions $\theta_1, \theta_2, \ldots, \theta_n$ and hence to the same point of the manifold V.

In that case it is convenient to adjoin to the inequalities (9) the further inequalities

$$\psi_\gamma > 0 \tag{10}$$

chosen so that different systems of values of the variables y which correspond to the same point of V always satisfy one and only one of the inequalities (9) and (10).

Take for example a torus, with equation

$$(x_1^2 + x_2^2 + x_3^2 + R^2 - r^2)^2 - 4R^2(x_1^2 + x_2^2) = 0.$$

We set

$$\begin{aligned} x_1 &= (R + r\cos y_1)\cos y_2, \\ x_2 &= (R + r\cos y_1)\sin y_2, \\ x_3 &= r\sin y_1, \end{aligned}$$

and we see that the same point of the torus corresponds to an infinity of systems of values of the y, comprised by the formulae

$$y_1 + 2K_1\pi, \quad y_2 + 2K_2\pi,$$

where the K are integers.

But if we constrain the y by the conditions

$$0 \leq y_1 < 2\pi, \quad 0 \leq y_2 < 2\pi,$$

we shall have only one system of values y corresponding to each point of the torus.

§4. Oppositely oriented manifolds

With each definition there is reason to make a distinction, the importance of which will be realized later.

Suppose firstly that we have a manifold V defined in the first manner, i.e., by relations

$$F_\alpha = 0, \quad \varphi_\beta > 0.$$

We take account of the order in which the equations $F_\alpha = 0$ are arranged; if two of the equations are permuted it will be convenient to say that the system of relations represents, not the manifold V, but the *oppositely oriented* manifold.

We can now replace the equations

$$F_\alpha = 0 \quad (\alpha = 1, 2, \ldots, p)$$

by the following:

$$\Phi_1 = A_{11}F_1 + A_{12}F_2 + \cdots + A_{1p}F_p = 0$$
$$\Phi_2 = A_{21}F_1 + A_{22}F_2 + \cdots + A_{2p}F_p = 0$$
$$\cdots\cdots\cdots\cdots\cdots\cdots\cdots$$
$$\Phi_p = A_{p1}F_1 + A_{p2}F_2 + \cdots + A_{pp}F_p = 0,$$

where the A_{ik} are any functions of the x.

If the determinant Δ of the coefficients A_{ik} does not vanish in the domain considered, the equations $\Phi_\alpha = 0$ will be equivalent to the equations $F_\alpha = 0$ and consequently, if we adjoin to them a certain number of inequalities, they will again represent the manifold V or the oppositely oriented manifold.

We shall agree to say that if Δ (which always has the same sign, since I have assumed it does not vanish) is positive then these equations represent the manifold V, and if Δ is negative, they represent the oppositely oriented manifold.

If we imagine replacing the y by m analytic functions of m new variables $z_1, z_2, \ldots, z_m$, so that we have

$$x_i = \theta_i'(z_1, z_2, \ldots, z_m),$$

then the new equations again represent V or the oppositely oriented manifold.

We have to assume that the Jacobian Δ of the y with respect to the z never vanishes, so that a system of values of y corresponds to only a single system of values of z. It then will always have the same sign.

We shall agree to say that if Δ is positive the new equations again represent V, and if it is negative they represent the oppositely oriented manifold.

We now see what happens when we pass from one definition to the other.

Let a manifold V be defined by

$$F_1 = F_2 = \cdots = F_p = 0$$

and certain inequalities.

We adjoin to these p equations the following:

$$y_1 = F_{p+1}, \quad y_2 = F_{p+2}, \quad \ldots, \quad y_{n-p} = F_n,$$

where $F_{p+1}, y_2 = F_{p+2}, \ldots, y_{n-p} = F_n$ are any $n-p$ functions of the x.

We have seen that if the Jacobian Δ of the n functions $F_1, F_2, \ldots, F_n$ does not vanish we can solve the n equations for the x and thus find n equations

$$x_i = \theta_i(y_1, y_2, \ldots, y_{n-p})$$

which represent a manifold of $n-p$ dimensions. But we can ask whether this represents V or the oppositely oriented manifold.

We agree to say that it represents V if Δ is positive, and the oppositely oriented manifold if Δ is negative.

Consider the manifold of $n-p$ dimensions

$$F_\alpha = 0, \ldots, \varphi_\beta > 0$$

which I call V. We have seen that the manifold

$$F_\alpha = 0, \quad \varphi_\gamma = 0, \quad \varphi_\beta > 0 \quad (\beta \neq \gamma)$$

of $n-p-1$ dimensions, which I call v, forms part of the boundary of V.

However, it is important to arrange the equations which define v in the following order

$$F_1 = 0, \quad F_2 = 0, \quad \dots, \quad F_p = 0, \quad \varphi_\gamma = 0,$$

since if two of them are permuted, we have agreed to say that these equations no longer represent the boundary of V, or a part of it, but a manifold oppositely oriented to that boundary.

§5. Homologies

Consider a manifold V of p dimensions; now let W be a manifold of q dimensions ($q \leq p$) which is a part of V. We suppose that the boundary of W is composed of λ manifolds of $q-1$ dimensions

$$v_1, \quad v_2, \quad \dots, \quad v_\lambda.$$

We express this fact by the notation

$$v_1 + v_2 + \dots + v_\lambda \sim 0.$$

More generally, the notation

$$k_1 v_1 + k_2 v_2 \sim k_3 v_3 + k_4 v_4,$$

where the k are integers and the v are manifolds of $q-1$ dimensions, will denote that there exists a manifold W of q dimensions forming part of V, the boundary of which is composed of k_1 manifolds similar to v_1, k_2 manifolds similar to v_2, k_3 manifolds similar to v_3 but oppositely oriented, and k_4 manifolds similar to v_4 but oppositely oriented.

Relations of this form will be called homologies.

Homologies can be combined like ordinary equations. Thus we employ the following notation: assuming we have

$$k_1 v_1 + k_2 v_2 + \dots + k_p v_p \sim w_1 + w_2 + \dots + w_p,$$

and that the manifolds $w_1, w_2, \dots, w_p$ form part of the boundary of V, we shall occasionally write

$$k_1 v_1 + k_2 v_2 + k_p v_p \sim \varepsilon.$$

§6. Betti numbers

We say that the manifolds

$$v_1, \quad v_2, \quad \dots, \quad v_\lambda$$

which have the same number of dimensions and form part of V are *linearly independent* if they are not connected by any homology with integral coefficients.

If there exist $P_m - 1$ *closed* manifolds of m dimensions which are linearly independent and form part of V, but not more than $P_m - 1$, then we shall say that the connectivity of V with respect to manifolds of m dimensions is equal to P_m.

Thus for a manifold V of m dimensions there are $m-1$ numbers which I call

$$P_1, \quad P_2, \quad \dots, \quad P_{m-1}$$

and which are the connectivities of V with respect to manifolds of

$$1, \quad 2, \quad \dots, \quad m-1$$

dimensions.

I shall call this the sequence of *Betti numbers* in what follows.

The definitions may be clarified by an example:

Let D be a domain forming part of ordinary space and bounded by n closed surfaces

$$S_1, \quad S_2, \quad \ldots, \quad S_n$$

which do not intersect.

This domain is a manifold of three dimensions. It then admits two Betti numbers, P_1 and P_2.

This manifold is defined by the inequalities

$$\varphi_1 > 0, \quad \varphi_2 > 0, \quad \ldots, \quad \varphi_n > 0$$

if the equations to the n surfaces S are

$$\varphi_1 = 0, \quad \varphi_2 = 0, \quad \ldots, \quad \varphi_n = 0.$$

Since the surfaces do not intersect, there are no values x_1, x_2, x_3 which simultaneously satisfy two of these equations

$$\varphi_i = 0, \quad \varphi_k = 0.$$

Since the surfaces $S_1, S_2, \ldots, S_n$ are two-dimensional, they each have only a single Betti number, which will be the Riemann connectivity; let

$$2Q_1 + 1, \quad 2Q_2 + 1, \quad \ldots, \quad 2Q_n + 1$$

be the connectivities (which are odd, because the surfaces are closed) of the n surfaces

$$S_1, \quad S_2, \quad \ldots, \quad S_n.$$

Then we have

$$P_2 = n, \quad P_1 = Q_1 + Q_2 + \cdots + Q_n + 1.$$

Thus, for the region inside a sphere

$$P_2 = 1, \quad P_1 = 1,$$

for the region inside two spheres

$$P_2 = 2, \quad P_1 = 1,$$

for the region inside a torus

$$P_2 = 1, \quad P_1 = 2,$$

for the region inside two tori

$$P_2 = 2, \quad P_1 = 2.$$

§7. The use of integrals

Consider a manifold V which we may represent by the equations and inequalities (8), (9) and (10) in such a way that all the conditions enunciated above are satisfied.

We know then what we mean by an m-tuple multiple integral

$$\int F dy_1 dy_2 \cdots dy_m$$

over the manifold V; F of course denotes a given function of the y. Integration can be effected with respect to the m variables successively, and the limits of integration are defined by the inequalities (9) and (10).

That being given, I am going to define the following integral

$$\int \sum X_{\alpha_1\alpha_2\ldots\alpha_m} dx_{\alpha_1} dx_{\alpha_2} \cdots dx_{\alpha_m}. \tag{11}$$

The differentials $dx_{\alpha_1}, dx_{\alpha_2}, \ldots, dx_{\alpha_m}$ are any m of the n differentials $dx_1, dx_2, \ldots, dx_n$. The functions $X_{\alpha_1\alpha_2\ldots\alpha_m}$ are given functions of $x_1, x_2, \ldots, x_n$ having all possible combinations of the indices

$$\alpha_1, \alpha_2, \ldots, \alpha_m,$$

that is to say, all combinations of n letters m at a time. We make the convention that the function X is zero if two of the indices are equal, and changes sign when two of the indices are permuted.

That being given, the integral (11) by definition will equal the integral of order m

$$\int \sum X_{\alpha_1\alpha_2\ldots\alpha_m} \frac{\partial(x_{\alpha_1}, x_{\alpha_2}, \ldots, x_{\alpha_m})}{\partial(y_1, y_2, \ldots, y_m)} dy_1 dy_2 \cdots dy_m.$$

Now if the manifold V is not susceptible to representation by relations of the form (8), (9) and (10) satisfying all the conditions previously enunciated, we decompose the manifold V into partial manifolds small enough to be susceptible to that mode of representation, and the integral (11), understood to be over the total manifold V will, by definition, be the sum of the integrals (11), understood to be over the various partial manifolds.

This definition may nevertheless still be ambiguous.

In fact, if we permute two of the letters y_1 and y_2 the integral changes sign; it is important then to order these letters and make a permutation of two of these letters equivalent to a change of the sense of integration in the study of simple integrals. I shall speak then of the sense of integration in connection with the order in which we find it convenient to arrange the letters $y_1, y_2, \ldots, y_m$.

I had occasion to deal with an analogous question in a memoir *Sur les résidus des intégrales doubles* in vol. IX of *Acta Mathematica*, in particular in the paragraph 3 of that memoir entitled: *Conditions d'intégrabilité.*

I investigated under what circumstances these *conditions of integrability* are complete, i.e., under what circumstances the integral (11) is always zero when taken over a closed manifold.

Here is what I found: writing

$$(\alpha_1, \alpha_2, \ldots, \alpha_m)$$

in place of $X_{\alpha_1\alpha_2\ldots\alpha_m}$ and $[\alpha_p]$ in place of x_{α_p}, the conditions of integrability may be written

$$\left\{\begin{aligned} &\frac{\partial(\alpha_1,\alpha_2,\ldots,\alpha_m)}{\partial[\alpha_{m+1}]} \pm \frac{\partial(\alpha_2,\alpha_3,\ldots,\alpha_{m-1})}{\partial[\alpha_1]} \\ &\pm \frac{\partial(\alpha_3,\alpha_4,\ldots,\alpha_m,\alpha_{m+1},\alpha_1)}{\partial[\alpha_2]} \pm \cdots \pm \frac{\partial(\alpha_{m+1},\alpha_2,\ldots,\alpha_{m-1})}{\partial[\alpha_m]} = 0. \end{aligned}\right. \tag{12}$$

Here the following law governs the choice of the signs $\pm$. We always take the sign $+$ if m is even, and alternate $+$ and $-$ if m is odd.

We have as many equations (12) as there are systems of indices

$$\alpha_1, \quad \alpha_2, \quad \ldots, \quad \alpha_m, \quad \alpha_{m+1},$$

i.e., since the indices may be chosen from the letters

$$1, \quad 2, \quad \ldots, \quad n$$

as many as the number of combinations of n letters $m+1$ at a time.

We suppose now that the conditions (12), instead of being satisfied for all possible values of the n variables

$$x_1, \quad x_2, \quad \ldots, \quad x_n$$

are satisfied only for certain values of these variables. For example, consider a manifold V defined by the conditions

$$F_\alpha = 0, \quad \varphi_\beta > 0.$$

Next, let a domain D be defined containing all points near to V, for example, by the conditions

$$-\varepsilon < F_\alpha < \varepsilon, \quad \varphi_\beta > -\varepsilon,$$

where ε is a small positive number.

We assume that the conditions (12) hold for all points of the domain D.

By repeating the reasoning of the memoir cited, with suitable modifications, we can show the following. Let a manifold V' of $m+1$ dimensions form part of V (the number $m+1$ must then be less than or equal to the number of dimensions of V). We assume that the boundary of V' is composed of k manifolds of m dimensions

$$W_1, \quad W_2, \quad \ldots, \quad W_k$$

so that $W_1 + W_2 + \cdots + W_k \sim 0$.

Then if the integral (11) satisfies the conditions (12) in the domain D, the algebraic sum of the integrals (11) taken over the manifolds $W_1, W_2, \ldots, W_k$ is zero. It is necessary, of course, to pay attention to the sense of integration for each of them.

The conditions (12) are sufficient for this to happen, but they are not necessary; these conditions, as we have seen, are equal in number to the combinations of n letters $m+1$ at a time; it would suffice for the integral (11) to satisfy, at all points of V, certain conditions equal in number to the combinations of $n-p$ letters $m+1$ at a time, where p is the number of dimensions of V.

These conditions are quite easy to construct, but that would take me too far away from my subject.

So if the integral (11) satisfies the conditions (12) in the domain D, the various values of that integral taken over various closed manifolds of m dimensions forming part of V will be linear combinations, with integral coefficients, of a certain number of them, which we could call the periods of the integral (11).

The maximum number of periods is equal to $P_m - 1$, since, if we consider P_m closed manifolds of m dimensions there will always be a manifold of $m+1$ dimensions which has these P_m manifolds as its boundary, or even some subset of them. Then there will always be a linear relation with integral coefficients between the P_m corresponding integrals. Moreover, we can see that there always exist integrals of the form (11) for which the maximum period is attained. This means of elucidating the definition of Betti numbers was employed by Betti himself for the first and the last of these numbers, i.e., for P_1 and P_{m-1}; but we have come to see that it is easy to do the same for the other Betti numbers.

§8. Orientable and nonorientable manifolds

We consider a manifold V defined in the second manner, i.e., in the form of a chain or network of partial manifolds each of which is defined by relations of the form (8) and (9).

Let v_1 be a partial manifold defined by the conditions

$$x_i = \theta(y_1, y_2, \dots, y_m), \qquad |y_k| < \beta_k.$$

Let v_2 be another partial manifold defined by the conditions

$$x_i = \theta_i'(z_1, z_2, \dots, z_m), \qquad |z_k| < \gamma_k.$$

Suppose that these two manifolds have a common part v' forming a connected manifold. I claim that in the interior of this manifold the Jacobian

$$\Delta = \frac{\partial(y_1, y_2, \dots, y_n)}{\partial(z_1, z_2, \dots, z_n)}$$

always has the same sign.

In fact, it cannot change sign without vanishing or becoming infinite. We have

$$\Delta = \frac{\partial(x_1, x_2, \dots, x_m)}{\partial(z_1, z_2, \dots, z_m)} \div \frac{\partial(x_1, x_2, \dots, x_m)}{\partial(y_1, y_2, \dots, y_m)}$$

so that Δ is itself the quotient of two Jacobians; since these two Jacobians are essentially finite, Δ cannot be zero unless we have

$$\frac{\partial(x_1, x_2, \dots, x_m)}{\partial(z_1, z_2, \dots, z_m)} = 0,$$

and since nothing distinguishes the first m variables $x_1, x_2, \dots, x_m$ from the $n - m$ others, $x_{m+1}, x_{m+2}, \dots, x_n$, it will be necessary for the Jacobian of any m of the x with respect to the z to be zero.

All the Jacobians of the functions θ' will then vanish simultaneously, contrary to hypothesis. Δ then cannot vanish, and we see in exactly the same way that it cannot become infinite.

Thus Δ is always of the same sign, and we can choose the order of the variables z in such a way that this sign is positive.

A difficulty may occur in certain cases; suppose that the common part of v_1 and v_2, instead of reducing to a single connected manifold v', is composed of several connected manifolds v', v'', v'''; in each of them the sign of Δ remains constant, but it may change in passing from one to the other. In that case we say that the manifold V is *nonorientable.*

We assume that this circumstance does not occur, and consider a sequence of partial manifolds forming a closed chain

$$v_1, \quad v_2, \quad \dots, \quad v_q, \quad v_1.$$

Let

$$y_1^i, \quad y_2^i, \quad \dots, \quad y_m^i$$

be the m variables which play the role of $y_1, y_2, \dots, y_m$ in relation to v_i.

Suppose that v_1 and v_2 have a common part v_1', v_2 and v_3 a common part $v_2', \dots, v_{q-1}$ and v_q a common part v_{q-1}' and finally v_q and v_1 a common part v_q'.

Let Δ_i be the Jacobian of

$$y_1^{i+1}, \quad y_2^{i+1}, \quad \dots, \quad y_m^{i+1}$$

with respect to

$$y_1^i, \quad y_2^i, \quad \ldots, \quad y_m^i.$$

This determinant will be defined in the interior of v_i'. In the interior of v_q' I also define the Jacobian Δ_q of

$$y_1^1, \quad y_2^1, \quad \ldots, \quad y_m^1$$

with respect to

$$y_1^q, \quad y_2^q, \quad \ldots, \quad y_m^q.$$

From what we have seen, we know we can always choose the order of variables so that

$$\Delta_1, \quad \Delta_2, \quad \ldots, \quad \Delta_{q-1}$$

are always positive. On the other hand, though D_q is always of the same sign, is this sign $+$ or $-$?

If the sign is $-$, then I say that the manifold V is *nonorientable.* I could also say that the manifolds $v_1, v_2, \ldots, v_q$ form a nonorientable chain.

Suppose now that we have constructed a certain connected network of partial manifolds

$$(4) \qquad v_1, \quad v_2, \quad \ldots, \quad v_q$$

such that each point of V is in the interior (exclusive of the boundary) of one or more of the manifolds (4). If the determinant Δ is positive in the common part of any two of the manifolds (4), I say that the manifold is orientable. If this is not the case then it is clear that we could always form a nonorientable chain with some of the manifolds (4), and that the manifold V is nonorientable. I could also say that the network of manifolds (4) forms an orientable system.

But to justify our definition completely, we have to see that V cannot be orientable and nonorientable at the same time. It is clear, first of all, that in the orientable system (4) we cannot find a nonorientable chain.

It remains to show that the system (4) remains orientable when we adjoin any manifold v_{q+1} forming part of V.

Let v_i' be the common part of v_i and v_{q+1}; each point of v_{q+1} belongs to at least one of the manifolds v_i', and, since v_{q+1} is continued, if I consider two points M_1 and M_k of v_{q+1} belonging to v_1' and v_k', respectively, we can find intermediate manifolds which I can call (since the numbering remains arbitrary)

$$v_1', \quad v_2', \quad \ldots, \quad v_k'$$

and which form a chain.

I can always choose the order of the variables, analogous to the y, defining the manifold v_{q+1} in such a way that the determinant analogous to Δ between v_1 and v_{q+1} is positive in v_1'; I shall call it Δ_1.

Likewise, we let Δ_i be the analogue of Δ between v_i and v_{q+1}; Δ_i is defined in the interior of v_i'.

Then let Δ' be the determinant between v_1 and v_2; it will be defined throughout the common part of the two manifolds and, in particular, in the common part of v_1' and v_2'; it will be positive because the system (4) is orientable.

Now, in the common part of v_1' and v_2', since Δ_1 and Δ' are positive, it follows that their ratio

$$\Delta_2 = \frac{\Delta_1}{\Delta'}$$

is positive, and since it always has the same sign, it will be positive for all points of v_2'.

Step by step, we show that $\Delta_3, \ldots, \Delta_k$ are likewise positive.

The system thus remains orientable with the adjunction of v_{q+1}, it also remains so if the variables $y_1, y_2, \ldots, y_m$ in the equations for one of the manifolds v_i are replaced by holomorphic functions of new variables $y_1', y_2', \ldots, y_m'$ with nonvanishing Jacobian (this is necessary in order that a system of values of the y correspond to a single system of values of the y'). It is necessary, of course, to choose the order of the new variables y' in such a way that this Jacobian is positive.

The system always remains orientable, we cannot construct a nonorientable chain, so that a manifold cannot be orientable and nonorientable at the same time. Q.E.D.

Everyone knows the example of a one-sided (nonorientable) surface obtained from a rectangle of paper $ABCD$ (where the pairs of opposite sides are AB, CD and BC, DA) by joining the sides AB, CD so that A coincides with C and B with D.

Examples of orientable manifolds are easy to construct. Thus in the space of n dimensions:

1° Every n-dimensional domain is orientable.
2° Every 1-dimensional curve is orientable.
3° Every *closed* hypersurface of $n-1$ dimensions is orientable.

But we can go further.

Consider a manifold V defined *in the first manner*, i.e., by equations and inequalities of the form (1). I claim that this will always be orientable.

In fact, let

$$F_1 = F_2 = \cdots = F_p = 0$$

be the p equations which, together with some inequalities we shall not write, define V. Let V be decomposed into a certain number of partial manifolds v, defined by relations of the form (8) and (9). Let

$$v_1, \quad v_2, \quad \ldots, \quad v_q, \quad v_1$$

be a certain number of partial manifolds which form a connected chain, i.e., such that each has a part in common with its successor. I claim that this chain is always orientable.

In fact, we have assumed above that the Jacobians of the p functions F relative to any p of the variables x are never simultaneously zero. I may then assume that the manifolds v are sufficiently small that in the interior of v_1 one of the Jacobians (which I shall call Δ_1) does not vanish; in the interior of v_2 another of the Jacobians (which I shall call Δ_2) does not vanish, and so on.

If this is not already the case, we can attain it by subdividing each of the manifolds v into sufficiently small manifolds.

Suppose, for example, that

$$\Delta_1 = \frac{\partial(F_1, F_2, \ldots, F_p)}{\partial(x_{\alpha_1}, x_{\alpha_2}, \ldots, x_{\alpha_p})}$$

$$\Delta_2 = \frac{\partial(F_1, F_2, \ldots, F_p)}{\partial(x_{\beta_1}, x_{\beta_2}, \ldots, x_{\beta_n})}$$

$$\cdots\cdots\cdots\cdots\cdots$$

Since the order of the letters $x_{\alpha_1}, x_{\alpha_2}, \ldots, x_{\alpha_p}$ remains arbitrary and Δ_1 does not change sign in the interior of v_1, I can always assume that Δ_1 is positive in the interior of v_1. Likewise, Δ_2 will be positive in the interior of v_2; and so on.

Now in the interior of v_1 we set

$$\left\{\begin{array}{llll} F_1 = 0, & F_2 = 0, & \ldots, & F_p = 0 \\ y_1^1 = x'_{\alpha_1}, & y_2^1 = x'_{\alpha_2}, & \ldots, & y_m^1 = x'_{\alpha_m}. \end{array}\right. \tag{13}$$

Here $m = n - p$; the variables $x^1_{\alpha_i}$ are the $m = n - p$ variables x which remain when we remove the p variables x_{α_i}.

In the interior of v_2 we set

$$\begin{array}{llll} F_1 = 0, & F_2 = 0, & \ldots, & F_p = 0, \\ y_1^2 = x'_{\beta_1}, & y_2^2 = x'_{\beta_2}, & \ldots, & y_m^2 = x'_{\beta_m}. \end{array} \tag{14}$$

The variables x'_{β_i} are the $m = n - p$ variables which remain after we remove the p variables x_{β_i}.

And so on.

I assume that the order of the variables x'_{α_i} has been chosen in such a way that we can pass from the normal order of the variables x, i.e.,

$$x_1, \quad x_2, \quad \ldots, \quad x_n,$$

to the order

$$x_{\alpha_1}, \quad x_{\alpha_2}, \quad \ldots, \quad x_{\alpha_p}, \quad x'_{\alpha_1}, \quad x'_{\alpha_2}, \quad \ldots, \quad x'_{\alpha_m}$$

by a substitution of the alternating group.

Then, in solving the equations (13) we see that in the interior of v_1 the x are holomorphic functions of

$$y'_1, \quad y'_2, \quad \ldots, \quad y'_m$$

and so on. In general, in the interior of v_i the x are holomorphic functions of

$$y_1^i, \quad y_2^i, \quad \ldots, \quad y_m^i.$$

It is then a question of calculating the determinant

$$\frac{\partial(y_1^2, y_2^2, \ldots, y_m^2)}{\partial(y_1^1, y_2^1, \ldots, y_m^1)}$$

which we shall abbreviate

$$\frac{\partial y_i^2}{\partial y_i^1}$$

in the interior of the common part of v_1 and v_2.

To do this, we replace the equations (13) and (14) by the more general equations

$$\left\{\begin{array}{llll} F_1 = \lambda_1, & F_2 = \lambda_2, & \ldots, & F_p = \lambda_p, \\ & & y_i^1 = x'_{\alpha_i}, & \end{array}\right. \tag{13'}$$

and

$$F_k = \lambda_k, \quad y_i^2 = x'_{\beta_i}. \tag{14'}$$

We later put $\lambda_k = 0$.

Solving the equations (13′), we obtain the x as functions of the λ_k and the y'_i holomorphic in the interior of v_1 and the Jacobian of

$$x_1, \quad x_2, \quad \ldots, \quad x_n$$

with respect to

$$\lambda_1, \quad \lambda_2, \quad \ldots, \quad \lambda_p, \quad y_i^1, \quad y_i^2, \quad \ldots, \quad y_i^m$$

is evidently $\frac{1}{\Delta_1}$.

Likewise solving the equations (14′) we obtain the x as functions of the λ_k and y_i^2 holomorphic in the interior of Δ_2, and the Jacobian is $\frac{1}{\Delta_2}$.

It is then true for points common to v_1 and v_2 that

$$\frac{\partial y_i^2}{\partial y_i'} = \frac{\partial(\lambda_1, \lambda_2, \ldots, \lambda_p, y_1^2, y_2^2, \ldots, y_m^2)}{\partial(\lambda_1, \lambda_2, \ldots, \lambda_p, y_1', y_2', \ldots, y_m')} = \frac{\Delta_1}{\Delta_2},$$

and it only remains to make $\lambda_k = 0$ in the expressions for the x.

We similarly find that in the part common to v_k and v_{h+1}

$$\frac{\partial y_i^{h+1}}{\partial y_i^h} = \frac{\Delta_h}{\Delta_{h+1}},$$

and in the part common to v_p and v_1

$$\frac{\partial y_i^1}{\partial y_i^q} = \frac{\Delta_q}{\Delta_1}.$$

We have already seen that we can always assume that Δ_i is positive in the interior of v_i. All these quotients are then positive and the chain is orientable. Q.E.D.

Thus all varieties which satisfy the first definition are orientable, and, since I cited an example of a nonorientable manifold satisfying the second definition we conclude that there are manifolds satisfying the second definition which do not satisfy the first. This is what I claimed originally.

Any nonorientable manifold is, by definition, *opposite to itself.*

§9. Intersection of two manifolds

Let V and V' be two manifolds[6] defined in the second manner, one having p dimensions, the other $n-p$, and let M_0 and M_0' be two points belonging to V and V' respectively with respective coordinates

$$x_1^0, \quad x_2^0, \quad \ldots, \quad x_n^0$$

and

$$x_1'^0, \quad x_2'^0, \quad \ldots, \quad x_n'^0.$$

Around the point M_0 we construct a partial manifold v on V, analogous to those envisioned in paragraph 3, in such a way that for a point $x_1, x_2, \ldots, x_n$ of v we have

$$x_i = \theta_i(y_1, y_2, \ldots, y_p) \quad (i = 1, 2, \ldots, n).$$

Likewise, around M_0' we construct a manifold v' on V' so that for a point $x_1', x_2', \ldots, x_n'$ of v' we have

$$x_i' = \theta_i'(y_1', y_2', \ldots, y_{k-p}') \quad (i = 1, 2, \ldots, n).$$

[6]The first supplement to *Analysis situs* takes this up again. (Translator's note.)

Consider the determinant

$$\begin{vmatrix} \frac{\partial x_1}{\partial y_1} & \frac{\partial x_2}{\partial y_1} & \cdots & \frac{\partial x_n}{\partial y_1} \\ \frac{\partial x_1}{\partial y_2} & \frac{\partial x_2}{\partial y_2} & \cdots & \frac{\partial x_n}{\partial y_2} \\ \cdots & \cdots & \cdots & \cdots \\ \frac{\partial x_1}{\partial y_p} & \frac{\partial x_2}{\partial y_p} & \cdots & \frac{\partial x_n}{\partial y_p} \\ \frac{\partial x'_1}{\partial y'_1} & \frac{\partial x'_2}{\partial y'_1} & \cdots & \frac{\partial x'_n}{\partial y'_1} \\ \cdots & \cdots & \cdots & \cdots \\ \frac{\partial x'_1}{\partial y'_{n-p}} & \frac{\partial x'_2}{\partial y'_{n-p}} & \cdots & \frac{\partial x'_n}{\partial y'_{n-p}} \end{vmatrix}.$$

This is a function of $y_1, y_2, \dots, y_p, y'_1, y'_2, \dots, y'_{n-p}$; thus it depends on the position of the points M and M' with coordinates $x_1, x_2, \dots, x_n$ and $x'_1, x'_2, \dots, x'_n$ on the manifolds v and v'. I therefore call it $f(M, M')$.

I can change the variables by replacing $y_1, y_2, \dots, y_p$ by holomorphic functions of p variables $z_1, z_2, \dots, z_p$ chosen in such a way that a system of values of the y corresponds to a single system of values of the z. For this it is necessary that the Jacobian of the y with respect to the z never vanishes, and I can always assume that the z are arranged in an order which makes the determinant positive.

The function $f(M, M')$ is then multiplied by this Jacobian, and consequently retains its sign.

It is the same when we make an analogous change of variables on the manifold v'.

Consider now another manifold v_1, analogous to v, constructed on V, and a manifold v'_1, analogous to v', constructed on V'.

Let M_1 be a point of v_1 and M'_1 a point of v'_1. We can construct a function analogous to $f(M, M')$ which I shall call $f_1(M_1, M'_1)$.

Suppose now that the manifolds v_1 and v (likewise the manifolds v'_1 and v') have a common part and that the points M and M' are identified, likewise the points M' and M'_1.

We compare the two functions

$$f(M, M'), \quad f_1(M, M').$$

Suppose that the two manifolds V and V' are orientable. We may then assume (by suitable choice of the order of the variables y relative to v_1) that the determinant analogous to that we called Δ in §8, relative to v and v_1, is positive on the part common to v and v_1; likewise, the determinant analogous to Δ relative to v' and v'_1 will be equally positive.

Then f and f_1 have the same sign.

Then let $S(M, M')$ be a function which is equal to $+1$, -1 or 0 according as $f(M, M')$ is positive, negative or zero. This function is well defined, it does not depend on the position of the points M and M' on V and V'; nor does it depend on the manner in which the manifolds v and v' have been constructed around M and M'.

This is no longer true if one of the manifolds V and V' is nonorientable.

Suppose in particular that the points M and M' are the same, so that the point M is a point of intersection of V and V'; consideration of the function $S(M, M')$ is then of great interest.

Imagine that we construct the function $S(M, M)$ for all points of intersection M of V and V', and take the sum of all the functions S so obtained; I shall denote this sum by $N(V, V')$.

This definition of $N(V, V')$ is applicable when the manifolds V and V' have been defined as in §3. But we can simplify it when V and V' have been defined as in §1.

Let

$$F_\alpha = 0, \quad \varphi_\beta > 0 \ \ (\alpha = 1, 2, \ldots, p; \beta = 1, 2, \ldots, q)$$

be the conditions which define V and let

$$F'_\gamma = 0, \quad \varphi'_\delta > 0 \ \ (\gamma = 1, 2, \ldots, n-p; \delta = 1, 2, \ldots, q')$$

be those which define V'. V has $n-p$ dimensions and V' has p.

Consider a point M of V with coordinates $x_1, x_2, \ldots, x_n$ and a point M' of V' with coordinates $x'_1, x'_2, \ldots, x'_n$ and form the determinant

$$\begin{vmatrix} \frac{\partial F_1}{\partial x_1} & \frac{\partial F_2}{\partial x_1} & \cdots & \frac{\partial F_p}{\partial x_1} & \frac{\partial F'_1}{\partial x'_1} & \frac{\partial F'_2}{\partial x'_1} & \cdots & \frac{\partial F'_{n-p}}{\partial x'_1} \\ \cdots & \cdots & \cdots & \cdots & \cdots & \cdots & \cdots & \cdots \\ \frac{\partial F_1}{\partial x_n} & \frac{\partial F_2}{\partial x_n} & \cdots & \frac{\partial F_p}{\partial x_n} & \frac{\partial F'_1}{\partial x'_n} & \frac{\partial F'_2}{\partial x'_n} & \cdots & \frac{\partial F'_{n-p}}{\partial x'_n} \end{vmatrix}$$

which I shall call $\psi(M, M')$. What is the relationship between $\psi(M, M')$ and $S(M, M')$?

To give an account of this, we set

$$x'_i = x_i + h_i \quad (i = 1, 2, \ldots, n)$$

and allow the x'_i and x_i to simultaneously vary, but in such a way that their difference h_i remains constant.

We set

$$F_\alpha(x_i) = y_\alpha,$$
$$F'_\gamma(x_i + h_i) = y_\gamma.$$

If the determinant $\psi(M, M')$ is never zero we can solve these equations for the x_i and obtain

$$x_i = \theta_i(y_\alpha, y'_\gamma).$$

The Jacobian of the θ_i with respect to the y_α and the y'_γ will then be $\frac{1}{\psi(M,M')}$. If we set $y_\alpha = 0$ we shall have

$$x_i = \theta_i(0, y'_\gamma),$$

and this will be the equation of a manifold of $n-p$ dimensions which forms part of V; likewise if we set

$$x'_i = h_i + \theta_i(y_\alpha, 0),$$

we have the equation of a variety of p dimensions which forms part of V'.

If we define it in the manner of the manifolds v and v' at the beginning of this section, we see that the Jacobian of the θ_i is none other than $f(M, M')$, whence

$$f(M, M')\psi(M, M') = 1.$$

Suppose now that the two manifolds V and V' are parts of a manifold W of a greater number of dimensions. Suppose, for example, that

$$(1) \qquad F_\alpha = 0, \quad \varphi_\beta > 0 \ \ (\alpha = 1, 2, \ldots, p; \beta = 1, 2, \ldots, q)$$

are the relations which define W, which then has $n-p$ dimensions.

Let

(2) $$F'_\gamma = 0, \quad \varphi'_\delta > 0 \quad (\gamma = 1, 2, \ldots, p; \delta = 1, 2, \ldots, q')$$

be the relations which, together with (1), define V, which then has $n - p - p'$ dimensions.

Finally let

(3) $$F''_\epsilon = 0, \quad \varphi''_\zeta > 0 \quad (\epsilon = 1, 2, \ldots, p'', \zeta = 1, 2, \ldots, q'')$$

be the relations which, together with (1), define V', which then has $n - p - p''$ dimensions. I assume that $p + p' + p'' = n$.

Let M be a point of intersection of V and V'.

I form the Jacobian of the F_α, the F'_γ and F''_ϵ with respect to the x, and this will be the determinant I call $\psi(M)$.

Then by definition, $S(M)$ will equal $+1$ or -1 according as $\psi(M)$ is positive or negative.

Next, I denote by $N(V, V')$ the sum of all the quantities $S(m)$ relative to points of intersection of V and V'.

It is clear that if one of the manifolds V or V' is replaced by the oppositely oriented manifold, then the expression $N(V, V')$ changes sign.

Now let V be a closed one-dimensional manifold and let W be a domain of n dimensions, the boundary of which is composed of manifolds of $n - 1$ dimensions

$$V_1, \quad V_2, \quad \ldots, \quad V_k$$

so that

$$V_1 + V_2 + \cdots + V_k \sim 0.$$

I claim that

$$N(V, V_1) + N(V, V_2) + \cdots + N(V, V_k) = 0.$$

In fact, the manifold V will be composed of a certain number of connected one-dimensional manifolds, i.e., a certain number of closed lines. The equations of these lines take the form

$$x_i = \theta_i(t),$$

where t is a variable we can make increase from $-\alpha$ to $+\infty$.

Since the line is closed we have

$$\theta_i(-\infty) = \theta_i(+\infty).$$

Imagine now a point M of intersection of one of these lines with the boundary of W; if, as t increases, this line passes from the interior of W to the exterior, we have

$$S(M) = 1.$$

If it passes from exterior to interior, $S(m)$ will equal -1. But since the line is closed it returns to its starting point, so that the sum of all the $S(M)$ will be zero. Q.E.D.

Suppose now that all the manifolds we consider, in particular V, W, V_1, V_2, $\ldots$, V_k form part of a manifold U, and that this is a manifold for which we know the connectivities and the homologies in relation to them.

(What I mean to say is that when I defined homologies in §5 I had a certain manifold V which played an important role in that definition; well, U plays that role here.)

I assume that U has h dimensions, V one dimension, W h dimensions, and $V_1, V_2, \ldots, V_k$ $h-1$ dimensions.

Without any change in the preceding, we see that if V is closed and one-dimensional and if

$$V_1 + V_2 + \cdots + V_k \sim 0,$$

then

$$N(V, V_1) + N(V, V_2) + \cdots + N(V, V_k) = 0.$$

I claim that this is again true if the one-dimensional manifold V is no longer closed, but if its two extremities are on the boundary of U and if W has no points on the boundary of U.

In fact, U can be decomposed into two regions, connected or otherwise, namely: W and the region R exterior to W. By hypothesis, the boundary of U is contained entirely in R.

Since the line V is not closed, the initial point corresponding to $t = -\infty$ does not coincide with the final point corresponding to $t = +\infty$; however, by hypothesis both these points belong to the boundary of V, and hence to R.

Our line then passes from R to W exactly as often as from W to R; so that the sum is zero. Q.E.D.

I now want to establish the converse.

Let $V_1, V_2, \ldots, V_k$ be k closed manifolds of $h-1$ dimensions situated in U, where U has no point in common with the boundary of V_i. Suppose that we do not have

$$V_1 + V_2 + \cdots + V_k \sim 0;$$

I claim that we can always find a line V situated in U which is closed, which has two extremities on the boundary of U, and which is such that

$$N(V, V_1) + N(V, V_2) + \cdots + N(V, V_k) \neq 0.$$

In fact, suppose first of all that the manifolds $V_1, V_2, \ldots, V_k$ do not decompose U into different regions. Then we can go from any point of U to any other point of U without encountering any of the manifolds V_i.

Consider then a small line meeting V_1 at a point M and not encountering any of the other manifolds V_i; we can connect the two extremities of that line by another continuous line which does not meet any of the manifolds V_i; we can thus construct a closed line which does not meet any of the manifolds V_i except at a single point. The sum of the $S(M)$ then cannot be zero.

Suppose now that the manifolds V_i decompose U into several regions R, while nevertheless being linearly independent. In that case the boundary of any of the regions R cannot consist of part of the manifolds V_i, otherwise we would have a homology between these manifolds, contrary to hypothesis.

The boundary of R is composed then of some of the manifolds V_i and a part of the boundary of U. It follows that at any point of R we can go to the boundary of U without leaving R and without meeting any of the manifolds V_i.

Now consider one of the manifolds V_i, V_1 for example, and imagine a small line cutting V_1 at a point M; let R and R' be the regions to which the two extremities μ and μ' of that small line belong. We can go from μ to the boundary of U by a continuous line without leaving R, and from μ' by a third continuous line to the boundary of U without leaving R'.

The union of these three lines then goes continuously from the boundary of V to the boundary of U, and if we call it V, we have

$$N(V, V_1) = 1, \quad N(V, V_i) = 0 \quad (i > 1).$$

Thus we can choose V in such a way that all the $N(V, V_i)$ are zero, except one which is equal to 1.

Suppose finally that we have a certain number of homologies between the V_i, say three, for example

$$(\alpha) \qquad \Sigma k_i V_i \sim \Sigma k_i' V_i \sim \Sigma k_i'' V_i \sim 0.$$

Then among the k manifolds V_i we can find $k-3$ which are linearly independent, say $V_1, V_2, \ldots, V_{k-3}$.

We can choose V in such a way that all the $N(V, V_i)$ where $i = 1, 2, \ldots, k-3$ are zero with the exception of one of them which equals 1.

The values of

$$N(V, V_{k-2}), \quad N(V, V_{k-1}), \quad N(V, V_k)$$

may be deduced with the aid of the relations

$$(\beta) \qquad \Sigma k_i N(V, V_i) = \Sigma k_i' N(V, V_i) = \Sigma k_i'' N(V, V_i) = 0$$

which are a necessary consequence of our homologies (α).

But we cannot have

$$(\gamma) \qquad \Sigma N(V, V_i) = 0,$$

regardless of which of the $k-3$ quantities

$$N(V, V_i) \quad (i = 1, 2, \ldots, k-3)$$

is equal to 1. This would mean, in fact, that the equation (γ) was a necessary consequence of the equations (β); and then the homology

$$(\delta) \qquad \Sigma V_i \sim 0$$

would be a necessary consequence of the homologies (α). But we have assumed at the beginning that the homology (δ) does not hold.

Thus we can always choose V in such a way that the equation (γ) is not valid. Q.E.D.

We shall give the name *cut* to each manifold contained in U, whether closed or not, if its boundary forms part of the boundary of U.

We can then enunciate the following theorem which summarizes the preceding discussion:

The necessary and sufficient condition (if the manifolds V_i are closed and of $h-1$ dimensions) *for the existence of a cut V such that the equation*

$$\Sigma k_i N(V, V_i) = 0$$

does not hold, is that the homology

$$\Sigma k_i V_i \sim 0$$

does not hold.

We now try to extend this theorem to the case where V has p dimensions and $v_1, v_2, \ldots, v_k$ have $h-p$.

Let

$$F_\alpha = 0, \quad \varphi_\beta > 0 \quad (\alpha = 1, 2, \ldots, n-k)$$

be the equations of U, and let

$$F_\alpha = 0, \quad F'_\gamma = 0, \quad \varphi_\beta > 0 \quad (\gamma = 1, 2, \ldots, h-p)$$

be those of V.

As far as $V_1, V_2, \ldots, V_k$ are concerned, we define them in the following manner. We can always find $p-1$ equations

$$\Phi_1 = \Phi_2 = \cdots = \Phi_{p-1} = 0$$

satisfied by all the points of $V_1, V_2, \ldots, V_k$; this is so if $h = 3, p = 2$, if U is ordinary space and if $V_1, V_2, \ldots, V_k$ are curves, for we can always make a surface pass through the curves.

To define V_i we adjoin a pth equation

$$F''_i = 0.$$

Then let U' be the manifold of $h - p + 1$ dimensions

$$F_\alpha = 0, \quad \Phi_v = 0, \quad \varphi_\beta > 0$$

and V' the one-dimensional manifold

$$F_\alpha = F'_\gamma = \Phi_v = 0, \quad \varphi_\beta > 0.$$

We have firstly that

$$N(V', V_i) = N(V, V_i).$$

On the other hand, if the homology

$$\Sigma k_i V_i \sim 0$$

holds with respect to U', it holds with respect to U. It is true that the converse is false, and that this homology can hold with respect to U without holding with respect to U'; however, if it holds with respect to U we can always find a manifold U' of $h-p+1$ dimensions with respect to which it holds, by suitable choice of the functions Φ.

We must then conclude that the theorem is again true when V has more than one dimension.

Then if V_1 and V_2 are two closed manifolds of $h - p$ dimensions, and if for all cuts V of p dimensions we have

$$N(V, V_1) = N(V, V_2),$$

then we also have

$$V_1 \sim V_2$$

and conversely.

We confine ourselves to the case where U is closed. Then U has no boundary and all the cuts are closed.

The number of linearly independent cuts of p dimensions is $P_p - 1$.

Let

$$C_1, \quad C_2, \quad \ldots, \quad C_\lambda \qquad (\lambda = P_p - 1)$$

be the cuts.

Next, let

$$V_i \quad (i = 1, 2, \ldots, \mu)$$

be μ closed manifolds of $h - p$ dimensions.

The necessary and sufficient condition for a homology

$$\Sigma k_i V_i \sim 0$$

is that we have

$$\Sigma k_i N(C_1, V_i) = \Sigma k_i N(C_2, V_i) = \cdots = \Sigma k_i N(C_\lambda, V_i) = 0.$$

But, if the number μ of the V_i is greater than λ, we can always find integers k_i satisfying these conditions, because we have μ numbers k_i and λ conditions to satisfy. If the V_i are linearly independent, then we have

$$\mu \leq \lambda.$$

Then

$$P_{h-p} \geq P_p,$$

but, changing p into $h-p$, we find

$$P_p \leq P_{h-p},$$

so

$$P_p = P_{h-p}.$$

Consequently, for a closed manifold the Betti numbers equally distant from the ends of the sequence are equal.

This theorem has not, I believe, been announced previously; nevertheless it was known to various people who made applications of it.

We now look at what happens to the middle number $P_{h/2}$ when h is even. Suppose that h is a multiple of $4+2$, in such a way that $\frac{h}{2}$ is odd.

We know that when we make an odd number of permutations of lines in a determinant, it changes sign. When we permute V and V_i, which have $\frac{h}{2}$ dimensions, the determinant $f(M, M)$ will change sign; we then have

$$N(V, V_1) = -N(V_1, V)$$

from which we deduce

$$N(V, V) = 0.$$

The symbol $N(V, V)$ in itself has no meaning, because with two coincident manifolds V and V the number of points of intersection is infinite; we therefore rectify the definition by setting

$$N(V, V) = N(V, V')$$

where

$$V \sim V'.$$

That being given, I claim that $P_{h/2}$ is odd. Suppose in fact that it is even, and let

$$V_1, \quad V_2, \quad \ldots, \quad V_\mu$$

be μ linearly independent manifolds of $\frac{h}{2}$ dimensions where

$$\mu = P_{h/2} - 1$$

is odd.

We form the determinant where the ith term in the kth column is $N(V_i, V_k)$. This determinant will be skew symmetric, i.e., the terms on the principal diagonal will be zero, and terms symmetric with respect to the diagonal will be equal but of opposite sign. Since the number of columns will be odd, this determinant will be zero. It then follows, contrary to hypothesis, that the manifolds $V_1, V_2, \ldots, V_\mu$ are not linearly independent.

Thus $P_{h/2}$ is odd.

This will no longer be true when h is a multiple of 4, nor when the manifold U is nonorientable; since all our arguments assume an orientable manifold. We shall see examples later.

§10. Geometric representation

There is a manner of representing manifolds of three dimensions situated in a space of four dimensions which considerably facilitates their study.

Consider a certain number of polyhedra in ordinary space

$$P_1, \quad P_2, \quad \ldots, \quad P_n.$$

We may assume that in the space of four dimensions there are three-dimensional manifolds

$$Q_1, \quad Q_2, \quad \ldots, \quad Q_n$$

homeomorphic to the $P_1, P_2, \ldots, P_n$, respectively.

Let F_1 be a face of the polyhedron P_1 and Φ the set of points on the boundary of Q_1 which correspond to the points of F_1. Likewise, let F_2 be a face of P_2 and Φ_2 the image of that face on the boundary of Q_2.

It can happen that Φ_1 coincides with Φ_2. In that case the two manifolds Q_1 and Q_2 are contiguous, and we pass from the interior of one to the interior of the other by crossing Φ_1.

In that case we say that the faces F_1 and F_2 are *conjugate*.

It can happen that the faces F_1 and F_2 belong to the *same* polyhedron P_1. Then the two-dimensional manifold Φ_1, which is the same as the two-dimensional manifold Φ_2, separates two portions of the manifold Q_1.

We can understand this better in terms of an example in ordinary space. Consider a rectangle $ABCD$ and a torus on which we draw two cuts, namely, latitudinal and longitudinal circles; let H be their point of intersection. The surface of the torus will then be homeomorphic to the rectangle; the two sides of the cut formed by the longitudinal circle correspond to the two sides AB and CD, while the two sides of the cut formed by the latitudinal circle correspond to the sides AD and BC. We see the analogy with the preceding: the rectangle corresponds to the polyhedron P_1, the torus to the manifold Q_1, the sides AB and CD to the two faces F_1 and F_2, the two sides of the longitudinal circle to the two manifolds Φ_1 and Φ_2, which, as we see, coincide.

That being given, imagine that among the faces of the n polyhedra P_i we have a certain number which are conjugate in pairs and a certain number which remain *free*.

Consider the total manifold V consisting of the set of manifolds Q_i. Since some of the manifolds Q_i are contiguous, it may happen that the total manifold V is connected: this is what I assume.

If there is none among the faces of the P_i which remains free, then the manifold V will be closed. In the contrary case, the points which correspond to free faces form the boundary of V.

We conceive that the knowledge of the polyhedra P_i and the mode of conjugation of their faces provides us with an image in ordinary space of the manifold V, sufficient for the study of its properties from the point of view of *analysis situs*.

We should comment on the method of defining conjugation of faces. It is clear first of all that for two faces to be conjugate they must have the same number of sides. Next, for the mode of conjugation to be completely known, it does not suffice to know merely which faces are conjugate, but also which vertices correspond.

Only then is the mode of conjugation completely defined.

Corresponding to two conjugate faces, there is, by definition, a two-dimensional manifold inside V. It may also happen that several edges of the polyhedra P correspond to the same line inside V, or that several vertices of the polyhedra correspond to the same point inside V. We then say that these edges or vertices belong to the same cycle.[7]

We note here that we can form cycles of edges and cycles of vertices.

Let A_1 be an edge, F'_1 one of the two faces that meet in A_1, F_2 the conjugate of F'_1, A_2 the edge of F_2 which corresponds to A_i, F'_2 the other face which meets A_2, F_3 the conjugate of F'_2, A_3 the edge corresponding to A_2, etc.

We stop when we arrive at a free face or when we return to the edge A_1. The edges $A_1, A_2, A_3, \ldots$ form a cycle.

The same is true for vertices. Let S_1 be a vertex, F'_1 one of the faces which meet at S_1, F_2 the conjugate of F'_1, S_2 the vertex corresponding to S_1, F'_2 *one* of the faces which meet at S_2, etc.

$S_1, S_2, S_3, \ldots$ belong to the same cycle.

However, in this case more than two faces meet at each vertex, so that F'_2, for example, can be chosen in several ways; we should not stop until we have exhausted all possible choices.

The analogy with the formation of cycles in the theory of Fuchsian groups is evident. It is even stronger if we assume there is only a single polyhedron P_1.

First example. The simplest example is that where we have a single polyhedron which is a cube $ABCDA'B'C'D'$, with vertices having the respective coordinates

$$\begin{array}{llll} A & \ldots\ 0\,0\,0 & A' & \ldots\ 0\,0\,1 \\ B & \ldots\ 0\,1\,0 & B' & \ldots\ 0\,1\,1 \\ C & \ldots\ 1\,0\,0 & C' & \ldots\ 1\,0\,1 \\ D & \ldots\ 1\,1\,0 & D' & \ldots\ 1\,1\,1 \end{array}$$

I suppose that the opposite faces are conjugate, in the following fashion:

$$\text{(1)} \qquad \begin{cases} ABDC & \equiv & A'B'D'C', \\ ACC'A' & \equiv & BDD'B', \\ CDD'C' & \equiv & ABB'A'. \end{cases}$$

Here is what I intend this notation to mean: the relation

$$ABDC \equiv A'B'D'C'$$

means

1º The faces $ABDC$ and $A'B'D'C'$ are conjugate.

2º The vertices on the first of these faces occur in the order $ABDC$.

[7]This appearance of the term "cycle" is a one-off, and should not be confused with Poincaré's systematic use of the term "cycle" in the homology of algebraic surfaces studied in Supplements 3 and 4. (Translator's note.)

3° The vertices A and A', B and B', D and D', C and C' correspond.

We observe in passing that the vertices $ABDC$ are encountered on the first of these faces in clockwise order; on the other hand, the vertices $A'B'C'D'$ are encountered on the second face in anti-clockwise order.

This condition must always be fulfilled *if we want the manifold V to be orientable.* If two faces F_1 and F_2 are conjugate and if a point describes the perimeter of F_1 clockwise, then the corresponding point on F_2 must describe the perimeter of that face anti-clockwise.

That being given, suppose that the mode of conjugation is defined by the relations (1). It is then easy to see that all the vertices form a single cycle and that there are three cycles of edges, comprising those edges which are parallel to the x axis, those parallel to the y axis, and those parallel to the z axis.

Second example. We retain our cube, but change the mode of conjugation to that defined by the relations

$$\tag{2} \begin{cases} ABDC & \equiv & B'D'C'A', \\ ABB'A' & \equiv & DD'C'C, \\ ACC'A' & \equiv & DD'B'B. \end{cases}$$

We then have two cycles of edges and two cycles of vertices, and I summarize the results in the following relations; I place the sign $\equiv$ between two edges (or two vertices) to denote that they are part of the same cycle.

Two cycles of edges

$$AB \equiv B'D' \equiv C'C \equiv B'A' \equiv AC \equiv DD',$$

$$AA' \equiv DC \equiv C'A' \equiv B'B \equiv C'D' \equiv DB.$$

Two cycles of vertices

$$A \equiv B' \equiv C' \equiv D,$$
$$B \equiv D' \equiv C \equiv A'.$$

Later we shall see why this mode of conjugation is inadmissible.

Third example. We retain our cube, with the following mode of conjugation.

$$\tag{3} \begin{cases} ABDC & \equiv & B'D'C'A', \\ ABB'A' & \equiv & C'CDD', \\ ACC'A' & \equiv & DD'B'B. \end{cases}$$

We then find

Four cycles of edges

$$AB \equiv B'D' \equiv C'C, \quad AA' \equiv C'D' \equiv DB,$$

$$AC \equiv DD' \equiv B'A', \quad CD \equiv BB' \equiv A'C'.$$

Two cycles of vertices

$$A \equiv B' \equiv C' \equiv D,$$
$$B \equiv D' \equiv A' \equiv C.$$

Fourth example. Now let

$$\tag{4} \begin{cases} ABDC & \equiv & B'D'C'A', \\ ABB'A' & \equiv & CDD'C', \\ ACC'A' & \equiv & BDD'B'. \end{cases}$$

We find:
Three cycles of edges:

$$AA' \equiv CC' \equiv BB' \equiv DD',$$
$$AB \equiv CD \equiv B'D' \equiv A'C',$$
$$AC \equiv BD \equiv D'C' \equiv B'A',$$

and a single cycle of vertices.

Fifth example. Consider a regular octahedron; it has six vertices of which four form a square $BCED$ (the letters are arranged in the order in which they are encountered in a circuit of the square's perimeter) and of which two, A and F, are outside the square. Let

$$(5) \qquad \begin{cases} ABC & \equiv & FED \\ ACE & \equiv & FDB \\ AED & \equiv & FBC \\ ADB & \equiv & FCE \end{cases}$$

be the mode of conjugation; we find six cycles of edges and three cycles of vertices; each edge forms a cycle with the opposite edge, i.e., the edge symmetric to it in relation to the center of symmetry of the octahedron, and each vertex forms a cycle with the opposite vertex.

It is unnecessary to multiply examples any longer; I now propose to survey all the modes of conjugation which are admissible.

Let a *star* be the figure formed by a certain number of solid angles with the same vertex and arranged around that vertex in such a way that each point of space belongs to exactly one of the solid angles.

I could suppose that the edges of the faces are extended indefinitely, or else that they end on the surface of a sphere with its center at the vertex of the solid angles. Then the various solid angles are cut by the sphere in a certain number of spherical polygons, so that the surface of the sphere is subdivided into spherical polygons. The surface subdivided in this way can be regarded as homeomorphic to a convex polyhedron, the faces of which correspond to the spherical polygons just defined.

Then the faces of the polyhedron correspond to the solid angles of the star, its edges correspond to the faces of the star, and its vertices to the edges.

Let S, F and A be the number of solid angles, faces and edges of the star. Since the polyhedron must satisfy Euler's theorem, we must have

$$S - F + A = 2.$$

We now return to the polyhedra

$$P_1, \quad P_2, \quad \ldots, \quad P_n$$

and imagine a cycle of vertices; all the vertices of this cycle correspond to the same point of V, which I call M. Among the manifolds

$$Q_1, \quad Q_2, \quad \ldots, \quad Q_n,$$

there will be a certain number which have the point M on their boundary; I call these the manifolds Q_α, and they are those that correspond to the polyhedra P_α to which various vertices of the cycle belong.

Now consider two manifolds Q_α which are contiguous and the two-dimensional manifold which is their common boundary. I call the two-dimensional manifolds defined in this way Φ_α; they correspond to those faces of the polyhedra P_α to which the various vertices of the cycle belong.

Now imagine finally the one-dimensional manifolds which are the intersections of two manifolds Φ_α, I call these L_α. They correspond to those edges of the polyhedra P_α to which the various vertices of the cycle belong.

We consider the figure formed by the manifolds $Q_\alpha, \Phi_\alpha, L_\alpha$ or rather, the points of that manifold which satisfy the inequality

$$(x_1 - x_1^0)^2 + (x_2 - x_2^0)^2 + (x_3 - x_3^0)^2 + (x_4 - x_4^0)^2 < \varepsilon^2, \tag{6}$$

where ε is very small and $x_1^0, x_2^0, x_3^0, x_4^0$ are the coordinates of the point M.

This figure is evidently homeomorphic to a star with faces and edges bounded by a sphere. Let A be that star.

Consider any of the manifolds Q_α and, in addition, the manifold W consisting of those of its points which satisfy the inequality (6). This manifold will be connected or not. If it is not, I decompose it into a number of connected manifolds, and I call the connected manifolds formed in this way the Q'_α (thus the number of Q'_α will be greater than the number of Q_α if one of the manifolds W is not connected).

I similarly define the Φ'_α and the L'_α.

Let $q_\alpha, \varphi_\alpha, \ell_\alpha$ be the numbers of Q'_α, Φ'_α, and L'_α, respectively; these will also be the numbers of solid angles, faces, and edges of the star A, whence we reach the conclusion:

For a mode of conjugation to be admissible, it is necessary that for each cycle of vertices we have

$$q_\alpha - \varphi_\alpha + \ell_\alpha = 2.$$

We now see how to calculate the numbers $q_\alpha, \varphi_\alpha, \ell_\alpha$ for any cycle α of vertices:

1° q_α will be the number of vertices of the cycle.

2° To obtain φ_α we have to form the sum of the numbers of faces which occur at the various vertices of the cycle and divide it by two. If, for example, the cycle α consists of the vertices a, b, c belonging, respectively, to polyhedra P_1, P_2 and P_3, if the point a is the vertex of a trihedral angle, so that three faces of P_2 pass through a, if four faces of P_2 pass through b, and five faces of P_3 through c, then we have

$$\varphi_\alpha = \frac{3+4+5}{2} = 6.$$

3° To obtain ℓ_α we enumerate the edges which meet the various vertices of the cycle in the following way: all the edges of the same cycle of edges are counted once if one of the extremities belongs to the cycle α; they are counted twice if two extremities belong to the cycle α.

If we apply these rules to the six examples dealt with above, then we arrive at the following table.

Example	q_α	φ_α	ℓ_α
1st	8	12	6
2nd	4	6	2
3rd	4	6	4
4th	8	12	6
5th	2	4	4

It should be remarked that in the examples 2, 3, and 5 there are several cycles of vertices, but we arrive at the same three numbers $q_\alpha, \varphi_\alpha, \ell_\alpha$ for each cycle in the same example.

The table shows that the relation

$$q_\alpha - \varphi_\alpha + \ell_\alpha = 2$$

is satisfied for all examples except the second. The mode of conjugation in the second example must therefore be rejected.

§11. Representation by a discontinuous group

Here is another mode of representation which can also be applied in certain cases.

Let (x, y, z) be a point of ordinary space; consider a series of substitutions which change x, y, z, respectively, into

$$\begin{gathered}
\varphi_1(x,y,z),\ \psi_1(x,t,z),\ \chi_1(x,y,z)\\
\varphi_2(x,y,z),\ \psi_2(x,t,z),\ \chi_2(x,y,z)\\
\dots\dots\dots,\ \dots\dots\dots,\ \dots\dots\dots\\
\varphi_n(x,y,z),\ \psi_n(x,t,z),\ \chi_n(x,y,z)\\
\dots\dots\dots,\ \dots\dots\dots,\ \dots\dots\dots
\end{gathered}$$

Suppose that the set of these substitutions forms a properly discontinuous group. The space is then found to be divided into an infinity of domains

$$D_0, \quad D_1, \quad D_2, \quad \dots$$

such that each domain D_i corresponds to a substitution S_i of the group which changes D_0 into D_i.

Consider a surface Σ which forms that part of the boundary of D_0 which separates D_0 from D_i; the substitution S_i^{-1} inverse to S_i changes D_i into D_0, and since the points of Σ belong to the boundary of D_i, the transform[8] of the surface Σ will be another part of the boundary of D_0.

The boundary of D_0 is thus divided into pieces of surface which are conjugate in pairs, in such a way that each of them is transformed into its conjugate by a substitution of the group.

The domain D_0, with its boundary subdivided in this way, will be homeomorphic to a polyhedron, the faces of which are conjugate in pairs, as in the preceding section. Then, as in the preceding section, we could make this polyhedron, and hence the domain D_0, correspond to a closed three-dimensional manifold situated in the space of four dimensions, obtained by transporting D_0 to this space, then deforming it and *gluing together* the conjugate portions of its boundary.

The analogy with the theory of Fuchsian groups is too evident to need stressing; I shall confine myself to a single example:

Sixth example. Consider the group generated by the three substitutions

$$(1) \qquad \begin{cases} (x,y,z; & x+1, y, z),\\ (x,y,z; & x, y+1, z),\\ (x,y,z; & \alpha x+\beta y, \gamma x+\delta y, z+1), \end{cases}$$

[8]On page 50 Poincaré introduces the word "transform" in another sense, to denote what we now call the group-theoretic conjugate. (Translator's note.)

where $\alpha, \beta, \gamma, \delta$ are integers such that

$$\alpha\delta - \beta\gamma = 1.$$

I call this group the group $(\alpha, \beta, \gamma, \delta)$. I claim that it is properly discontinuous.

To justify this, let us find how the transforms of a single point

$$x = a, \quad y = b, \quad z = c$$

are distributed in space.

First of all, all the transforms of this point by any combination of the first two substitutions are covered by the formula

$$x = a + k, \quad y = b + k_1, \quad z = c, \tag{2}$$

where k and k_1 are any two integers; it is easy to see that these substitutions commute with each other.

If we now transform the set of points (2) by the third substitution we find

$$x = \alpha(a + k) + \beta(b + k_1), \quad y = \gamma(a + k) + \delta(b + k_1), \quad z = c + 1. \tag{3}$$

If we set

$$\alpha a + \beta b = a_1, \quad \gamma a + \delta b = b_1,$$

then the point (a_1, b_1) is the transform of the point (a, b) under the substitution

$$(x, y;\ \alpha x + \beta y,\ \gamma x + \delta y)$$

which I shall call s.

We can then replace the equations (3) by the following:

$$x = a_1 + k', \quad y = b_1 + k_1', \quad z = c_1 + 1, \tag{4}$$

where k' and k_1' are two new integers; applying the first two substitutions to these points we always recover the same points.

We now apply the third substitution to them.

Let

$$a_2 = \alpha a_1 + \beta b_1, \quad b_2 = \gamma a_1 + \delta b_1$$

so that the point a_2, b_2 is the transform of the point a_1, b_1 by s.

Then the transforms of the points (4) by the third substitution are covered by the formula

$$x = a_2 + k'', \quad y = b_2 + k_1'', \quad z = c + 2 \tag{5}$$

where k'' and k_1'' are two integers.

In general, suppose that the successive transforms of the point (a, b) by the substitution s are $a_1, b_1;\ a_2, b_2,\ \ldots;\ a_n, b_n;\ \ldots$ and the successive transforms by the inverse substitution are $a_{-1}, b_{-1};\ a_{-2}, b_{-2};\ \ldots$.

Then all the transforms of the point (a, b, c) by substitutions of the group (1) are given by the formulae

$$x = a_n + k, \quad y = b_n + k_1, \quad z = c + n,$$

where n, k and k_1 are three arbitrary integers.

Moreover, we see easily that the substitutions generated by the first two in the group commute with the third.

The fundamental domain D_0 is a cube with side of length 1 and bounded by the six points

$$x, y, z = 0, 1.$$

The most simple case is that where

$$\alpha = \delta = 1, \quad \beta = \gamma = 0,$$

so that our three substitutions reduce to

$$(x, y, z;\ x+1, y, z;\ x, y+1, z;\ x, y, z+1).$$

Each of these changes one face of the cube into an opposite face, so we have simply recovered our first example.

However, the manner in which the surface of the cube D_0 is divided into conjugate regions is not always so simple.

Suppose for example

$$\alpha = \beta = \delta = 1, \quad \gamma = 0.$$

Each of the faces parallel to the z axis will then be conjugate to the opposite face; but for the faces $z = 0, z = 1$ perpendicular to the z axis it will be more complicated.

Suppose that the points $ABCD, A'B'C'D'$ have the same coordinates as in our first example. Each of the faces $ABCD, A'B'C'D'$ must be divided into two triangles, namely: ABC and BCD on the one hand, $A'D'C'$ and $A'B'D'$ on the other, and the rule of conjugation of the faces is expressed by the relations

$$\begin{aligned} ACC'A' &\equiv BDD'A', \\ CDD'C' &\equiv ABB'A', \\ ABC &\equiv A'D'C', \\ BCD &\equiv B'A'D'. \end{aligned}$$

More generally, the faces parallel to the z axis remain conjugated in pairs, but the faces perpendicular to the z axis must be divided into polygons small enough to be as numerous as the numbers $\alpha, \beta, \gamma, \delta$ are large, and conjugate in pairs by a law which is more or less complicated.

A simple case is that of

$$\alpha = \delta = 0, \quad \beta = 1, \quad \gamma = -1.$$

In this case the mode of conjugation is the same as that of our fourth example.

§12. The fundamental group

We are led in this way to the notion of fundamental group of a manifold. Let

$$F_1, \quad F_2, \quad \ldots, \quad F_\lambda$$

be λ functions of the coordinates $x_1, x_2, \ldots, x_n$ of a point M of the manifold V defined by the relations

$$f_\alpha = 0, \quad \varphi_\beta > 0.$$

I do not assume that the functions F are uniform. When the point M leaves its initial position M_0 and returns to that position after traversing an arbitrary path, it may happen that the functions F do not return to their initial values.

To better fix ideas, without omitting any of the essentials, we shall assume that the functions F are defined in the following manner. They must satisfy differential equations of the form

$$dF_i = X_{i,1}dx_1 + X_{i,2}dx_2 + \cdots + X_{i,n}dx_n, \tag{1}$$

where the coefficients $X_{i,k}$ are given functions of the x_k and the F_i. These functions must be uniform, finite and continuous, and have derivatives for all values of F and all points sufficiently close to V.

I likewise suppose that for all points sufficiently close to V the equations (1) satisfy conditions of integrability, which can be written

$$\begin{aligned}&\frac{\partial X_{i,k}}{\partial x_q} + \frac{\partial X_{i,k}}{\partial F_1} X_{1,q} + \frac{\partial X_{i,k}}{\partial F_2} X_{2,q} + \cdots + \frac{\partial X_{i,k}}{\partial F_\lambda} X_{\lambda,k} \\ &= \frac{\partial X_{i,q}}{\partial x_k} + \frac{\partial X_{i,q}}{\partial F_1} X_{1,k} + \frac{\partial X_{i,q}}{\partial F_2} X_{2,k} + \cdots + \frac{\partial X_{i,q}}{\partial F_\lambda} X_{\lambda,k}.\end{aligned}$$

Then if the point M describes an infinitely small contour on the manifold V, the functions F return to their original values. It will again be the case if the point M describes a *hairpin bend*, i.e., travels from M_0 to M_1 by any path M_0BM_1, describes an infinitely small contour, then returns from M_1 to M_0 by the *same* path M_1BM_0.

But they need not be the same if it describes a closed finite contour.

Suppose for example that we take ordinary space and let our manifold be the torus. It is evident that the functions F will return to their original values when the point M describes a hairpin bend on the torus, but this need not be the case if M describes a longitudinal or latitudinal circle.

The final values of the functions F, when the mobile point M leaves an initial point M_0 and returns after describing a closed contour, naturally depend on the initial values.

So let F_a^0 and F_a' be the initial and final values of F_a. The F_a' are functions of the F_a^0 or, in other words, the functions F undergo a certain substitution when M describes the closed contour considered, which changes F_a^0 into F_a'.

The substitutions undergone by the functions F when the point M describes all the closed contours that can be traced on the manifold V from an initial point M_0 evidently form a group, which I call g.

Now consider a closed contour M_0BM_0 traced on V from an initial point M_0. If that closed contour reduces to a hairpin bend I shall write

$$M_0BM_0 \equiv 0.$$

Now if M_0AM_1, M_0BM_1, M_0CM_1 are three different paths traced on V from M_0 to M_1 I shall write

$$M_0AM_1CM_0 \equiv M_0AM_1BM_0 + M_0BM_1CM_0.$$

It is important to remark that $M_0AM_1CM_0$ is not the same as $M_0CM_1AM_0$, nor is $M_0AM_1BM_0 + M_0BM_1CM_0$ the same as

$$M_0BM_1CM_0 + M_0AM_1BM_0;$$

we cannot change the order of terms in a sum.

It follows from this convention that we have

$$M_0BM_0 \equiv 0$$

if the closed contour M_0BM_0 constitutes a boundary of a two-dimensional manifold forming part of V; and in fact the closed contour can then be decomposed into a large number of hairpin bends.[9]

[9]This paragraph is manifestly false; H. Poincaré ultimately rectified it (Fifth supplement).

We are thus led to consider relations of the form

$$k_1C_1 + k_2C_2 \equiv k_3C_3 + k_4C_4,$$

where the k are integers and the C are closed contours traced on V, leaving from M_0. These relations, which I call *equivalences*, resemble the homologies studied above. They differ in that

1° With homologies, the contours can leave from any initial point.
2° With homologies, we can change the order of terms of a sum.

We can add two equivalences to each other if the order of terms is respected; thus if

$$A \equiv B \text{ and } C \equiv D$$

we can conclude

$$A + C \equiv B + D$$

but not

$$C + A \equiv B + D.$$

If we have

$$2A \equiv 0$$

we do not have the right to conclude

$$A \equiv 0.$$

This being given, it is clear that we can envisage a group G satisfying the following conditions:

1° Each closed contour M_0BM_0 corresponds to a substitution S of the group.
2° The necessary and sufficient condition for S to reduce to the identity substitution is that

$$M_0BM_0 \equiv 0.$$

3° If S and S' correspond to contours C and C' and if

$$C'' \equiv C + C',$$

then the substitution corresponding to C'' will be SS'.

The group G is called the *fundamental group* of the manifold V.

We compare it to the group g of substitutions undergone by the functions F.

The group g will be isomorphic to G.[10]

The isomorphism can be holoedric, but it will not be if there is a closed contour M_0BM_0 indecomposable into hairpin bends on which the functions F return to their original values.

[10]Here, and in the next line, Poincaré is using the 19th-century terminology of "holoedric isomorphisms" and "meriedric isomorphisms," which are what we call "isomorphisms" and "homomorphisms", respectively. Thus we would say that there is a homomorphism $G \to g$. (Translator's note.)

§13. Fundamental equivalences

The group G will be generated by a certain number of principal substitutions $S_1, S_2, \dots, S_p$. Each of them corresponds to a closed contour, so that we have p *fundamental closed contours* $C_1, C_2, \dots, C_p$, and any other closed contour is equivalent to a combination of fundamental contours in a certain order.

These fundamental contours are not, in general, independent, and there are certain relations between them that I call *fundamental equivalences.*[11]

Suppose for example that we have the equivalence

$$k_1C_1 + k_2C_2 + k_1'C_1 + k_3C_3 \equiv 0.$$

This signifies that the substitution $S_1^{k_1}S_2^{k_2}S_1^{k_1'}S_3^{k_3}$ of the group G reduces to the identity.

The fundamental equivalences enable us to know the structure of the group G.

Suppose that the manifold V has been defined by the mode of representation of §10 and that we have only a single polyhedron P_1. It is clear that we obtain all the fundamental contours in the following manner. Let M_0 be a point inside P, A a point on one of the faces of P_1, A' the corresponding point on the conjugate face. If we go from A to M_0, then from M_0 to A', without leaving P_1, then the corresponding path on the manifold V will be closed.

Thus there are as many fundamental contours as there are pairs of faces.

Now here is how we find the fundamental equivalences:

Consider a cycle of edges. Take for example an edge which is the intersection of faces F_1 and F'_μ, and which I call for that reason the *edge* F_1, F'_μ; let F'_1 be the face conjugate to F_1 and $F_2F'_1$ be the edge on that face corresponding to $F_1F'_\mu$; let F'_2 be the face conjugate to F_2 and $F_3F'_2$ the edge on that face corresponding to $F_2F'_1$; and so on until we return to the face F'_μ and the edge $F_1F'_\mu$.

We remark that in carrying out this process, we may return several times to the same face.

Let A_i be a point of F_i and A'_i the corresponding point of F'_i; let C_i be the fundamental contour

$$M_0A_i + A'_iM_0.$$

Then we have the fundamental equivalence

$$C_1 + C_2 + \cdots + C_\mu \equiv 0,$$

and there are as many fundamental equivalences as there are cycles of edges.

When we have formed the fundamental equivalences in this way, we realize that the fundamental homologies are none other than the results of letting the order of terms become immaterial. The knowledge of these homologies immediately yields the Betti number P_1.

We apply these principles to the examples above, and remark that all the manifolds cited are closed and three-dimensional, so $P_1 = P_2$.

---

[11]The fundamental contours are what we now call *generators* of the fundamental group, and the fundamental equivalences are what we call *defining relations.* (Translator's note.)

First example:

$$(C_1) \qquad ABDC \equiv A'B'D'C',$$
$$(C_2) \qquad ABB'A' \equiv CDD'C',$$
$$(C_3) \qquad ACC'A' \equiv BDD'B'.$$

Here is what I mean by the notation

$$(C_1) \qquad ABDC \equiv A'B'D'C'.$$

I want to say that the face $ABDC$ is conjugate to $A'B'D'C'$ and that if α denotes a point of $ABDC$ and α' a point of $A'B'D'C'$, then the fundamental contour $M_0\alpha + \alpha' M_0$ is denoted by C_1.

Fundamental equivalences:

$$C_1 + C_2 \equiv C_2 + C_1, \quad C_1 + C_3 \equiv C_3 + C_1, \quad C_2 + C_3 \equiv C_3 + C_2.$$

The fundamental homologies reduce to the identity

$$P_1 = P_2 = 4.$$

I now pass immediately to the third example, since we have seen that the second must be rejected.

Third example:

$$(C_1) \qquad ABDC \equiv B'D'C'A',$$
$$(C_2) \qquad ABB'A' \equiv C'CDD',$$
$$(C_3) \qquad ACC'A' \equiv DD'B'B.$$

Fundamental equivalences:

$$C_1 + C_3 + C_2 \equiv 0, \qquad C_1 - C_3 - C_2 \equiv 0,$$
$$C_2 - C_1 - C_3 \equiv 0, \qquad C_3 - C_2 - C_1 \equiv 0,$$

which can also be written

$$2C_1 \equiv 2C_2 \equiv 2C_3, \qquad 4C_1 \equiv 0.$$

Fundamental homologies:

$$C_1 \sim C_2 \sim C_3 \sim 0$$

whence

$$P_1 = P_2 = 1.$$

We can give this result a simple geometric interpretation.

The group G is of finite order and consists of only eight distinct substitutions, corresponding to the following contours:

$$0, \quad C_1, \quad C_2, \quad C_3, \quad 2C_1, \quad 3C_1, \quad 3C_2, \quad 3C_3.$$

The group is isomorphic to the following group:

$$\begin{aligned}(x, y, z, t;\ & z, -t, x, -t, z;\\ & -t, -z, y, x;\ z, -t, -x, y;\\ & \quad -x, -y, -z, -t;\ y, -x, t, -z\\ & \qquad t, z, -y, -x;\ -z, t, x, -y).\end{aligned}$$

This group, which transforms into itself the four-dimensional hypercube with faces defined by

$$x = \pm 1, \quad y = \pm 1, \quad z = \pm 1, \quad t = \pm 1,$$

may be called the *hypercubic* group.

Fourth example:

(C_1) $$ABDC \equiv B'D'C'A',$$

(C_2) $$ABB'A' \equiv CDD'C',$$

(C_3) $$ACC'A' \equiv BDD'B'.$$

Fundamental equivalences:

$$C_2 + C_3 \equiv C_3 + C_2, \quad C_3 + C_1 \equiv C_1 + C_2, \quad -C_2 + C_1 \equiv C_1 + C_3.$$

Fundamental homologies:

$$C_2 \sim C_3 \sim 0$$

whence

$$P_1 = P_2 = 2.$$

Fifth example:

(C_1) $$ABD \equiv FED,$$

(C_2) $$ACE \equiv FDB,$$

(C_3) $$AED \equiv FBC,$$

(C_4) $$ADB \equiv FCE.$$

Fundamental equivalences:

$$C_1 \equiv C_2 \equiv C_3 \equiv C_4, \quad 2C_1 \equiv 0$$

whence

$$C_1 \sim C_2 \sim C_3 \sim C_4 \sim 0$$

and

$$P_1 = P_2 = 1.$$

The group G reduces to two substitutions corresponding to the contours 0 and C_1.

Sixth example. The group G is evidently isomorphic to the group $(\alpha, \beta, \gamma, \delta)$.

The three substitutions

$$(x, y, z;\ x + 1, y, z),$$
$$(x, y, z;\ x, y + 1, z),$$
$$(x, y, z;\ \alpha x + \beta y, \gamma x + \delta y, z + 1),$$

correspond to the fundamental contours C_1, C_2 and C_3, respectively.

First we have the fundamental equivalences

$$C_1 + C_2 \equiv C_2 + C_1,$$
$$C_1 + C_3 \equiv C_3 + \alpha C_1 + \delta C_2,$$
$$C_2 + C_3 \equiv C_3 + \beta C_1 + \delta C_2,$$

whence it follows first of all that any combination of fundamental contours can be expressed in the form

$$m_3 C_3 + m_1 C_1 + m_2 C_2,$$

where the m are integers. Since such an expression cannot be equivalent to 0 unless the three integers m are zero, it follows that we possess all the fundamental equivalences.

Fundamental homologies:

$$(\alpha - 1)C_1 + \gamma C_2 \sim 0,$$
$$\beta C_1 + (\delta - 1)C_2 \sim 0.$$

If these two homologies are not distinct we have

$$C_1 \sim C_2 \sim 0,$$

whence

$$P_1 = P_2 = 2,$$

which is what happens in the general case, and in our fourth example in particular.

If the determinant of these homologies is zero, i.e., if

$$(\alpha - 1)(\delta - 1) - \beta\gamma = 0$$

or

$$\alpha + \delta = 2,$$

we have

$$P_1 = P_2 = 3$$

except for the case where the two homologies reduce to the identity. This is what happens for

$$\alpha = \delta = 1, \ \beta = \gamma = 0,$$

i.e., for our first example; we then have

$$P_1 = P_2 = 4.$$

§14. Conditions for homeomorphism

We know that two closed two-dimensional manifolds with the same Betti numbers are homeomorphic. This follows, for example, from the study of the periods of abelian functions. We consider a Riemann surface R and let z be the corresponding imaginary variable; we can introduce a new imaginary variable t such that z is a Fuchsian function of t and t, as a function of z, has no singular point on the surface R. All the Fuchsian groups corresponding to Riemann surfaces with the same connectivity are isomorphic.

Moreover, it is evident that the Fuchsian group is none other than the fundamental group g of the surface R, considered as a two-dimensional manifold.

We remark that not all Fuchsian groups arise in this way from a closed two-dimensional manifold. Consider the Fuchsian fundamental polygon R_0 and, if the Fuchsian function exists over the whole plane, it is necessary to adjoin its mirror image R_0' in the real axis; but then the domain R_0+R_0' will not be simply connected. To each point of the closed manifold V there corresponds a point of R_0 (or of $R_0 + R_0'$), and conversely. Suppose that there exists one or more cycles of vertices and that the sum of the angles of this cycle is zero or $\frac{2\pi}{n}$ where n is an integer greater than 1. Then let M be the point of V corresponding to this cycle of vertices and imagine an infinitely small loop around M. By definition of the group g, this loop must correspond, in g, to the identity substitution, but in the Fuchsian group it corresponds to a nonidentity substitution. Thus the Fuchsian group cannot be isomorphic to g.

This leaves Fuchsian groups of the first family such that the sum of the angles of each cycle is 2π, and those of the third family. But the latter must likewise be rejected. In fact, if the group is of the third family then the domain $R_0 + R'_0$ is not simply connected. Let C be a closed contour traced in this domain such that we do not have

$$C \sim 0.$$

Corresponding to this contour, in the Fuchsian group, we have the identity substitution (because the variable z returns to its point of departure). But in the group g we have a nonidentity substitution. Here again, the Fuchsian group cannot be isomorphic to g.

Thus we are left with the groups of the first family for which the sum of the angles of each cycle is 2π.

All of the groups of the same genus are isomorphic, and it is for this reason that all the closed two-dimensional manifolds with the same Betti number are homeomorphic.

Is it the same when the number of dimensions is greater? Are two closed manifolds of dimension $h > 2$ with the same Betti numbers homeomorphic?

We shall see that they are not, and that this is why questions of *analysis situs* become more complicated when the number of dimensions increases.

It is clear first of all that, if two manifolds are homeomorphic, then their groups are isomorphic.

Now we return to our sixth example and enquire whether two of the groups $(\alpha, \beta, \gamma, \delta)$ can be isomorphic.

Let $(\alpha, \beta, \gamma, \delta)$ and $(\alpha', \beta', \gamma', \delta')$ be the two groups, and let C_1, C_2, C_3; C'_1, C'_2, C'_3 be the three fundamental contours of each, so

$$(1) \qquad \begin{cases} C_1 + C_2 & \equiv & C_2 + C_1 \\ C_1 + C_3 & \equiv & C_3 + \alpha C_1 + \gamma C_2 \\ C_2 + C_3 & \equiv & C_3 + \beta C_1 + \delta C_2 \end{cases}$$

$$(1') \qquad \begin{cases} C'_1 + C'_2 & \equiv & C'_2 + C'_1 \\ C'_1 + C'_3 & \equiv & C'_3 + \alpha' C'_1 + \gamma' C'_2 \\ C'_2 + C'_3 & \equiv & C'_3 + \beta' C'_1 + \delta' C'_2 \end{cases}$$

are the fundamental equivalences of the two groups.

Suppose that the two groups are isomorphic and let

$$a_3 C_3 + a_1 C_1 + a_2 C_2,$$

$$b_3 C_3 + b_1 C_1 + b_2 C_2,$$

$$c_3 C_3 + c_1 C_1 + c_2 C_2$$

be the contours of the first group that correspond, respectively, to the contours C'_1, C'_2, C'_3 of the second group. The a, b, and c are integers, and we have seen above that each contour of the first group can be expressed in this form.

If we are to have an isomorphism we must recover the equivalences (1) when $a_3 C_3 + a_1 C_1 + a_2 C_2, \ldots$ are substituted in place of C'_1, C'_2, C'_3 in the equivalences $(1')$.

Thus it is necessary, first of all, that the substitutions (which I denote by the same symbols as the corresponding contours)

$$a_3 C_3 + a_1 C_1 + a_2 C_2,$$

$$b_3C_3 + b_1C_1 + b_2C_2$$

commute. To make writing simpler, I shall use the following notation: I put

$$a_3 = h,\ b_3 = k,\ a_1C_1 + a_2C_2 = S_0; \quad b_1C_1 + b_2C_2 = T_0,$$

so that our first two substitutions reduce to $hC_3 + S_0$, $kC_3 + T_0$.

I let S_1 denote the result of applying the linear substitution $(\alpha, \beta, \gamma, \delta)$ to the coefficients of S_0, that is, replacing a_1 and a_2 by

$$\alpha a_1 + \beta a_2 \text{ and } \gamma a_1 + \delta a_2.$$

S_2 will be the result of applying the same substitution to the coefficients of S_1, and so on; likewise, $T_1, T_2, \ldots$ are the successive transforms of T_0.

That being so, the equivalences (1) give

$$S_0 + hC_3 \equiv hC_3 + S_h.$$

For our substitutions to commute, it is therefore necessary that

$$hC_3 + S_0 + kC_3 + T_0 \equiv kC_3 + T_0 + hC_3 + S_0$$

or

$$(k+h)C_3 + S_k + T_0 \equiv (k+h)C_3 + T_h + S_0$$

or

$$(\lambda) \qquad S_k + T_0 \equiv T_h + S_0.$$

Suppose first of all that h and k are equal and not 0. The preceding equivalence can then be replaced by the equations

$$(\alpha_h - 1)(a_1 - b_1) + \beta_h(a_2 - b_2) = 0,$$

$$\gamma_h(a_1 - b_1) + (\delta_h - 1)(a_2 - b_2) = 0,$$

where I let

$$\begin{pmatrix} \alpha_h & \beta_h \\ \gamma_h & \delta_h \end{pmatrix}$$

denote the coefficients of the hth power of the linear substitution

$$\begin{pmatrix} \alpha & \beta \\ \gamma & \delta \end{pmatrix}.$$

These equations can be satisfied in two ways:

1° Certainly if

$$a_1 = b_1, \quad a_2 = b_2,$$

in which case the two substitutions

$$hC_3 + S_0, \quad kC_3 + T_0$$

will be identical.

2° Or else, if the determinant

$$\begin{vmatrix} \alpha_h - 1 & \beta_h \\ \gamma_h & \delta_h - 1 \end{vmatrix}$$

is zero. But the latter cannot happen unless we have

$$s^h = 1,$$

where s is one of the roots of the equation

$$\begin{vmatrix} \alpha - s & \beta \\ \gamma & \delta - s \end{vmatrix} = 0, \tag{2}$$

which means, if the substitution $(\alpha, \beta, \gamma, \delta)$ is *elliptic*, that the roots of equation (2) for s are imaginary (in which case each must equal an hth root of unity) and if it is *parabolic* it means they are equal.

Suppose then that $(\alpha, \beta, \gamma, \delta)$ is *hyperbolic*, which means that the roots of the equation for s are real, and suppose no longer that $h = k$.

Then the kth power of

$$hC_3 + S_0$$

and the hth power of

$$kC_3 + T_0$$

must commute. Let

$$h'C_3 + S_0',$$
$$K'C_3 + T_0'$$

be these two powers. Since we have

$$h' = k' = hk,$$

these two powers must be identical, by what we have said above. In the isomorphic group $(\alpha', \beta', \gamma', \delta')$ the kth power of C_1' must be equal to the hth power of C_2'. This cannot be so unless

$$h = k = 0,$$

that is,

$$a_3 = b_3 = 0.$$

We pass to the case of elliptic substitutions, among which we include the substitution

$$\begin{pmatrix} -1 & 0 \\ 0 & -1 \end{pmatrix}.$$

Then the determinant

$$\begin{vmatrix} \alpha_h - 1 & \beta_h \\ \gamma_h & \delta_h - 1 \end{vmatrix}$$

can vanish for a certain value of h I call ν.

But then the substitution

$$\begin{pmatrix} \alpha_\nu & \beta_\nu \\ \gamma_\nu & \delta_\nu \end{pmatrix}$$

reduces to the identity substitution, that is, we have

$$\alpha_\nu = \delta_\nu = 1, \quad \beta_\nu = \gamma_\nu = 0.$$

Then it happens that νC_3 commutes with C_1 and C_2, and more generally the two substitutions

$$a_3C_3 + a_1C_1 + a_2C_2,$$
$$b_3C_3 + b_1C_1 + b_2C_2$$

commute, provided a_3 and b_3 are divisible by ν. The latter sufficient condition is also necessary, if we exclude, as above, the case where the powers of our two substitutions are identical.

To show this, I shall again employ an abbreviated notation. The symbol

$$m_1C_1 + m_2C_2$$

makes no sense unless m_1 and m_2 are integers, but the following

$$\mu(a_1'C_1 + a_2'C_2) + \rho(b_1'C_1 + b_2'C_2)$$

can be given a meaning when μ, ρ, the a' and the b' are not integers, provided

$$\mu a_1' + \rho b_1', \quad \mu a_2' + \rho b_2'$$

are integers. It is evident that we can permit this, because C_1 and C_2 commute.

So choose

$$\xi_0 = a_1'C_1 + a_2'C_2,$$
$$\eta_0 = b_1'C_1 + b_2'C_2$$

so that

$$\xi_1 = \xi_0 s, \quad \eta_1 = \eta_0 s^{-1},$$

where ξ_1 and η_1 are, following our conventions, the transforms of ξ_0 and η_0 by the linear substitution $(\alpha, \beta, \gamma, \delta)$; s is one of the roots of equation (2). It follows from the theory of linear substitutions that we can always choose the numbers a' and b' (which are generally irrational or even imaginary) in this fashion. Then we put

$$S_0 = \mu\xi_0 + \rho\eta_0, \quad T_0 = \mu'\xi_0 + \rho'\eta_0.$$

We have

$$\mu\xi_0 + \rho\eta_0 + hC_3 \equiv hC_3 + \mu s^h\xi_0 + \rho s^{-h}\eta_0$$

and

$$k(hC_3 + \mu\xi_0 + \rho\eta_0) \equiv khC_3 + \mu\frac{s^{kh}-1}{s^h-1}\xi_0 + \rho\frac{s^{-kh}-1}{s^{-h}-1}\eta_0.$$

That being so, the equivalence (λ) can be written

$$\mu(s^k - 1) = \mu'(s^h - 1),$$
$$\rho(s^{-k} - 1) = \rho'(s^{-h} - 1).$$

If k and h are not divisible by ν, these relations are not satisfied identically, and we can put

$$\mu = (s^h - 1)\varepsilon, \quad \mu' = (s^k - 1)\varepsilon, \quad \rho = (s^{-h} - 1)\zeta, \quad \rho' = (s^{-k} - 1)\zeta.$$

But then

$$k(hC_3 + S_0) \equiv khC_3 + \varepsilon(s^{kh} - 1)\xi_0 + \zeta(s^{-kh} - 1)\eta_0,$$
$$h(kC_3 + T_0) \equiv khC_3 + \varepsilon(s^{kh} - 1)\xi_0 + \zeta(s^{-kh} - 1)\eta_0,$$

which shows that the kth power of our first substitution is identical with the hth power of the second. Since we have excluded this case, it follows that we must have

$$a_3 \equiv b_3 \equiv 0 \pmod{\nu}.$$

But we can go further: if C_1' and C_2' are our first two substitutions, we can replace them by

$$\omega_1C_1' + \omega_2C_2', \quad \omega_1'C_1' + \omega_2'C_2',$$

where the ω are any integers such that $\omega_1\omega_2' - \omega_2\omega_1' = 1$. We can then always suppose that $b_3 = 0$ (by replacing C_1' and C_2' by $\omega_1 C_1' + \omega_2 C_2'$ and $\omega_1' C_1' + \omega_2' C_2'$ and choosing the numbers ω so as to make the new b_3 vanish).

But now the subgroup generated by C_1' and C_2', thanks to the isomorphism with $(\alpha', \beta', \gamma', \delta')$, must commute with all substitutions in the group, in particular, with C_3.

We can therefore say that the substitution

$$-C_3 + C_2' + C_3$$

belongs to the subgroup generated by C_1' and C_2'. We have

$$C_2' = \mu'\xi_0 + \rho'\eta_0,$$
$$-C_3 + C_2' + C_3 = \mu' s\xi_0 + \rho' s^{-1}\eta_0.$$

For the latter substitution to belong to the group it is necessary, if a_3 is nonzero, for it to be a multiple of C_2'.

But it is evident that this can happen only if $s = s^{-1} = -1$.

If we leave this case aside, we must have

$$a_3 = b_3 = 0.$$

Then if we leave aside the case where

$$\alpha + \delta = \pm 2$$

(parabolic substitutions and the substitution $(-1, 0, 0, -1)$), we must have

$$a_3 = b_3 = 0.$$

I add that c_3 must be equal to 1, otherwise the combinations of the three fundamental substitutions

$$\begin{aligned} a_1 C_1 + a_2 C_2 &\equiv C_1', \\ b_1 C_1 + b_2 C_2 &\equiv C_2', \\ c_3 C_3 + c_1 C_1 + c_2 C_2 &\equiv C_3', \end{aligned}$$

could not generate all the substitutions

$$m_3 C_3 + m_1 C_1 + m_2 C_2$$

in the group $(\alpha, \beta, \gamma, \delta)$, but only those where the integer m_3 is divisible by c_3.

Now each substitution in group $(\alpha, \beta, \gamma, \delta)$ can be put in the form

$$m_3 C_3' + m_1 C_1' + m_2 C_2'.$$

For C_1 to be expressed in this form it is first of all necessary that m_3 be zero, and then that we have identically

$$C_1 = m_1(a_1 C_1 + a_2 C_2) + m_2(b_1 C_1 + b_2 C_2),$$

and likewise that

$$C_2 = m_1'(a_1 C_1 + a_2 C_2) + m_2'(b_1 C_1 + b_2 C_2).$$

Since the m and m' are integers, it follows that the determinant

$$a_1 b_2 - a_2 b_1 = 1.$$

But I have said above that we can replace C_1' and C_2' by

$$\omega_1 C_1' + \omega_2 C_2', \quad \omega_1' C_1' + \omega_2' C_2'.$$

If $a_1b_2 - a_2b_1$ equals 1 we can choose the ω in such a way that

$$\omega_1 C_1' + \omega_2 C_2' = C_1, \quad \omega_1' C_1' + \omega_2' C_2' = C_2,$$

that is, we can always suppose that

$$a_1 = b_2 = 1, \quad a_2 = b_1 = 0.$$

But then the equivalence (1′)

$$C_1' + C_3' \equiv C_3' + \alpha' C_1' + \gamma' C_2'$$

becomes

$$C_1 + C_3 \equiv C_3 + \alpha' C_1 + \gamma' C_2,$$

whence

$$\alpha = \alpha', \quad \gamma = \gamma'.$$

We similarly find that

$$\beta = \beta', \quad \delta = \delta'.$$

We must therefore conclude that the two groups $(\alpha, \beta, \gamma, \delta)$ and $(\alpha', \beta', \gamma', \delta')$ cannot be isomorphic unless we can pass from one to the other by changing C_1' and C_2' into

$$\omega_1 C_1' + \omega_2 C_2', \quad \omega_1' C_1' + \omega_2' C_2'.$$

This can be stated in another way.

Let

$$S = \begin{pmatrix} \alpha & \beta \\ \gamma & \delta \end{pmatrix}$$

be a linear substitution with integer coefficients such that

$$\alpha\delta - \beta\gamma = 1.$$

Let

$$T = \begin{pmatrix} \omega_1 & \omega_2 \\ \omega_1' & \omega_2' \end{pmatrix}$$

be another linear substitution with integer coefficients such that

$$\omega_1 \omega_2' - \omega_1' \omega_2 = 1.$$

The substitution $T^{-1}ST$, which is called the transform[12] of S by T, is also linear with integer coefficients and of determinant 1.

If two linear substitutions S and S' with integer coefficients and determinant 1 are transformed into each other by a substitution T, I say that S and S' belong to the same class.

It is clear first of all that S and S' cannot belong to the same class unless the sum $\alpha + \delta$ has the same value for each, but this condition is not sufficient, and $\alpha + \delta$ can have the same value for several classes of linear substitutions, just as the same determinant can correspond to several classes of quadratic forms.

Thus the necessary and sufficient condition for the two groups $(\alpha, \beta, \gamma, \delta)$ and $(\alpha', \beta', \gamma', \delta')$ to be isomorphic is that the two linear substitutions $(\alpha, \beta, \gamma, \delta)$ and $(\alpha', \beta', \gamma', \delta')$ belong to the same class.

---

[12]It would be nice to call this substitution the *conjugate* of S by T, to avoid confusion with Poincaré's previous use of the word "transform," but of course this would clash with Poincaré's use of the word "conjugate" to denote the pairing between identified edges of faces. (Translator's note).

We have so far left aside the case where

$$\alpha + \delta = \pm 2.$$

If $\alpha + \delta = 2$, the substitution $(\alpha, \beta, \gamma, \delta)$ will be in the same class as $(1, h, 0, 1)$; the group $(\alpha, \beta, \gamma, \delta)$ will be isomorphic to $(1, h, 0, 1)$.

The latter contains a remarkable substitution C_2, which is not a multiple of any other and which commutes with all substitutions in the group. Moreover, we see without difficulty that if h is nonzero, C_2 is the only substitution with this property.

We can also leave aside the case where h is zero, since the group $(1, 0, 0, 1)$, whose substitutions all commute with each other, evidently cannot be isomorphic to any other group $(\alpha, \beta, \gamma, \delta)$.

If $\alpha + \delta = -2$, the group $(\alpha, \beta, \gamma, \delta)$ will be isomorphic to $(-1, h, 0, -1)$. The latter contains a remarkable substitution C_2, which is not a multiple of any other, does not commute with all substitutions in the group, but whose *double* commutes with all these substitutions. If h is not zero, C_2 is the only substitution that enjoys the latter property. If, on the other hand, h is zero, then there are infinitely many such elements, which proves already that $(-1, 0, 0, -1)$ cannot be isomorphic to $(-1, h, 0, -1)$.

Likewise, the presence of this remarkable substitution C_2 in $(1, h, 0, 1)$ and the absence of any substitution enjoying the same property in $(-1, h', 0, -1)$ shows that these two groups cannot be isomorphic.

It remains to resolve two questions:

1° Can the groups $(1, h, 0, 1)$ and $(1, h', 0, 1)$, where $|h| \neq 0$, $|h'| \neq 0$, $|h| \neq |h'|$, be isomorphic?

2° The same question for the groups $(-1, h, 0, -1)$, $(-1, h', 0, -1)$.

I shall begin with the first question.

Observe first of all that, since C_2 is the only substitution in the first group with the characteristic property stated above, we must have

$$C_2' \equiv C_2$$

(or $C_2' \equiv -C_2$, but then we change C_1' and C_2' into $-C_1'$ and $-C_2'$).

Then, in order to recover all substitutions in the first group by combining

$$\begin{aligned} C_2' &\equiv a_3 C_3 + a_1 C_1 + a_2 C_2, \\ C_2' &\equiv C_2, \\ C_3' &\equiv c_3 C_3 + c_1 C_1 + c_2 C_2, \end{aligned}$$

it is necessary that

$$a_3 c_1 - c_3 a_1 = 1.$$

One then proves easily that

$$C_1' + C_3' \equiv C_3' + C_1' + h C_2',$$

which shows that the two groups cannot be isomorphic unless

$$h = \pm h'.$$

We pass to the second question.

We know, from a moment ago, that we must have

$$C'_2 \equiv C_2,$$

$$a_3c_1 - c_3a_1 = 1.$$

It follows that

$$C'_1 + C'_3 \equiv (a_3 + c_3)C_3 + (c_1 + a_1\varepsilon)C_1 + (c_2 + a_2\varepsilon - a_1c_3h\varepsilon)C_2,$$

where

$$\varepsilon = (-1)^{c_3}$$

and

$$C'_3 - C'_1 \equiv (c_3 - a_3)C_3 + (c_1 - a_1)\varepsilon' C_1 + [(c_2 - a_2)\varepsilon' + (a_1 + c_1)a_3\varepsilon' h]C_2,$$

where

$$\varepsilon' = (-1)^{a_3}.$$

If we want

$$C'_1 + C'_2 \equiv C'_3 - C'_1 + h'C'_2,$$

then we need

$$a_3 = 0, \quad \varepsilon' = 1, \quad \varepsilon = -1, \quad a_1c_3 = 1, \quad h = h'.$$

Therefore our two groups cannot be isomorphic unless

$$h = \pm h'.$$

The result stated above is therefore general and it extends to parabolic substitutions.

If the two linear substitutions $(\alpha, \beta, \gamma, \delta)$, $(\alpha', \beta', \gamma', \delta')$ are not in the same class, then the two corresponding groups cannot be isomorphic.

The result of this long discussion is that the different groups $(\alpha, \beta, \gamma, \delta)$ give rise to an infinity of different—that is, nonhomeomorphic—closed manifolds V. However, the number P_1 can take only one of the three values 2, 3 or 4.

Thus for two closed manifolds to be homeomorphic, it does not suffice for them to have the same Betti numbers.

This is shown equally well by our other examples.

In the third example, the group G reduces to eight substitutions and, in the fifth example, to only two.

On the other hand, the hypersphere

$$x_1^2 + x_2^2 + x_3^2 + x_4^2 = 1$$

is a manifold whose group G consists of only a single substitution, the identity.

Thus we have three manifolds whose groups are of finite order, but nonisomorphic, so the manifolds cannot be homeomorphic. Nevertheless, they have the same Betti numbers

$$P_1 = P_2 = 1.$$

It seems natural to restrict the meaning of the term *simply connected* to manifolds whose group G reduces to a single substitution. Then a closed manifold of more than two dimensions can have a group G of finite order without being simply connected.

This does not happen with two-dimensional manifolds: the group G of such a manifold cannot be finite without reducing to a single substitution.[13]

We have seen why a Fuchsian group which admits a vertex cycle with angle sum $\frac{2\pi}{n}$ $(n > 1)$ cannot be the group G of a closed two-dimensional manifold.

For the same reason, the group of such a manifold cannot be of finite order, or, more generally, it cannot contain a substitution of finite order.

It would be interesting to treat the following questions.

1° Given a group G defined by a certain number of fundamental equivalences, can it give rise to a closed n-dimensional manifold?

2° How may we construct such a manifold?

3° Are two manifolds of the same number of dimensions, and with the same group G, always homeomorphic?

These questions will demand lengthy and difficult study. I shall not say more about them here.

However, I want to draw attention to one point.

Riemann has related the study of algebraic curves to that of two-dimensional manifolds from the point of view of *analysis situs*.

Similarly, the study of algebraic surfaces is related to that of four-dimensional manifolds. These manifolds have three Betti numbers:

$$P_1 = P_3 \text{ and } P_2.$$

Picard has shown that, if the algebraic surface is the most general of its degree, then the number P_1 reduces to 1; it takes a value greater than 1 only in certain special cases. Thus multiple connectivity presents itself as a special case of simple connectivity.

This result seems paradoxical, but it is a little less paradoxical to us now; a group G can be much more complex than another group G' and still have a smaller value for the number P_1.

§15. Other modes of generation

It is possible to present manifolds by other definitions which are, so to speak, intermediate between the first two.

For example, we can view a manifold V as the set of points satisfying certain inequalities and the equations

$$\left.\begin{cases} F_1(x_1, x_2, \ldots, x_n; y_1, y_2, \ldots, y_q) \\ F_2(x_1, x_2, \ldots, x_n; y_1, y_2, \ldots, y_q) \\ \ldots\ldots\ldots\ldots\ldots\ldots\ldots\ldots \\ F_p(x_1, x_2, \ldots, x_n; y_1, y_2, \ldots, y_q) \end{cases}\right\} = 0, \tag{1}$$

where the x are the coordinates of a point and the y are arbitrary parameters. The dimension of V is then $n - p + q$.

We can adjoin to the equations (1) the relations

$$\varphi_\alpha(y_1, y_2, \ldots, y_q) = 0 \quad (\alpha = 1, 2, \ldots, \lambda),$$

between the y. The y are then still entirely arbitrary parameters and the dimension of V becomes equal to $n - p + q - \lambda$.

[13]Poincaré is evidently considering only orientable manifolds, since the projective plane is a counterexample among nonorientable two-dimensional manifolds. (Translator's note.)

Again, we can view a manifold V as defined by certain inequalities and by the relations

$$x_i = \theta_i(y_1, y_2, \ldots, y_p) \quad (i = 1, 2, \ldots, n), \tag{2}$$

where the y are parameters connected by λ equations

$$\varphi_\alpha(y_1, y_2, \ldots, y_p) = 0 \quad (\alpha = 1, 2, \ldots, \lambda).$$

The dimension of V is then $p - \lambda$.

Suppose for a moment that $y_1, y_2, \ldots, y_p$ are the coordinates of a point P in a space of p dimensions. The equations

$$\varphi_\alpha = 0$$

then define a certain manifold W in that p-dimensional space.

Each point of W corresponds to a point of V, because the equations (2) express the x as functions of the y.

The simplest case is that where, conversely, each point of V corresponds to a single point of W. But another case, also very interesting, is the following.

Suppose that the manifold W remains unaltered when the y undergo substitutions from a certain group G. Let P be a point of W, and let $P_1, P_2, \ldots, P_h$ be the images of P under the substitutions in G.

The points $P, P_1, P_2, \ldots, P_h$ form what I call a *system of points.*

If the functions θ_i are not altered by the substitutions in G, it is clear that the various points in the same system correspond to the same point of V.

The interesting case is where a point of V corresponds to a single system of points of W.

Given a manifold W and a group G which does not alter it, we can always construct a manifold V in such a way that each point of V corresponds to a system of points of W, and exactly one.

For V to be orientable, it is necessary and sufficient that W be orientable and that all substitutions in G have the following property.

Let $y_1, y_2, \ldots, y_p$ be the coordinates of P, and let $y'_1, y'_2, \ldots, y'_p$ be those of its image; then the Jacobian of the y' with respect to the y must be positive.

Seventh example. Let

$$y_1^2 + y_2^2 + y_3^2 = 1$$

be the equation of the manifold W, which is therefore a sphere in ordinary space.

This sphere is not altered when we change y_1, y_2, y_3 to $-y_1, -y_2, -y_3$. This will be our group G.

Then if we put, for example,

$$x_1 = y_1^2, \quad x_2 = y_2^2, \quad x_3 = y_3^2,$$

$$x_4 = y_2 y_3, \quad x_5 = y_1 y_3, \quad x_6 = y_1 y_2,$$

the x do not change when the y change their sign, and we have defined a two-dimensional manifold V in six-dimensional space.

This manifold will be closed; it will be *nonorientable.*

In fact, let P be a point of W, with coordinates y_1, y_2, y_3. To define the position of this point on the sphere W it suffices to know two of the coordinates, for example y_1 and y_2, since the equation of the sphere gives us y_3 as a function of y_1 and y_2.

Its image P' with coordinates $-y_1, -y_2$ and $-y_3$ is diametrically opposite. But now it does not suffice to define the position of the point P from y_1 and y_2, because y_3 is not a uniform function of these two variables. Rather we must put

$$\begin{aligned} y_1 &= \cos\varphi \sin\theta, \\ y_2 &= \sin\varphi \sin\theta, \\ y_3 &= \cos\theta. \end{aligned}$$

The coordinates of the point P in the new system are φ and θ, those of P' are $\varphi + \pi$ and $\pi - \theta$, and now we see that

$$\frac{\partial(\varphi + \pi, \pi - \theta)}{\partial(\varphi, \theta)} = -1 < 0.$$

The manifold V is therefore nonorientable.

Eighth example. Let $y_1, y_2, \ldots, y_q; z_1, z_2, \ldots, z_q$ be $2q$ parameters connected by the relations

$$(3) \qquad \begin{cases} y_1^2 + y_2^2 + \cdots + y_q^2 &= 1, \\ z_1^2 + z_2^2 + \cdots + z_q^2 &= 1. \end{cases}$$

If we regard these $2q$ parameters as the coordinates of a point in $2q$-dimensional space, then the equations (3) represent a closed manifold W of $2q - 2$ dimensions.

If we regard $y_1, y_2, \ldots, y_q$ and $z_1, z_2, \ldots, z_q$ as the coordinates of two points Q and Q' in q-dimensional space, these two points will both be found on the *hypersphere* S with equation

$$y_1^2 + y_2^2 + \cdots + y_q^2 = 1,$$

which is a closed manifold of $q - 1$ dimensions.

Thus each pair of points of S corresponds to a single point of W, and conversely, *provided we regard the two pairs QQ' and $Q'Q$ as distinct.*

Now consider the $\frac{q(q+3)}{2}$ combinations

$$y_i + z_i, \quad y_i z_i, \quad y_i z_k + y_k z_i \qquad (i, k = 1, 2, \cdots, q),$$

and set them equal to the $\frac{q(q+3)}{2} = n$ variables

$$x_1, \quad x_2, \quad \ldots, \quad x_n.$$

We have now defined a manifold V of $2q-2$ dimensions in n-dimensional space.

When we change y_i into z_i and z_i into y_i (that is, when we permute the two points Q and Q'), the $\frac{q(q+3)}{2}$ combinations do not change.

Thus each pair of points on the hypersphere corresponds to a single point on the manifold V, and conversely, but *under the condition that the two pairs QQ' and $Q'Q$ not be considered distinct.*

Is this manifold V closed? I wish to show that it is not for $q = 2$, but it is for $q > 2$.

In the former case we have

$$x_1 = y_1 + z_1, \quad x_2 = y_1 z_1, \quad x_3 = y_2 + z_2,$$
$$x_4 = y_2 z_2, \quad x_5 = y_1 z_2 + y_2 z_1.$$

Thus for the y and z to be real we must have

$$x_1^2 > 4x_2, \quad x_3^2 > 4x_4.$$

We have analogous inequalities for $q > 2$; but in the latter case do the inequalities define a true boundary of our manifold V?

To make the situation clearer, I want to treat a simpler example first.

Suppose that, in ordinary space, we have the circle

$$x^2 + y^2 = 1, \quad z = 0.$$

If we confine ourselves to the points of the circle for which y is positive, then we have the following relations:

$$x^2 + y^2 = 1, \quad z = 0, \quad y > 0,$$

which define a one-dimensional manifold (in this case, a semicircle).

This manifold is not closed; it has two boundary points:

$$x = \pm 1, \quad y = z = 0.$$

On the other hand, consider the following surface:

$$x^2 + y^2 - z^2 + (x^2 + y^2 + z^2)^2 = 0. \tag{4}$$

It is the surface generated by revolving a lemniscate about its major axis. It consists of two distinct sheets N_1 and N_2 with a common conical point, which is the origin. One passes from one sheet to the other through the origin. Thus if we adjoin to equation (4) the inequality

$$z > 0, \tag{5}$$

the relations (4) and (5) define a two-dimensional manifold which is none other than the sheet N_1. The latter manifold can be regarded as closed; it is homeomorphic to a sphere, and there is no need to regard the conical point

$$x = y = z = 0$$

as a true boundary point.

In general, if a p-dimensional manifold is not closed, its boundary will consist of one or more $(p-1)$-dimensional manifolds. If the set of points we suspect to be boundary points forms one or more manifolds of less than $p-1$ dimensions, then they cannot be true boundary points, and the given manifold is closed.

But in our case we obtain the suspected boundary points by supposing the points Q and Q' to become equal, that is

$$y_1 = z_1, \quad y_2 = z_2, \quad \ldots, \quad y_q = z_q.$$

This gives a manifold of $q-1$ dimensions. Thus the boundary of V will have $q-1$ dimensions, whereas V has $2q-2$. V will therefore be closed unless

$$2q - 2 = (q-1) + 1 \text{ or } q = 2.$$

To understand this better, we compare the two examples $q = 2$ and $q = 3$.

First suppose $q = 2$ and consider our manifold in the neighborhood of the point

$$y_1 = z_1 = 0, \quad y_2 = z_2 = 1,$$

that is, the point

$$x_1 = 0, \quad x_2 = 0, \quad x_3 = 2, \quad x_4 = 1, \quad x_5 = 0.$$

We notice that, for small values of x_1 and x_2, the other three variables x_3, x_4 and x_5 can be expanded in powers of x_1 and x_2; it therefore suffices to study the variations of x_1 and x_2.

Next we see that x_1 and x_2 can take all values such that

$$x_1^2 > 4x_2.$$

The (x_1, x_2)-plane is therefore divided into two regions by the line

$$x_1^2 = 4x_2,$$

which is a true boundary line.

We obtain the same result if we study the manifold V in the neighborhood of any other boundary point. This manifold is therefore not closed.

Now suppose $q = 3$ and let

$$x_1 = y_1 + z_1, \quad x_2 = y_1 z_1, \quad x_3 = y_2 + z_2, \quad x_4 = y_2 z_2, \quad x_5 = y_1 z_2 + y_2 z_1,$$

$$x_6 = y_3 + z_3, \quad x_7 = y_3 z_3, \quad x_8 = y_1 z_3 + y_3 z_1, \quad x_9 = y_2 z_3 + y_3 z_2.$$

We study the manifold V in the neighborhood of the point P_0, which is such that

$$y_1 = z_1 = y_2 = z_2 = 0, \quad y_3 = z_3 = 1,$$

whence

$$x_1 = x_2 = x_3 = x_4 = x_5 = x_8 = x_9 = 0, \quad x_6 = 2, \quad x_7 = 1.$$

We see that, in the neighborhood of this point, x_6, x_7, x_8, x_9 can be expanded in powers of x_1, x_2, x_3, x_4, and x_5, so that it suffices to study the variations of the latter five variables.

In order to have only three variables, and make a geometric representation possible, I cut my manifold V by the plane manifold

$$x_1 = 0, \quad x_3 = 0,$$

so that the intersection will be a two-dimensional manifold V'.

Let x_2, x_4, and x_5 be the coordinates of a point of V'. We can regard these three variables as the coordinates of a point in ordinary space, and thus we have a geometric representation of the manifold V', or rather the portion of this manifold in the neighborhood of P_0.

We then find

$$y_1 = -z_1, \quad y_2 = -z_2,$$

because x_1 and x_3 are assumed to be zero and therefore

$$x_2 = -y_1^2, \quad x_4 = -y_2^2, \quad x_6 = -2y_1 y_2,$$

whence

$$4x_2 x_4 - x_5^2 = 0.$$

The latter equation is a cone of second degree, but only a single sheet of this cone appears, since we must have

$$x_2 < 0, \quad x_4 < 0.$$

The portion of the cone that appears is therefore separated from the portion that does not by the vertex, which cannot be regarded as a true boundary point. Thus the manifold V', and likewise V, is again closed.

We obtain a similar result by studying V in the neighborhood of another boundary point, or when we cut V by other plane manifolds.

I could not have asked for an example better suited to clarify the preceding reasoning.

In summary, *the manifold V is closed if $q > 2$ and not if $q = 2$.*

Now, is the manifold V orientable or nonorientable?

I propose to show that it is orientable if q is odd and nonorientable if q is even.

We put

$$\begin{array}{ll}
y_1 = \cos\theta_1, & z_2 = \cos\theta_1', \\
y_2 = \sin\theta_1 \cos\theta_2, & z_2 = \sin\theta_2', \\
y_3 = \sin\theta_1 \sin\theta_2 \cos\theta_3, & z_3 = \sin\theta_1' \sin\theta_2' \sin\theta_3', \\
\cdots\cdots\cdots & \cdots\cdots\cdots \\
y_{q-1} = \sin\theta_1 \sin\theta_2 \cdots \sin\theta_{q-2} \cos\theta_{q-1}, & z_{q-1} = \sin\theta_1' \sin\theta_2' \cdots \sin\theta_{q-2}' \cos\theta_{q-1}', \\
y_q = \sin\theta_1 \sin\theta_2 \cdots \sin\theta_{q-2} \sin\theta_{q-1}, & z_q = \sin\theta_1' \sin\theta_2' \cdots \sin\theta_{q-2}' \sin\theta_{q-1}'.
\end{array}$$

The position of a point on W is defined by the $2q-2$ coordinates

$$\theta_1, \theta_2, \ldots, \theta_{q-1}; \theta_1', \theta_2', \ldots, \theta_{q-1}'.$$

On the other hand, the group G consists (apart from the identity substitution) of the single substitution that exchanges θ_i with θ_i'. For the manifold V to be orientable, it is therefore necessary and sufficient for the Jacobian

$$\frac{\partial(\theta_i, \theta_i')}{\partial(\theta_i', \theta_i)} \quad (i = 1, 2, \cdots, q-1)$$

to be positive. But the latter is equal to $(-1)^{q-1}$, and hence it is $+1$ if q is odd and -1 if q is even. Therefore

V is orientable if q is odd,
V is nonorientable if q is even.

Q.E.D.

Now we shall be concerned with determining the Betti number, P_{q-1}.

We first determine the Betti numbers of W.

We can construct two $(q-1)$-dimensional manifolds on W in the following way. We know that each point of W corresponds to a pair of points QQ' on the hypersphere S. The pairs Q_0Q' for fixed Q_0 and Q' varying over the hypersphere therefore form a closed $(q-1)$-dimensional manifold U_1 which forms part of W. Similarly, the pairs QQ_0, where Q varies over the hypersphere and Q_0 is fixed, form a closed $(q-1)$-dimensional manifold U_3 which forms part of W.

These two manifolds are linearly independent (from the point of view of homology).

To see this, consider the integral of order $q-1$

$$J = \int \sin^{q-1}\theta_1 \sin^{q-2}\theta_2 \cdots \sin^2\theta_{q-2} \sin\theta_{q-1} d\theta_1 d\theta_2 \cdots d\theta_{q-1}$$

and let J' be the integral of order $q-1$ obtained from J by replacing the θ_i by the θ_i'.

Now consider the integral

$$J + \lambda J',$$

where λ is an irrational number.

We know the x as functions of the θ and θ'; conversely, we can calculate the θ and θ' as functions of the x, so that the integral $J + \lambda J'$ takes the form

$$\int \Sigma X \delta,$$

where X is an entire function of the x, and δ is the product of the $(q-1)$ differentials dx_i.

I observe, at the outset, that the integral $J + \lambda J'$ is zero when taken over an infinitely small manifold forming part of W.

When taken over U_2, it is equal to σ (surface of the hypersphere S); when taken over U_1, it is equal to $\lambda\sigma$; and since there is no linear relation between σ and $\lambda\sigma$, U_1 and U_2 are linearly independent.

The Betti number P_{q-1} therefore equals at least 3.

To go further, we consider the manifold W' obtained by removing from W all points of the two manifolds U_1 and U_2; I propose to show that W' is simply connected.

Consider, similarly, the manifold S' obtained by removing the point Q_0 from the hypersphere S. To each pair of points QQ' in the manifold S' there corresponds a point of W', and conversely.

But S' is homeomorphic to the domain D,

$$\eta_1^2 + \eta_2^2 + \cdots + \eta_{q-1}^2 < 1,$$

in $(q-1)$-dimensional space, where the coordinates are denoted by $\eta_1, \eta_2, \ldots, \eta_{q-1}$ (just as the surface of an ordinary sphere, with one point removed, is homeomorphic to the interior of a disc).

To each pair of points in the domain D there corresponds a point of W', whence it follows that W' is homeomorphic to the domain Δ defined by the inequalities

$$\eta_1^2 + \eta_2^2 + \cdots + \eta_{q-1}^2 < 1,$$

$$\zeta_1^2 + \zeta_2^2 + \cdots + \zeta_{q-1}^2 < 1$$

and situated in $(2q-2)$-dimensional space, where the coordinates are denoted by η_i and ζ_i.

But the domain Δ is simply connected because it is convex. In fact, let v be any closed manifold of any number of dimensions which forms part of Δ. I say that we can reduce it to a point by a continuous deformation without leaving V. We construct a manifold v' by multiplying the coordinates of all points of v by a positive factor k less than 1 (the manifold v' will be homothetic to v, with the center of homothety being the origin, and the homothety ratio being k). The manifold v' will be entirely inside Δ and, when k decreases from 1 to 0, it begins as v and ends as a single point. Q.E.D.

Thus Δ, and consequently W, is simply connected. We now determine the Betti numbers of W, beginning with the P_h for $h < q-1$. If we consider a closed h-dimensional manifold in W, it can always be taken as homologous to a closed h-dimensional manifold disjoint from U_1 and U_2. This means it will be homologous to zero in W' (because W' is simply connected) and *a fortiori* in W.

Thus P_h is equal to 1.

On the other hand, since the Betti numbers equally distant from the extremes are equal, P_h will also be equal to 1 when h is greater than $q-1$.

It remains to determine P_{q-1}.

Consider a closed $(q-1)$-dimensional manifold v contained in W.

If it is disjoint from U_1 and U_2 it will be homologous to zero; indeed, it will be part of W', hence homologous to zero in W' and *a fortiori* in W.

Suppose that it meets U_1 and U_2.

The points of W corresponding to pairs Q_1Q', where Q_1 is fixed and Q' describes the whole hypersphere, form a manifold U_1' homologous to U_1. But each point of W belongs to exactly one manifold U_1'.

The points of W corresponding to pairs QQ_2, where Q_2 is fixed and Q describes the whole hypersphere, form a manifold U_2' homologous to U_2.

Each point of W belongs to exactly one manifold U_2'.

Two manifolds U_1' and U_2' have exactly one point in common (namely, the point corresponding to the pair Q_1Q_2). On the other hand, two manifolds U_1' (or two manifolds U_2') are disjoint.

Having established this, we return to the manifold v and suppose, to fix ideas, that it meets U_1 in two points, M' and M'', and U_2 in one point N'. Through the point M' I then take a manifold U_2', through the point M'' an analogous manifold U_2'', and through the point N' a manifold U_1'.

Next, by a small deformation of the manifold v, I arrange that it does not meet U_1 and U_2 except at the points M', M'', and N', and that it has around M' a small part u_2' in common with U_2', around M'' a small part u_2'' in common with U_2'', and around N' a small part u_1' in common with U_1'.

Then the manifolds

$$U_1' - u_1', \quad U_2' - u_2', \quad U_2'' - u_2'', \quad -v + u_1' + u_2' + u_2'',$$

together form a closed manifold which meets neither U_1 nor U_2 and which will therefore be homologous to zero. We then have

$$U_1' - u_1' + U_2' - u_2' + U_2'' - u_2'' \sim v - u_1' - u_2' - u_2'',$$

whence

$$v \sim U_2' + U_1' + U_2'' \sim U_1 + 2U_2.$$

It could also happen, for example, that the number we have called $S(M)$ above does not have the same value for the point M' considered as the point of intersection of v and U_1 as for the point M' considered as the point of intersection of U_2' and U_1. In that case U_2' must be replaced by the opposite manifold, and we have

$$v \sim U_1' - U_2' + U_2'' \sim U_1.$$

In either case, v, U_1 and U_2 are not linearly independent, and we have

$$P_{q-1} = 3.$$

Finally, we determine the Betti numbers for V.

Pairs QQ' and $Q'Q$ correspond to the same point of V. It follows that U_1 and U_2 correspond to the same manifold in V, so I can write

$$U_1 \sim U_2.$$

On the other hand, the integral $J + J'$ is nonzero over that manifold, which shows that we do not have

$$U_1 \sim 0.$$

The number P_{q-1} therefore equals at least 2.

Each closed manifold v contained in V corresponds to a manifold w contained in W, but there are two cases. We know that to each point of V there correspond two points of W, and I shall say that these two points are *symmetric* because we pass from one to the other by exchanging the y_i with the z_i.

We construct w by taking, for each point of v, *one* of the two points corresponding to it. Then w may or may not be closed, but its boundary consists of two symmetric parts.

First consider the case where w is closed.

If the number of dimensions is different from $q-1$, then w (and consequently v) can be reduced to a point by continuous deformation and we have

$$v \sim 0.$$

If the number of dimensions equals $q-1$ we have, with respect to W,

$$w \sim mU_1 + nU_2,$$

where m and n are integers; but U_1 is homologous to U_2 with respect to V.

We therefore have

$$v \sim (m+n)U_1$$

with respect to V.

Now consider the case where w is not closed. We suppose that the number of dimensions is less than or equal to $q-1$; the case where the number is greater than $q-1$ follows easily, because the Betti numbers are equal in pairs. The boundary f of w therefore has less than $q-1$ dimensions.

Let H be the $(q-1)$-dimensional manifold, contained in W, which consists of the points symmetric to themselves, that is, the points

$$y_i = z_i.$$

The boundary f can be continuously deformed, without its points ceasing to be symmetric pairs, until the deformed boundary becomes part of H.

Then w can be continuously deformed, without ceasing to correspond to a closed manifold v, so that w becomes closed at the end of the deformation.

Thus v is always homologous to a manifold corresponding to a closed manifold w.

It follows that the possible cases are

$$v \sim 0, \quad v \sim (m+n)U_1.$$

Hence all the Betti numbers of V are equal to 1, except P_{q-1} which is equal to 2.

This has two consequences:

1° If q is odd, V will be orientable; whence it follows that, for an orientable manifold of $4k$ dimensions, the number P_{2k} is not necessarily odd.

2° If q is even, V will be nonorientable; whence it follows that, for a nonorientable manifold of $4k+2$ dimensions. The number P_{2k+1} is not necessarily odd, although it must be for an orientable manifold.

These are the results I announced at the end of §9.

§16. The theorem of Euler

We all know the theorem of Euler, according to which, if S, A and F are the numbers of vertices, edges and faces of a convex polyhedron,

$$S - A + F = 2.$$

This theorem has been generalized by M. de Jonquieres to nonconvex polyhedra. If a polyhedron forms a closed two-dimensional manifold with Betti number P_1, then we have

$$S - A + F = 3 - P_1.$$

The fact that the faces are planes is evidently of no importance; the theorem applies just as well to curvilinear polyhedra. It also applies to a subdivision of any closed surface into simply connected regions. These regions correspond to the faces of the polyhedron, their boundary lines correspond to the edges, and the extremities of these lines to the vertices.

I now propose to generalize these results to an arbitrary space.

Suppose then that V is a p-dimensional manifold. We subdivided it into a certain number of p-dimensional manifolds v_p; the manifolds v_p are not closed, and their boundaries consist of a certain number of $(p-1)$-dimensional manifolds v_{p-1}. The boundaries of the v_{p-1} in turn consist of a certain number of $(p-2)$-dimensional manifolds v_{p-2}, and so on; I finally arrive at a certain number of one-dimensional manifolds v_1, bounded by a certain number of isolated points or zero-dimensional manifolds I call v_0.

The manifold V can have arbitrary Betti numbers, but I assume that the manifolds $v_p, v_{p-1}, \dots, v_1$ are simply connected.

I let $\alpha_p, \alpha_{p-1}, \dots, \alpha_1$ and α_0 denote the numbers of the v_p, the $v_{p-1}, \dots$, the v_1 and the v_0.

The figure formed by all these manifolds may be called a polyhedron, since the analogy with ordinary polyhedra is evident. An ordinary polyhedron is in fact a closed two-dimensional manifold V, subdivided into a certain number of manifolds v_2, which are the faces. The faces are bounded by a certain number of manifolds v_1, which are the edges and which are bounded by a certain number of manifolds v_0 called *vertices.*

I propose to calculate the number

$$N = \alpha_p - \alpha_{p-1} + \alpha_{p-2} - \cdots \mp \alpha_1 \pm \alpha_0.$$

Here I introduce some new terminology, not very well justified perhaps, but convenient.

If two polyhedra are obtained by subdividing the same manifold V, I shall say they are *congruent.*

Suppose now that the polyhedron P is formed from the manifold V, by regions v_p and their successive boundaries $v_{p-1}, \dots, v_1, v_0$.

If we subdivide the v_p into smaller regions v'_p, then the boundaries of the v'_p will consist of a certain number of new regions v''_{p-2}, together with the regions v'_{p-1} obtained by subdividing the v_{p-1}. The boundaries of the v'_{p-1} and the v''_{p-1} consist of a certain number of new regions v''_{p-2} together with the regions v'_{p-2} obtained by subdividing the v_{p-2}, and so on. We finally arrive at v'_1 and v''_1 whose boundaries consist of a certain number of new points v''_0, together with the points v_0.

Let P' be the polyhedron formed by the regions $v'_p, v'_{p-1}, v''_{p-1}, v'_{p-2}, v''_{p-2}, \dots,$ v'_1, v''_1, v'_0, v''_0.

I then say that the polyhedron P' is *derived* from the polyhedron P.

I shall clarify this definition with an example from ordinary geometry. Consider a regular tetrahedron T. In each of its faces I join each vertex to the midpoint of the opposite side. Each face is thereby decomposed into six triangles; altogether there are twenty-four triangles. The polyhedron with twenty-four triangles obtained in this way is *derived* from T.

Now let P and P' be two congruent polyhedra, that is, obtained from the same manifold V by two different decompositions. Then there always exists a polyhedron P'' derived from both P and P', and which we obtain by combining the two decompositions. Thus if we let v_p, v'_p and v''_p denote the subdivisions of V under the three decompositions corresponding to the three polyhedra P, P' and P'', the necessary and sufficient condition for two points to belong to the same region v''_p is that they both belong to the same region v_p and the same region v'_p.

I propose to establish that the number N is the same for two congruent polyhedra and, since we have seen that two congruent polyhedra have a common derived polyhedron, it suffices to show that the number N is the same for a polyhedron and all those derived from it.

If we consider one of the regions v_{p-1} in the polyhedron P, it always lies in exactly two regions v_p, which it separates from each other. On the other hand, a region v_{p-2} can lie in more than two regions v_p and more than two regions v_{p-1}. This is the case in ordinary polyhedra, where an edge always separates two faces, but a vertex in general lies in more than two faces and in more than two edges.

However, we do not exclude the case where one region v_{p-2} belongs to only two regions v_{p-1}. Thus for an ordinary polyhedron we do not exclude the case where the midpoint of an edge is regarded as a vertex and where the edge is, consequently, regarded as two juxtaposed edges.

The regions v_{p-2}, which do not lie in more than two regions v_{p-1}, will be called *singular*. Now let v_{p-2} be a singular region that lies in two regions v_{p-1} I call v'_{p-1} and v''_{p-1}. It is clear that v'_{p-1} will separate *the same two regions* v_p that are separated by v''_{p-1}, so v_{p-2} will also lie in no more than two regions v_p.

Similarly, I say that the manifold v_q is *singular* if it lies in only two manifolds v_{q+1}, in which case the two v_{q+1} containing v_q will lie in the same v_{q+2}, the same $v_{q+3}, \ldots$, the same v_p, and the suppression of v_q and mutual annexation of the two v_{q+1} will change nothing in the v_{q+2}, $v_{q+3}, \ldots, v_p$.

Now consider a manifold v_h. This manifold contains a certain number of manifolds v_{h-1}; if one of them is singular I say that the manifold v_h is *irregular*. In the contrary case it will be *regular*.

Consider then the polyhedron P with regions v_p, $v_{p-1}, \ldots$ and the derived polyhedron P' with regions v'_p, $v'_{p-1}, \ldots$. We shall try to reconstruct the polyhedron P from the polyhedron P'. Take two regions v'_p I call α and β. I suppose that they are separated from each other by a region v'_{p-1} I call γ. Consequently, they are contiguous and form parts of the same region v_p. (Since α and β are parts of the same region v_p, the region γ is nothing but a subdivision of one of the regions v_{p-1} that separate the regions v_p from each other; γ is therefore one of the regions I have called v''_{p-1} in the definition of derived polyhedron, but here I do not make the distinction and I denote both the manifolds I previously called v''_{p-1} and those I previously called v'_{p-1} by the same notation v'_{p-1}.)

This being so, we suppress the region γ which bounds α and β and *annex* the region α to the region β. We have therefore suppressed one region v'_p and one region v'_{p-1}. On the other hand, we have not suppressed any region v'_{p-2} if γ is regular. If any of the regions v'_{p-2} is nonsingular, it will lie in at least three regions v'_{p-1} and, after the suppression of γ, it will still lie in at least two regions v'_{p-1}. Similarly, each region v'_q (where $q < p-2$) forms a part of γ lying in at least three regions v'_{p-1} and, after the suppression of γ, it will still lie in at least two regions v'_{p-1}. The suppression of γ therefore does not suppress any of the regions v'_q; *it therefore does not change the value of the number* N.

If, on the other hand, the region γ is irregular, we have all the more right to suppress it, since there then exists a region v'_{p-2} which lies only in γ and one other region v'_{p-1}. After the suppression of γ it will lie in at most a single region v'_{p-1}, which is inadmissible.

What should we do then? The region γ separates two regions v_p I have called α and β, but it does not form the whole boundary between α and β. Indeed, since γ is irregular, there is a singular region v'_{p-2} I call δ and which lies in γ and another region v'_{p-1} I shall call γ'. The latter region γ', from what we have seen above, separates the same regions as γ, namely α and β.

If the region δ is regular, we can suppress it and annex γ to γ'. The region $\gamma+\gamma'$ will separate α from β. We have therefore diminished α_{p-1} and α_{p-2} by one, without changing the other numbers α_i. Therefore N does not change.

If δ is irregular there will be a region v'_{p-3} I call ε which separates it from another region δ'. Then we suppress ε and annex δ to δ', and so on.

We can therefore suppress a region v'_q which separates two regions v'_{q+1}, and annex the two regions v'_{q+1} to each other, under two conditions:

1° If q is less than $p-1$ and the region v'_q is singular;
2° And, in any case, if it is regular.

This being so, here is how we order our operations:

I want to reconstruct the polyhedron P from the polyhedron P'. I can suppose without inconvenience that the polyhedron P has no singular region, although the polyhedron P' and the intermediate polyhedra may have them.

By a series of suppressions and annexations, we reconstruct P from P', passing through a series of intermediate polyhedra I call

$$P_0 = P', \quad P_1, \quad P_2, \quad \ldots, \quad P_{m-1}, \quad P_m.$$

How do we pass from the polyhedron P_i to the polyhedron P_{i+1}?

If P_i contains a singular v'_0, I suppress it. If not, then all the v'_1 are regular; if there is a singular v'_1, I suppress it.

If there is no singular v'_1, all the v'_2 are regular; if there is a singular v'_2, I suppress it.

And so on.

Finally, if there is no singular v'_{p-2}, then all the v'_{p-1} will be regular and we have the right to suppress any one of them. If one of the regions v_p is subdivided into several regions v'_p, I choose two of these v'_p that are contiguous and separated by a region v'_{p-1}, which is their common boundary. I annex these v'_p to each other by suppressing this common boundary.

None of these operations alter the number N.

We are not stopped until there are no more singular regions and none of the regions v_p are subdivided into regions v'_p. But then we have arrived at the polyhedron P.

None of the operations alter the number N.

This number is therefore the same for P and P'. Q.E.D.

This proof gives rise to certain objections, since it may be asked whether all the regions remain simply connected during the series of operations. However, before modifying the proof so as to meet these objections, I want to determine the value of N for a simply connected polyhedron.

If our theorem is true, the number N must have the same value for two polyhedra obtained by subdividing homeomorphic manifolds; it therefore has the same value for any two simply connected polyhedra.

It therefore suffices to make the determination for an arbitrarily chosen simply connected polyhedron.

I shall choose the generalized tetrahedron.

I give that name to the polyhedron bounding the domain

$$x_1 > 0, \quad x_2 > 0, \quad \ldots, \quad x_p > 0,$$

$$x_{p+1} > 0, \quad x_1 + x_2 + \cdots + x_p + x_{p+1} < 1.$$

We then have

$$\alpha_p = p + 2, \quad \alpha_{p-1} = \frac{(p+1)(p+2)}{2}, \quad \ldots,$$

$$\alpha_q = \frac{(p+2)!}{(q+2)!(p-q)!}, \quad \ldots, \quad \alpha_1 = \frac{(p+1)(p+2)}{2}, \quad \alpha_0 = p + 2,$$

that is, the numbers α_q are the binomial coefficients. Therefore

$$(1-1)^{p+2} = 1 - \alpha_p + \alpha_{p-1} - \cdots \pm \alpha_1 \mp \alpha_0 \pm 1 = 1 - N \pm 1,$$

where the sign of the last term is + if p is even and − if p is odd.

We therefore have $N = 2$ if p is even and $N = 0$ if p is odd.

I will arrive at the same result by choosing the generalized cube. I give that name to the polyhedron bounding the domain

$$-1 < x_i < 1 \quad (i = 1, 2, \ldots, p+1).$$

We then have

$$\alpha_p = 2(p+1), \quad \alpha_{p-1} = 2^2 \frac{p(p+1)}{2}, \quad \ldots, \quad \alpha_q = 2^{p-q+1} \frac{(p+1)}{(q+1)!(p-q)!}, \quad \ldots,$$

$$\alpha_1 = 2^p(p+1), \quad \alpha_0 = 2^{p+1},$$

whence

$$(1-2)^{p+1} = 1 - \alpha_p + \alpha_{p-1} - \cdots \pm \alpha_1 \mp \alpha_0 = 1 - N,$$

whence

$$N = 1 - (-1)^{p+1},$$

that is,

$$N = 2 \quad \text{if } p \text{ is even,}$$
$$N = 0 \quad \text{if } p \text{ is odd.}$$

Thus, for a simply connected polyhedron, the number N is equal to 2 if p is even and to 0 if p is odd.

This being so, I shall now establish our theorem in a complete and rigorous fashion by supposing it true for all manifolds of less than p dimensions.

Consider our polyhedron P and a q-dimensional region v_p contained in it. This region v_q will be part of a certain number of regions v_{q+1}, a certain number of regions v_{q+2}, ..., a certain number of regions v_p. The set of all these regions forms what I shall call the *star* of v_p.

I let γ_h denote the number of regions v_h $(h > q)$ that form part of the star of v_q.

Let $(x_1^0, x_2^0, \ldots, x_n^0)$ be a point of v_q. We consider the hypersphere S with equation

$$(x_1 - x_1^0)^2 + (x_2 - x_2^0)^2 + \cdots + (x_n - x_n^0)^2 = \varepsilon^2,$$

where ε is very small.

Let Π be the plane manifold defined by the q equations

$$A_i^1(x_1 - x_1^0) + A_i^2(x_2 - x_2^0) + \cdots + A_i^n(x_n - x_n^0) = 0 \quad (i = 1, 2, \cdots, q),$$

where the A_i^k are any constants.

The intersection of S, Π and V will be a $(p-q-1)$-dimensional manifold I call W, and it will be simply connected.

For this polyhedron the number α_h will equal γ_{h+q+1}; since it has less than p dimensions, our theorem will be applicable, so that we can write

$$\text{(A)} \qquad \gamma_p - \gamma_{p-1} + \cdots \pm \gamma_{q+2} \mp \gamma_{q+1} = 2 \text{ or } 0,$$

according as $p - q$ is even or odd.

We now define a polyhedron Q, formed by an operation we may call *cubing*.

Let V be a p-dimensional manifold in n-dimensional space. We construct an infinite number of plane manifolds defined by the equations

$$\text{(B)} \qquad \begin{aligned} &x_i = a_{k,l}, \\ &(i = 1, 2, \ldots, n;\ k = -\infty, \ldots, -1, 0, +1, +2, \ldots, +\infty). \end{aligned}$$

These plane manifolds decompose the space into an infinity of domains D_n analogous to rectangular parallelepipeds. The boundaries of the D_n will consist of a certain number of $(n-1)$-dimensional domains D_{n-1} forming part of the various plane manifolds $x_i = a_{k,i}$, and likewise analogous to rectangular parallelepipeds. The boundaries of the D_{n-1} consist of a certain number of domains D_{n-2} analogous to rectangular parallelepipeds in $(n-2)$-dimensional space, and so on.

The polyhedron Q is now defined as follows: the regions v_p are the intersections of V with the domains D_n, the regions v_{p-1} are the intersections of V with the domains D_{n-1}, and so on. Finally, the regions v_0 are the intersections of V with the domains D_{n-p}.

It follows from this definition that the polyhedron Q has no singular region.

I consider, in addition, an arbitrary polyhedron P congruent to Q, and a polyhedron P' derived from both P and Q.

I want to reconstruct P from P', on the one hand, and Q from P' on the other, and establish that N is unchanged under both these operations.

First we reconstruct P from P'.

Let $x_i = a$ be one of the plane manifolds defined by equation (B). We classify the regions v_q' of arbitrary dimension making up the polyhedron P' into four kinds.

Those of the first kind are those contained in the manifold

$$x_i = a.$$

Those of the second kind are those which include points such that

$$x_i = a + \varepsilon,$$

where ε is positive and very small.

Those of the third kind are those which include points such that

$$x_i = a - \varepsilon.$$

All the others are of the fourth kind.

Let $\delta_q, \delta'_q, \delta''_q$ be the numbers of q-dimensional manifolds that are, respectively, of the first, second, and third kind.

Every manifold of the second kind will be contiguous to a manifold of the third kind, and their common boundary will be a manifold of the first kind of one dimension less. The manifolds of the first three kinds are therefore in one-to-one correspondence and we have

$$\delta'_q = \delta''_q = \delta_{q-1}.$$

Moreover, it is clear that

$$\delta'_0 = \delta''_0 = \delta_p = 0.$$

If, in the set of plane manifolds (B) which constitute the cubing and which give rise to the polyhedra Q and P', we suppress the manifold $x_i = a$, then we obtain two polyhedra Q_1 and P_1 simpler than the originals. We compare P'_1 and P'.

When we suppress the plane manifold $x_i = a$, we suppress the manifolds of the first kind and we annex each manifold of the third kind to the corresponding manifold of the second kind. Therefore, in passing from P' to P'_1, the number α_q is diminished by

$$\delta''_q + \delta_q = \delta_q + \delta_{q-1}.$$

In particular, the numbers α_p and α_0 are diminished by δ_{p-1} and δ_0. It follows that the number N is diminished by

$$\delta_{p-1} - (\delta_{p-1} + \delta_{p-2}) + (\delta_{p-2} + \delta_{p-3}) - \cdots \pm (\delta_1 + \delta_0) \mp \delta_0 = 0.$$

Thus N does not change. Therefore, in suppressing the manifold $x_i = a$, we do not change N. However, in suppressing all the plane manifolds defined by (B), we recover the polyhedron P. The number N is therefore the same for P' and P.

Now we recover Q from P'.

Let $w_p, w_{p-1}, \ldots, w_1, w_0$ be the manifolds that make up the polyhedron Q. Similarly, let $v'_p, v'_{p-1}, \ldots, v'_1, v'_0$ be the manifolds that make up the polyhedron P'.

We divide the manifolds v'_p into $p+1$ classes.

Those of the first class are those that lie in one of the regions w_p without being in one of the regions w_{p-1}. Since the polyhedron P' is derived from Q, this first class includes all the manifolds v'_p (which are all subdivisions of the w_p), those manifolds v'_{p-1} that separate two manifolds v'_p in the same region w_p, and their intersections.

Those of the second class are those that lie in one of the regions w_{p-1}, without being in one of the regions w_{p-2}.

Those of the third class are those that lie in one of the regions w_{p-2}, without being in one of the regions w_{p-3}, etc.

Those of the pth class are those that lie in one of the regions w_1, without being in one of the points w_0.

Finally, the $(p+1)$-th class consists of the points w_0.

Here is how I proceed to recover W from P'. I commence by suppressing all manifolds in the first class that have at least p dimensions, which has the effect of uniting all the regions v'_p that are subdivisions of the same region w_p.

I claim that this operation does not change the number N.

Indeed, I can suppose that the mesh of the cubing that gives rise to Q is so fine that in the interior of one of the cells D_p, that is, in the interior of one of the regions w_p, we cannot find points belonging to two different manifolds v_{p-1}, except in the case where we find points belonging to the intersection of the two manifolds. (I always denote by v_p the manifolds which make up P.) More generally, I can suppose that w_p does not contain points from several manifolds v_q $(q < p)$ unless it contains points belonging to their intersection.

In a region w_p we can therefore have the points of a region v_q and all the regions v_h $(h > q)$ which make up the star of v_q, without having the points of a region v_{q-1} as well.

Then if I suppose that q is the least dimension of a region v_q with points in the interior of w_p, I shall have in w_p the points of single region v_q and its star.

Consider such a region w_p, containing the points of v_q and the regions in the star of v_q.

Let

$$\gamma_{q+1}, \quad \gamma_{q+2}, \quad \dots, \quad \gamma_p,$$

be the numbers we have defined above in defining the star.

It follows that, in the interior of w_p, we have

$$\begin{array}{c}1 \text{ region } v'_q \text{ of the first class}\\ \gamma_{q+1} \text{ regions } v'_{q+1} \text{ of the first class}\\ \gamma_{q+2} \text{ regions } v'_{q+2} \text{ of the first class}\\ \dots\dots\dots\dots\dots\dots\\ \gamma_p \text{ regions } v'_p \text{ of the first class.}\end{array}$$

In suppressing the regions of the first class with at least p dimensions and uniting the γ_p regions v'_p which make up w_p, we diminish α_p by $\gamma_p - 1, \alpha_{p-1}$ by $\gamma_{p-1}, \dots, \alpha_{q+1}$ by γ_{q+1}, α_q by 1. Therefore, by virtue of equation (A), the number N will not change.

Next we suppress all the manifolds of the second class with at least $p - 1$ dimensions, in uniting the manifolds v'_{p-1} of the second class which make up the same region w_{p-1}. Then we suppress all the manifolds of the third class with at least $p-2$ dimensions, in uniting the manifolds v'_{p-2} which make up the same region w_{p-2}, and so on.

We finally arrive at the polyhedron Q.

One shows as above that each of these operations leaves the number N unaltered.

The number N is therefore the same for P' and Q. It is therefore the same for P and Q, and consequently the same for any two congruent polyhedra. Q.E.D.

§17. The case where p is odd

I am now going to define, for an arbitrary polyhedron P, some remarkable new numbers I call $B_{\lambda,\mu}$.

First suppose $\lambda > \mu$. I consider all the manifolds v_λ; for each of them I take all the manifolds v_μ they contain, sum all the numbers relative to the various manifolds v_λ, and call the sum $\beta_{\lambda,\mu}$.

Since all the manifolds v_λ are simply connected by hypothesis, we have

$$\beta_{\lambda,\lambda-1} - \beta_{\lambda,\lambda-2} + \dots \pm \beta_{\lambda,1} \mp \beta_{\lambda,0} = 2\alpha_\lambda \quad \text{or} \quad 0,$$

according as λ is odd or even.

Now suppose $\lambda < \mu$. I consider all the manifolds v_λ; for each of them I take all the manifolds v_μ they are contained in (that is, take the number γ_μ relative to the star of v_λ), sum all the numbers relative to the different manifolds v_λ, and call the sum $\beta_{\lambda,\mu}$.

By virtue of equation (A) of the preceding section, we have

$$\beta_{\lambda,\beta} - \beta_{\lambda,p-1} + \cdots \pm \beta_{\lambda,\lambda+2} \mp \beta_{\lambda,\lambda+1} = 2\alpha_\lambda \text{ or } 0,$$

according as $p - \lambda$ is odd or even.

It follows from this definition that

$$\beta_{\lambda,\mu} = \beta_{\mu,\lambda}.$$

This being so, we form the following table

$$\begin{array}{rcr} +\beta_{p,p-1}, -\beta_{p,p-2}, +\beta_{p,p-3} & \ldots, & \pm\beta_{p,1}, \mp\beta_{p,0}, \\ +\beta_{p-1,p-2} - \beta_{p-1,p-3}, & \ldots, & \mp\beta_{p-1,1}, \pm\beta_{p-1,0}, \\ \cdots\cdots\cdots & \ldots, & \cdots\cdots\cdots \\ & & +\beta_{2,1}, -\beta_{2,0}, \\ & & +\beta_{1,0}. \end{array}$$

We see that, in each row and each column, each term $\beta_{\lambda,\mu}$ is affected alternatively by the $+$ sign and the $-$ sign, the $+$ sign when $\lambda - \mu$ is odd and the $-$ sign otherwise.

We form the sum of all terms in this table, which can be done in two ways—by rows and by columns.

The sums of the rows in the table, starting at the top, are

$$2\alpha_p, \quad 0, \quad 2\alpha_{p-2}, \quad 0, \quad \ldots, \quad 2\alpha_3, \quad 0, \quad 2\alpha_1$$

if p is odd, and

$$0, \quad 2\alpha_{p-1}, \quad 0, \quad 2\alpha_{p-3}, \quad \ldots, \quad 2\alpha_3, \quad 0, \quad 2\alpha_1$$

if p is even. The sum of the terms in the table is therefore

$$2\alpha_1 + 2\alpha_2 + \cdots + 2\alpha_p$$

if p is odd, and

$$2\alpha_1 + 2\alpha_3 + \cdots + 2\alpha_{p-1}$$

if p is even.

The sums of the columns in the table, starting at the left, are

$$2\alpha_{p-1}, \quad 0, \quad 2\alpha_{p-3}, \quad 0, \quad \ldots, \quad 2\alpha_2, \quad 0, \quad 2\alpha_0$$

if p is odd, and

$$2\alpha_{p-1}, \quad 0, \quad 2\alpha_{p-3}, \quad 0, \ldots, \quad 0, \quad 2\alpha_1, \quad 0$$

if p is even.

The sum of the terms in the table is therefore

$$2\alpha_0 + 2\alpha_2 + \cdots + 2\alpha_{p-1}$$

if p is odd, and

$$2\alpha_1 + 2\alpha_3 + \cdots + 2\alpha_{p-1}$$

if p is even.

Equating the two expressions for the sum, we obtain an identity if p is even, and the equation

$$N = 0$$

if p is odd.

This has the following consequence:

The number N is zero and independent of the Betti numbers if p is odd; however, it depends on the Betti numbers if p is even.

§18. Second proof

The second proof will show us how this works.

To assist understanding, I shall begin by explaining the case of an ordinary polyhedron with α_0 vertices, α_1 edges, and α_2 faces.

To each of the α_0 vertices I assign an arbitrary number, to each of the α_1 edges I assign a number δ equal to the difference of the numbers corresponding to its vertices.

We therefore have α_1 differences δ, but they cannot all be chosen arbitrarily. Indeed, they are determined when we know the α_0 numbers assigned to various vertices, and also when we know the excesses of $\alpha_0 - 1$ of these numbers over one of them. It follows that only $\alpha_0 - 1$ differences can be chosen arbitrarily.

This implies that there are

$$\alpha_1 - \alpha_0 + 1$$

linear relations between the differences δ.

It is clear that we can obtain all these linear relations in the following way. Consider a sequence of edges forming a closed contour. Then the algebraic sum of the differences corresponding to the various edges in the sequence will be zero.

We therefore proceed to construct closed contours consisting of edges.

First we have the polygonal contours of the faces; there are α_2 of them.

Then, if the polyhedron is not simply connected, we can trace on its surface $P_1 - 1$ linearly independent contours in the sense assigned to that word in the section on homologies. Let C be one of these contours, which successively traverses different faces. Let $a_1a_2, a_2a_3, \ldots, a_{n-1}a_n, a_na_1$ be the arcs of this contour in the successive faces.

We take the first of these arcs a_1a_2 and let F be the corresponding face.

The point a_1 and the point a_2 are on the perimeter of this face. We can therefore go from a_1 to a_2 following the perimeter, by a path I shall call

$$a_1m_1a_2 + a_2a_1 \sim 0,$$

that is, in the contour C we can replace the arc a_1a_2 by the arc $a_1m_1a_2$. Operating the same way on the other arcs of C we finally replace C by a homologous contour

$$a_1m_1a_2 + a_2m_2a_3 + \cdots + a_nm_na_1$$

I shall call C'. This contour C' consists of a certain number of edges and portions of edges. For example, the arc $a_1m_1a_2$ consists of an edge segment joining a_1 to the nearest vertex, then a certain number of complete edges, then a segment Sa_2 joining a_2 to the nearest vertex.

However, the segment Sa_2 is retraced in the arc $a_2m_2a_3$. The portions of edges in C' are therefore each traversed twice, in opposite senses, and hence we can suppress them to obtain a closed contour C'' consisting entirely of complete edges, and homologous to C or C'.

We have $P_1 - 1$ contours analogous to C'', which give $P_1 - 1$ relations between the δ.

We thus obtain $\alpha_2 + P_1 - 1$ closed contours consisting of edges. I claim that all possible closed contours are combinations of these.

First consider a closed contour, consisting of edges, which is homologous to zero. It separates the surface into two regions. If R is one of them, then it evidently consists of a certain number q of faces, because the contour consists entirely of edges. We can therefore replace the given contour by q partial contours which are the perimeters of these q faces.

Now if the given contour is not homologous to zero, we can always replace it by a combination of the contours C'' and a contour homologous to zero.

We therefore have

$$\alpha_2 + P_1 - 1$$

relations between the δ and no others, but are they all distinct?

To decide this, we have to see whether we can form a linear combination of these relations which reduces to an identity or, what comes to the same thing, whether we can form a combination of our $\alpha_2 + P_1 - 1$ contours, each edge of which is traversed twice, in opposite senses (or, if it is traversed more than twice, equally often in each sense).

Can we first form such a combination with the α_2 perimeters Π? To say that each edge is traversed twice in opposite senses is to say that the set of polygons whose perimeters are thereby traversed forms a closed polyhedron.

But we can evidently construct only one such closed polyhedron, which is the given polyhedron.

Thus with the α_2 relations corresponding to the α_2 perimeters Π we can form one linear identity, and only one.

Can we now form such a combination with the perimeters Π and the contours C''? If so, the set of polygons whose perimeters are thereby traversed form a nonclosed polyhedral surface whose boundary is a combination of the contours C''. But this is impossible, because the contours C'' are linearly independent.

We therefore form just one identity combination from our $\alpha_2 + P_1 - 1$ relations. It follows that there are

$$(\alpha_2 + P_1 - 1) - 1$$

distinct relations between the δ. The number of arbitrary δ is therefore

$$\alpha_1 - (\alpha_2 + P_1 - 1) + 1,$$

so that we have

$$\alpha_0 - 1 = \alpha_1 - (\alpha_2 + P_1 - 1) + 1$$

or

$$N = 3 - P_1.$$

We now extend this proof to the case of a polyhedron of three dimensions, where

the number of vertices v_0 is α_0,
the number of manifolds v_1 is α_1,
the number of manifolds v_2 is α_2,
the number of manifolds v_3 is α_3.

We assign a number to each v_0, and to each v_1 the difference δ of the numbers assigned to its vertices. We therefore have α_1 differences δ, $\alpha_0 - 1$ of which will be arbitrary.

We obtain the linear relations between the δ by constructing all the closed contours consisting exclusively of the v_1.

First we have the α_2 perimeters Π of the v_2. Now let C be any closed contour not homologous to zero. It traverses different v_3 in turn; let a_1a_2, a_2a_3, ... be the successive arcs of C in each of these v_3.

Consider the arc a_1a_2 in one of the v_3, which I call Φ, and whose extremities a_1 and a_2 are found in two of the v_2 making up the boundary of Φ, say a_1 in φ_1 and a_2 in φ_2.

Let b_1 be a vertex of φ_1 and let b_2 be a vertex of φ_2. We join a_1 and b_1 by any line a_1b_1, then b_1 to b_2 by a line b_1b_2 consisting entirely of v_1 in the boundary of Φ, and b_2 to a_2 by any line a_2b_2, so we have

$$a_1a_2 \sim a_1b_1 + b_1b_2 + b_2a_2,$$

and we can replace the arc a_1a_2 by the arc $a_1b_1b_2a_2$. We operate similarly on all the other arcs in C, so that C is replaced by the homologous contour

$$a_1b_1b_2a_2 + a_2b_2b_3a_3 + \cdots$$

which I call C'. The contour C' consists of a certain number of v_1 and the arcs a_1b_1, $a_2b_2, \ldots$, which are each traversed twice, in opposite senses. We can therefore suppress these arcs, leaving a contour C'' homologous to C and consisting exclusively of v_1.

There are $P_1 - 1$ contours C''.

I now claim that closed contour K formed from the v_1 is a combination of the Π and the C''. If $K \sim 0$, K will be the boundary of a certain two-dimensional region R. This region R can be subdivided into a certain number of manifolds r, each of which is the portion of R inside one of the v_3. We consider one of the manifolds r, and let φ be the region v_3 containing it. The boundary of r will be a closed one-dimensional manifold u, part of the boundary of φ. Since φ is simply connected, u will separate the boundary of φ into two regions. Let r' be one of these regions; it will consist of a certain number of complete v_2 and a certain number of portions of v_2 (because, among the v_2 which form the boundary of φ there may be some divided into two parts by u). We see that r is homologous to r', so we can replace r by r'. If we operate in the same way on all the regions r we obtain a manifold R' homologous to R and which consists of a certain number of complete v_2 and a certain number of portions which occur twice, with opposite senses. We can suppress these portions of v_2 and thus obtain a manifold R'' homologous to R and bounded by the same contour K. This manifold R'' can be decomposed into a certain number of polygons v_2, and hence the contour K can be replaced by the perimeters Π of these polygons.

If K is not homologous to zero we can replace it by a certain number of contours C'' and a contour homologous to zero.

We therefore have

$$\alpha_2 + P_1 - 1$$

relations between the δ, which I write

$$\varepsilon = 0,$$

and no others. But are they all distinct?

In other words, can we form a linear combination of the ε which is identically zero, or again, a combination of the Π and C'' in which each v_1 occurs twice, with opposite senses?

If a C'' occurs in the latter combination then the set of polygons v_2 whose perimeters Π occur in the combination will form a two-dimensional manifold which is not closed and has a certain number of the contours C'' as its boundary. But this is not possible, because the C'' are linearly independent.

It therefore remains to examine the combinations containing only the Π. The set of polygons v_2 whose perimeters Π are involved then form a closed manifold.

This leads us to examine the closed manifolds consisting exclusively of v_2.

We first have the boundaries that I call Φ and which are α_3 in number.

Now suppose D is any two-dimensional manifold, not homologous to zero. We treat it as we have treated R, and see that it is homologous to a closed two-dimensional manifold D'' formed exclusively from v_2.

The number of D'' is $P_2 - 1$.

I now claim that each closed manifold K formed from the v_2 is a combination of the Π and the D''. Indeed, if it is homologous to zero it will bound a three-dimensional region S composed of a certain number of v_3, because K consists of a certain number of v_2. We can therefore replace K by the boundaries Φ of these v_3. If K is not homologous to zero we can replace it by a certain number of D'' and a manifold homologous to zero and formed exclusively from the v_2.

We therefore have, between the ε,

$$\alpha_2 + P_2 - 1$$

linear relations which I write

$$\zeta = 0,$$

and no others. Are they distinct?

To form a linear combination of the ζ which is identically zero, it is necessary to form a combination of the Φ and the D'' such that each v_2 occurs twice, with opposite senses. We see, as above, that the D'' cannot figure in this combination and that the v_2, which figure in the boundaries Φ, must form a closed three-dimensional manifold. But we can construct only a single manifold of this kind: the given polyhedron itself.

It follows that there is only one linear combination of ζ which vanishes identically.

We therefore have

$$(\alpha_3 + P_2 - 1) - 1$$

distinct linear relations between the ε,

$$(\alpha_2 + P_1 - 1) - (\alpha_3 + P_2 - 1) + 1$$

distinct linear relations between the δ.

Therefore there are

$$\alpha_1 - (\alpha_2 + P_1 - 1) + (\alpha_3 + P_2 - 1) - 1$$

of the δ which remain arbitrary, so we find

$$\alpha_0 - 1 = \alpha_1 - (\alpha_2 + P_1 - 1) + (\alpha_3 + P_2 - 1) - 1.$$

And we find

$$\alpha_0 - 1 = \alpha_1 - (\alpha_2 + P_1 - 1) + (\alpha_3 + P_2 - 1) - (\alpha_4 + P_3 - 1) + 1$$

for polyhedra of four dimensions.

We therefore have

$$N = P_2 - P_1$$

for three-dimensional polyhedra and

$$N = 3 - P_1 + P_2 - P_3$$

for four-dimensional polyhedra.

In general we have

$$N = P_{p-1} - P_{p-2} + \cdots + P_2 - P_1$$

if p is odd, and

$$N = 3 - P_1 + P_2 - \cdots + P_{p-1}$$

if p is even.

If we now observe that the Betti numbers equally distant from the extremes are equal, we see that

$$N = 0$$

when P is odd, as we have already seen in the preceding paragraph.

Supplement to Analysis Situs

Rendiconti del Circulo Matematico di Palermo 13 (1899), pp. 285–343.

§I. Introduction

In the *Journal de l'École Polytechnique* (volume for the centenary of the founding of the school, 1894) I published a memoir entitled *Analysis situs*, or the study of manifolds in spaces of more than three dimensions and the properties of the Betti numbers. Since I shall have occasion to mention this memoir frequently in what follows, I shall use simply the title *Analysis situs*.

The following theorem is found in that memoir: *For any closed manifold, the Betti numbers equally distant from the extremes are equal.*

The same theorem was announced by M. Picard in his *Théorie des fonctions algébriques de deux variables.*

M. Heegaard returned to the same problem in his remarkable work, published in Danish under the title "*Forstudier til en topologisk teori for algebraiske Sammenhäng*" (Copenhagen, det. Nordiske Forlag Ernst Bojesen, 1898), according to which the theorem in question is inexact and its proof is without value.

After examining the objections of M. Heegaard, it is advisable to make a distinction. There are two ways of defining the Betti numbers.

Consider a manifold V which I shall assume, for example, to be closed; let $v_1, v_2, \ldots, v_n$ be n manifolds of p dimensions forming part of V. I assume that we cannot find a $(p+1)$-dimensional manifold forming part of V for which $v_1, v_2, \ldots, v_n$ is the boundary; but that if we adjoin an $(n+1)$-th manifold of p dimensions which I shall call v_{n+1}, and which forms part of V, then we can find a $(p+1)$-dimensional manifold forming part of V for which $v_1, v_2, \ldots, v_n, v_{n+1}$ constitutes the boundary, and *this is true whatever the choice of the* $(n+1)$*-th manifold* v_{n+1}. In that case we say that the Betti number is equal to $n+1$ for manifolds of p dimensions.

This is the definition adopted by Betti. However, we can give a second definition.

Suppose that we can find a $(p+1)$-dimensional submanifold of V for which $v_1, v_2, \ldots, v_n$ constitute the boundary; I express this fact by the following relation

$$v_1 + v_2 + \cdots + v_n \sim 0$$

which I call a *homology*.

It can happen that the same manifold v_1 appears several times in the boundary of our $(p+1)$-dimensional manifold; in that case the first member of the homology will appear with a coefficient, which must be an integer.

According to this definition, we can add and subtract homologies, and multiply them by integers.

We shall likewise make the *convention* that it is permissible to divide a homology by an integer provided all its coefficients are divisible by that integer. Consequently, if we have a $(p+1)$-dimensional manifold with boundary consisting of four times the manifold v_1, we shall agree to write not only the homology

$$4v_1 \sim 0$$

but also the homology

$$v_1 \sim 0$$

so that the latter homology signifies that there is a $(p+1)$-dimensional manifold, the boundary of which is a certain number of times v_1.

The homology

$$2v_1 + 3v_2 \sim 0$$

signifies that there is a $(p+1)$-dimensional manifold with boundary twice v_1 and thrice v_2, or four times v_1 and six times v_2, or six times v_1 and nine times v_2, etc.

Such are the conventions that I adopted in *Analysis situs* p. 15.

I shall say that several manifolds are independent if they are not connected by any homology with integer coefficients.

Then, if there are n independent p-dimensional manifolds, the Betti number according to the second definition is $n+1$.

This second definition, which is the one I adopted in *Analysis situs*, does not agree with the first.

The theorem enunciated above, and criticized by Heegaard, is true for the Betti numbers defined in the second manner, but false for Betti numbers defined in the first manner.

This is what happens with the example of Heegaard, p. 86:
If we adopt the first definition we have

$$P_1 = 2, \quad P_2 = 1$$

and consequently

$$P_2 < P_1.$$

If, on the contrary, we adopt the second definition, we find

$$P_1 = 1, \quad P_2 = 1$$

and consequently

$$P_2 = P_1$$

in conformity with the theorem enunciated.

This is likewise exhibited in an example I have cited myself in *Analysis situs*. This is the third example on p. 33.

We have formed the fundamental equivalences (p. 42), which are written in the following fashion:

$$2C_1 \equiv 2C_2 \equiv 2C_3 \equiv 0, \quad 4C_1 \equiv 0,$$

from which we deduce the homologies

$$4C_1 \sim 4C_2 \sim 4C_3 \sim 0.$$

Since, according to our convention, we can divide these homologies by 4, we arrive at the following system of fundamental homologies:

$$C_1 \sim C_2 \sim C_3 \sim 0.$$

Then if P_1 and P_2 are the Betti numbers, *defined in the second manner*, we find

$$P_1 = P_2 = 1.$$

But equality between the numbers P_1 and P_2 does not persist when we adopt the first definition, that of Betti; we always have $P_2 = 1$, but we no longer have $P_1 = 1$.

In fact, there is no two-dimensional manifold which has the closed line C_1 as boundary, because we do not have the equivalence $C_1 \equiv 0$.[14]

The only thing that is true is that there is a two-dimensional manifold which has boundary four times the line C_1. Then P_1 is not equal to 1.

Now, to return to the theorem according to which the Betti numbers equally distant from the extremes are equal.

The proof that I have given in *Analysis situs* seems to apply equally well to the two definitions of Betti number; therefore it must have a weak point, since the preceding examples show adequately that the theorem is not true for the first definition.

M. Heegaard has given a good account, but I do not believe that his first objection is warranted.

After having cited the fashion in which I define the manifolds $V_1, V_2, \ldots, V_p$ (*Analysis situs*, p. 29) by the equations $\Phi = 0, F_i'' = 0$, he adds (p. 70): "*Enhver af Mangfoldighederne V skulde altsaa kunne voere den fulstaendige Skoering mellem p Mangfoldighder af $h - 1$ Dimensioner iU.*"[15] This is not exact, since, as well as my equations, I have a certain number of inequalities which I introduced at the beginning of the memoir but neglected to describe afresh in what followed; my manifolds are then not complete intersections.

The second objection, however, is well founded. "*Naar omvendt*", says M. Heegaard, "*Homologien $\sum V_i \sim 0$ ikke finder Sted, saa i U' kan legges en lukket Kurve K' saa at*

$$\sum N(V', V_i) \neq 0$$

men det er ikke sikker at denne Kurve kan udskoeres of nogen Mangfoldighed V."[16] This is, indeed, the true weak point of the proof.

It is therefore necessary to return to this question, and this is the object of the present work.

In order to simplify the proofs, I shall often consider only the case of closed three-dimensional manifolds in a space of four dimensions. We can easily extend this to the general case.

I consider then, in what follows, a closed manifold V, but in order to calculate its Betti numbers I shall suppose that it is divided into smaller manifolds in the manner of forming a polyhedron.

---

[14]Here Poincaré repeats his error of p. 39, that the boundary of a surface is equivalent to 0, that is, null-homotopic. (Translator's note.)

[15]Each of the manifolds V must then be the complete intersection of the p $(h-1)$-dimensional manifolds in U.

[16]If conversely the homology $\sum_i V_i \sim 0$ does not hold, then we can trace a closed curve V' on it such that $\sum N(V', V) \neq 0$, but it is not certain that this curve is an intersection of the manifolds V.

§II. Schema of a polyhedron

We consider then, as in *Analysis situs*, a polyhedron of p dimensions, i.e., a manifold V of p dimensions, divided into manifolds v_p; the boundaries of the v_p are the v_{p-1}, those of the v_{p-1} are the $v_{p-2}, \dots$, those of the v_1 (edges) are the v_0 (vertices).

I let α_i be the number of v_i.

Let $a_1^q, a_2^q, \dots, a_{\alpha_q}^q$ be the different v_q.

Let a_1^q be one of the manifolds v_q and a_1^{q-1} one of the manifolds v_{q-1} which serve as its boundary. We study the connection between the a_1^q and the a_1^{q-1}.

Let

$$(1)\qquad F_1 = F_2 = \cdots = F_{n-q} = F_{n-q+1} = 0, \quad \varphi_j > 0,$$

be the equations and inequalities which define a_1^{q-1} according to the first definition of manifold (*Analysis situs*, p. 7).

The relations which define a_1^q can be arranged in the form

$$(2)\qquad F_1 = F_2 = \cdots = F_{n-q} = 0, \quad F_{n-q+1} > 0, \quad \varphi_j > 0.$$

In this case we say that the relation between a_1^q and a_1^{q-1} is *direct.*

This relation will become *inverse* if one of the two manifolds is replaced by its opposite; it remains direct if each of the two manifolds is replaced by its opposite.

We know that a manifold is replaced by its opposite (*Analysis situs*, p. 13) when we permute two of the functions F (which, being equal to zero, give the equations which define the manifold), or if we change the sign of one of them.

Thus the two manifolds

$$F_1 = F_2 = F_3 = 0; \quad F_1 = F_2 = 0, \quad F_3 > 0,$$
$$F_1 = F_2 = F_3 = 0; \quad F_1 = F_3 = 0, \quad F_2 < 0,$$
$$F_1 = F_2 = F_3 = 0; \quad F_2 = F_3 = 0, \quad F_1 > 0$$

are in a direct relation; whereas the two manifolds

$$F_1 = F_2 = F_3 = 0; \quad F_1 = F_2 = 0, \quad F_3 < 0,$$
$$F_1 = F_2 = F_3 = 0; \quad F_1 = F_3 = 0, \quad F_2 > 0$$

are in an inverse relation.

That being given, let $\varepsilon_{i,j}^q$ be a number which is equal to zero if a_j^{q-1} is not in the boundary of a_i^q, $+1$ if a_j^{q-1} is part of the boundary of a_i^q and in a direct relation to a_i^q; and finally -1 if a_j^{q-1} is a part of the boundary of a_i^q but in an inverse relation to a_i^q.

We agree to write the *congruence*

$$(3)\qquad a_i^q \equiv \sum \varepsilon_{i,j}^q a_j^{q-1}$$

to express the boundary of the a_i^q.[17]

The set of congruences (3) relative to the different $v_p, v_{p-1}, \dots, v_0$ of V constitute what may be called the *schema of a polyhedron.*

Two questions may be posed:

1° Given a schema, does there always exist a corresponding polyhedron?

[17]It is unfortunate that Poincaré chooses the symbol $\equiv$ to express the asymmetric relation "$\sum \varepsilon_{i,j}^q a_j^{q-1}$ is the boundary of a_i^q." (Translator's note.)

2° If two polyhedra have the same schema, are they homeomorphic?

Without dealing with these two questions for the moment, we seek some conditions a schema must satisfy in order to correspond to a polyhedron.

Consider one of the v_{p-1}, a_1^{p-1} for example; this manifold should separate two and only two of the v_p from each other; so that among the numbers $\varepsilon_{i,1}^p$ we have one which is equal to $+1$ and one which is equal to -1, and all the others are zero.

This is not all; consider any of the v_q, a_i^q for example, and any of the v_{q-2}, a_k^{q-2} for example.

There are two possibilities: first, where a_k^{q-2} does not belong to a_i^q, all the products

$$\varepsilon_{i,j}^q \varepsilon_{j,k}^{q-1} \tag{4}$$

must be zero, for if a_j^{q-1} does not belong to a_i^q the first factor is zero; while if a_j^{q-1} belongs to a_i^q the manifold a_k^{q-2} cannot belong to a_j^{q-1} (otherwise it would belong to a_i^q, contrary to hypothesis) and the second factor must be zero.

Or secondly, a_k^{q-2} could belong to a_i^q, but then we can argue about a_i^q as we have all along about the manifold, and conclude that a_k^{q-2} must separate two of the manifolds v_{q-1} from each other, and exactly two, belonging to a_i^q. Let them be a_1^{q-1} and a_2^{q-1}.

Among the products (4) we have two that are nonzero, namely

$$\varepsilon_{i,1}^q\, \varepsilon_{1,l}^{q-1}, \quad \varepsilon_{i,2}^q\, \varepsilon_{2,k}^{q-1}.$$

For all the others, in fact, either a_j^{q-1} does not belong to a_i^q or a_k^{q-2} does not belong to a_j^{q-1}.

Moreover, these two products are $+1$ and -1.

Then in all cases we have

$$\sum_j \varepsilon_{i,j}^q\, \varepsilon_{j,k}^{q-1} = 0. \tag{5}$$

We likewise have

$$\sum_i \varepsilon_{i,1}^p = 0$$

and more generally, for any k,

$$\sum_i \varepsilon_{i,k}^p = 0. \tag{5'}$$

The relation (5′) can be regarded, from a certain point of view, as a particular case of the relation (5).

Let P be the portion of $(p+1)$-dimensional space bounded by the polyhedron V; then the boundary of P is composed of various manifolds v_p, which, in their totality, form V; we can then write, in the sense of the congruence (3),

$$P \equiv \sum_i a_i^p \tag{3'}$$

or

$$P \equiv \sum_i \varepsilon_{0,i}^{p+1}\, a_i^p,$$

where the numbers $\varepsilon_{0,i}^{p+1}$ are all equal to 1 by definition.

On that account, the relation (5′), which can be written

$$\sum_i \varepsilon_{0,i}^{p+1}\,\varepsilon_{i,k}^{p} = 0$$

is no more than a particular case of the relation (5).

Next we have that each v_1 is bounded by two v_0, given by congruences (3) of the form

$$a_i^1 \equiv a_j^0 - a_k^0$$

and a relation analogous to (5) and (5′),

$$\sum_j \varepsilon_{i,j}^1 = 0,$$

which again takes the form (5) if we make the convention that all the ε_0 are equal to $+1$.

On the other hand, consider one of the a_i^q; all of the a_j^{q+1} which it bounds; all of the a_k^{q+2} which these a_j^{q+1} bound, and so on. The set of all these manifolds constitutes what we have called a *star* (*Analysis situs* p. 65).

We have seen (*loc. cit* p. 66) that the polyhedron which corresponds to a star must be simply connected. Thus one condition for a given schema to correspond to a polyhedron is that the polyhedra which correspond to different stars, according to the convention of page 66 of *Analysis situs*, must all be simply connected.

This is not a necessary condition for a schema to correspond to a polyhedron; it is simply one condition which, unless the contrary is stated, we shall suppose is satisfied.

To clarify these definitions by a few examples, let us see first of all what is the scheme of the generalized tetrahedron defined in p. 65 (*Analysis situs*).

The faces of the tetrahedron may be defined by the $n+1$ equations

$$(6)\qquad \begin{cases} x_1 = 0, \quad x_2 = 0, \quad \ldots, \quad x_n = 0, \\ \qquad x_1 + x_2 + \cdots + x_n = 1. \end{cases}$$

We obtain the a_i^q by suppressing $q+1$ of these equations; to define the sense of the manifold a_i^q we suppose that we suppress these $q+1$ equations without changing the order of the remaining $n-q$ equations.

That being given, we consider the relation between a_i^q and a_j^{q-1} and try to determine the number $\varepsilon_{i,j}^q$.

First of all, if a_i^{q-1} is to belong to a_i^q it is necessary that a_j^{q-1} be defined by the $n-q$ equations which define a_i^q, to which we must adjoin an $(n-q+1)$-th equation from among the equations (6). If this is not the case, then the number $\varepsilon_{i,j}^q$ is zero.

Suppose then that a_j^{q-1} is obtained by suppressing the q equations which occupy the positions

$$\alpha_1, \quad \alpha_2, \quad \ldots, \quad \alpha_q.$$

Suppose that a_i^q is obtained by suppressing, *in addition*, the βth equation; then the number $\varepsilon_{i,j}^q$, the absolute value of which is always equal to 1, has the same sign as the product

$$(\beta-\alpha_1)(\beta-\alpha_2)\cdots(\beta-\alpha_q).$$

It is then easy to verify that the relation (5) holds.

In fact, consider the manifold a_k^{q-2} obtained by suppressing the equations at positions $\alpha_1, \alpha_2, \ldots, \alpha_{q-1}$, and the manifold a_i^q obtained by suppressing, in addition, the equations at positions β and γ. (It is clear that if a_i^q is not obtained by suppressing the same equations as for a_k^{q-2}, plus two others, then all the products $\varepsilon_{i,j}^q \varepsilon_{j,k}^{q-1}$ will be zero.)

In that case all the products will again be zero except two,

$$\varepsilon_{i,1}^q \varepsilon_{i,2}^{q-1} \text{ and } \varepsilon_{i,2}^q \varepsilon_{2,k}^{q-1},$$

which correspond to the two manifolds a_1^{q-1} and a_2^{q-1} obtained by suppressing the equations at positions $\alpha_1, \alpha_2, \ldots, \alpha_{q-1}, \beta$ and $\alpha_1, \alpha_2, \ldots, \alpha_{q-1}, \gamma$.

Then the four numbers

$$\varepsilon_{i,1}^q, \quad \varepsilon_{1,k}^{q-1}, \quad \varepsilon_{i,2}^q, \quad \varepsilon_{2,k}^{q-1}$$

have respectively the same sign as

$$\begin{aligned}(\gamma-\beta)(\gamma-\alpha_1)(\gamma-\alpha_2 \quad &\cdots \quad (\gamma-\alpha_{q-1}),\\ (\beta-\alpha_1)(\beta-\alpha_2) \quad &\cdots \quad (\beta-\alpha_{q-1}),\\ (\beta-\gamma)(\beta-\alpha_1)(\beta-\alpha_2) \quad &\cdots \quad (\beta-\alpha_{q-1}),\\ (\gamma-\alpha_1)(\gamma-\alpha_2) \quad &\cdots \quad (\gamma-\alpha_{q-1}).\end{aligned}$$

Thus one verifies that the two products which are nonzero are equal but of opposite sign.[18] Q.E.D.

§III. Reduced Betti numbers

I am now going to find the Betti numbers of a polyhedron, but in order to avoid the doubts which I indicated above, I shall agree to define the numbers in the *second manner*, i.e., $P_q - 1$ will be the number of closed manifolds of q dimensions which we can trace on our polyhedron V and which are *linearly independent*, by which I mean there is no homology between them, in the sense of *Analysis situs*, p. 15.

However, I propose first of all to determine the number $P'_q - q$ of manifolds of q dimensions, closed and linearly independent, which can be traced on our polyhedron V but *restricted to those which are combinations of the variety v_q*.

The number P'_q is then what I called the *reduced Betti number.*

The manifolds of q dimensions in which combinations of the v_q can evidently be represented by $\sum_i \lambda_i a_i^q$, where the λ_i are integers and the letters a_i^q continue to represent the different manifolds v_q.

First of all, what is the condition for the manifold $\sum_i \lambda_i a_i^q$ to be closed?

To determine this we find the manifolds v_{q-1} which bound this manifold. To find them it evidently suffices to replace the a_i^q by their values given by the congruence (3).

This set of bounding manifolds will then be given by the formula

$$\sum_i \sum_j \lambda_i \varepsilon_{i,j}^q a_j^{q-1}.$$

Then for the manifold $\sum_i \lambda_i a_i^q$ to be closed it suffices that we have identically

$$\sum_i \sum_j \lambda_i \varepsilon_{i,j}^q a_j^{q-1} = 0,$$

[18]The polyhedron so defined, and hence all n-dimensional polyhedra bounded by $n+1$ hyperplanes, will be called the *generalized rectilinear tetrahedron.* I shall call each manifold homeomorphic to it a *generalized tetrahedron.*

i.e., for any j we have

$$\sum_i \lambda_i \varepsilon_{i,j}^q = 0.$$

In other words, the manifold $\sum_i \lambda_i a_i^q$ will be closed if

$$(7, q) \qquad \sum_i \lambda_i a_i^q \equiv 0$$

by virtue of the congruences (3, q), which is how I describe the congruences (3) that relate the a_i^q to the a_j^{q-1}.

We now find the homologies which can exist between the manifolds a_i^q. We obtain all these homologies by combining those obtainable in the following fashion.

Consider the congruence

$$(8) \qquad a_k^{q+1} \equiv \sum \varepsilon_{k,i}^{q+1} a_i^q$$

which, according to the convention we have just made, is a congruence $(3, q+1)$; replacing the sign $\equiv$ by $\sim$, and the first member by zero, we obtain

$$(9, q) \qquad \sum_i \varepsilon_{k,i}^{q+1} a_i^q \sim 0.$$

This homology evidently holds because, by definition and the congruence (8), it expresses the fact that the a_i^q form the boundary of a_k^{q+1}.

We shall show later (§VI) that there are no others.

I denote this homology by (9, q) to indicate that it holds between the a_i^q. I claim that if the homology (9, q) holds, the congruence

$$(10, q) \qquad \sum_i \varepsilon_{k,i}^{q+1} a_i^q = 0$$

will be a consequence of the congruences (3, q).

In fact, replacing the a_i^q by their values given by the congruences (3, q) we obtain

$$\sum_i \varepsilon_{k,i}^{q+1} a_i^q \equiv \sum_i \sum_j \varepsilon_{k,i}^{q+1} \varepsilon_{i,j}^q a_j^{q-1}.$$

The second (right-hand side) member is identically zero by virtue of the relations (5).

That being given, let α_q be the number of manifolds a_i^q; let α_q' be the number of those manifolds which remain distinct when we do not regard manifolds as distinct if connected by a homology of the form (9, q); let α_q'' be the number of manifolds which remain distinct when we do not regard manifolds as distinct if connected by a congruence of the form (7, q).

It follows from these definitions that

1° $\alpha_q - \alpha_q'$ is the number of distinct homologies of the form (9, q).
2° $\alpha_q - \alpha_q''$ is the number of distinct congruences of the form (7, q).
3° $\alpha_q' \geq \alpha_q''$ since, if several α_i^q are connected by a homology of the form $(9, q)$, they must likewise be connected by the corresponding congruence $(10, q)$.

Finally, the number sought, $P_q' - 1$, is equal to $\alpha_q' - \alpha_q''$ since the genuinely distinct closed manifolds of the form $\sum_i \lambda_i a_i^q$ are equal in number to the congruences (7, q), i.e., to the number $\alpha_q - \alpha_q''$.

The number $P'_q - 1$ is the number of those manifolds which remain distinct when those connected by a homology (9, q) are regarded as indistinct. But the number of these homologies is $\alpha_q - \alpha'_q$; we then have

$$P'_q - 1 = (\alpha_q - \alpha''_q) - (\alpha_q - \alpha'_q) = \alpha'_q - \alpha''_q.$$

Q.E.D.

Let $a^q_1, a^q_2, \dots, a^q_i$ be the manifolds v_q, i in number, and let

$$\begin{array}{ccc} a^q_1 & \equiv & \sum \varepsilon^q_{i,j} a^{q-1}_j \\ a^q_2 & \equiv & \sum \varepsilon^q_{2}, a^{q-1}_j \\ \dots & \dots & \dots\dots\dots \\ a^q_i & \equiv & \sum \varepsilon^q_{i,j}, a^{q-1}_j \end{array}$$

be the corresponding congruences (3). We form the corresponding homologies

$$\sum \varepsilon^q_{i,j} a^{q-1}_j \sim 0, \quad \sum \varepsilon^q_{2,j} a^{q-1}_j \sim 0, \quad \dots, \quad \sum \varepsilon^q_{i,j} a^{q-1}_j \sim 0.$$

The necessary and sufficient condition for these homologies to be distinct is that we do not have any congruence of the form

$$\lambda_1 a^q_1 + \lambda_2 a^q_2 + \cdots + \lambda_i a^q_i \equiv 0$$

between the i manifolds $a^q_1, a^q_2, \dots, a^q_j$. The number of distinct homologies is then equal to the number of distinct a^q_i, bearing in mind the congruences (7, q). Then

$$\alpha_{q-1} - \alpha'_{q-1} = \alpha''_q \quad \text{or} \quad \alpha_{q-1} = \alpha'_{q-1} + \alpha''_q.$$

On the other hand, we have

$$\alpha'_0 = 1.$$

If in fact we can go from any vertex a^0_1 to any other vertex a^0_i by following the edges (i.e., if the polyhedron is in one piece), we have the homology

$$a^0_1 \sim a^0_i,$$

i.e., we only have one distinct vertex when the homologies are taken into account.

Now consider the congruence (3′)

$$P = \sum a^p_i.$$

The corresponding homology is written

$$\sum a^p_i \sim 0,$$

and there is no other homology (9, p). Then

$$\alpha_p = \alpha'_p + 1.$$

In addition, since the polyhedron is a single piece, only one of the combinations $\sum \lambda_i a^p_i$ can be closed, this is the polyhedron in its entirety, represented by the formula $\sum a^p_i$.

We then will have a single congruence of the form (7, p)

$$\sum a^p_i \equiv 0.$$

Then

$$\alpha_p = \alpha''_p + 1, \quad \alpha'_p = \alpha''_p.$$

We then have the series of equations

$$
\begin{array}{ll}
\alpha'_0 = 0, & \\
\alpha_0 = \alpha'_0 + \alpha''_1, & \alpha'_1 - \alpha''_1 = P'_1 - 1, \\
\alpha_1 = \alpha'_1 + \alpha''_2, & \alpha'_2 - \alpha''_2 = P'_2 - 1, \\
\dots\dots\dots\dots, & \dots\dots\dots\dots\dots, \\
\alpha_{p-1} = \alpha'_{p-1} + \alpha''_p, & \alpha'_{p-1} - \alpha''_{p-1} = P'_{p-1} - 1, \\
\alpha_p = \alpha'_p + 1, & \alpha'_p - \alpha''_p = 0,
\end{array}
$$

whence we easily derive

$$\alpha_p - \alpha_{p-1} + \alpha_{p-2} - \cdots \pm \alpha_1 \mp \alpha_0 = 1 - (P'_{p-1} - 1) + \cdots \mp (P'_2 - 1) \pm (P'_1 - 1) \mp 1$$

entirely analogous to the formula

$$\alpha_p - \alpha_{p-1} - \cdots \pm \alpha_1 \mp \alpha_0 = 1 - (P_{p-1} - 1) + \cdots \mp (P_2 - 1) \pm (P_1 - 1) \mp 1$$

that we found in *Analysis situs* p. 74.

§IV. Subdivision of polyhedra

Consider a polyhedron V, of p dimensions, with its various manifolds

$$a_i^p, \quad a_i^{p-1}, \quad \dots, \quad a_0^1, \quad a_i^0.$$

Suppose that we subdivide each of these manifolds a_i^p into several others, which I shall call b_i^p; then let b_i^{p-1} be the manifolds of $p-1$ dimensions which bound the b_i^p; let b_i^{p-2} be the manifolds of $p-2$ dimensions which bound the b_i^{p-1}; and finally, let b_i^0 be the manifolds of zero dimensions (vertices) which bound the b_i^1 (edges).

We then have a new polyhedron V' which is derived from the polyhedron V, in the sense I attached to this word on p. 62 of *Analysis situs*.

We can assume, furthermore, that if a simply or multiply connected manifold v_{q-1} bounds two manifolds b_j^q and b_k^q, it is not necessarily a single one of the manifolds b_i^{q-1} but may itself be subdivided into several manifolds b_i^{q-1}. In that case, returning to the terminology of p. 63 of *Analysis situs*, the manifolds b_i^{q-1} are *irregular*, and the manifolds b_i^{q-2} which separate them are *singular*.

If a manifold b_i^q does not form part of one of the manifolds a_j^q it will form part of one of the manifolds a_j^{q+1}, or one of the manifolds a_j^{q+2}, ..., or at any rate one of the manifolds a_j^p.

Can it simultaneously be part of two manifolds a_j^m and a_k^m?

According to the assumed manner of subdivision, which always adjoins new boundaries without suppressing any, this cannot happen unless the two manifolds are contiguous and have a common boundary part a_k^{m-1}.

I assume then b_i^q forms part of a_j^h, and does not form part of any manifold a_k^m where $m < h$. The manifold a_j^h always exists and we have $h \geq q$; moreover, the manifold a_j^h is unique, i.e., b_i^q cannot simultaneously be part of two manifolds a_k^h and a_j^h.

Then if we agree to collect into the same class all the manifolds b_i^q which form part of the manifold a_j^h without forming part of any manifold a_k^m where $m < h$, then each manifold b_i^q will belong to exactly one class.

I may then represent b_i^q by a notation with four indices,

$$b_i^q = B(q, h, j, k),$$

where the index q indicates the number of dimensions of b_i^q, the indices h and j indicate that b_i^q belongs to the class a_j^h, and the index k serves to distinguish the various members of that class. We have $h \geq q$.

Then by the definition of the polyhedron V and its subdivision we will have:

1° The congruences (3, q) relative to the polyhedron V, which I shall write

(3,q,i) $\qquad a_i^q \equiv \sum_j \varepsilon_{i,j}^q a_j^{q-1}$.

2° The equations which give the subdivision of the manifold a_i^q

(1,q,i) $\qquad a_i^q = \sum_k B(q,q,i,k)$.

3° The congruences analogous to the congruences (3), but relative to the polyhedron V'; I shall write these

(2,q,h,j,k) $\qquad B(q,h,j,k) = \sum \zeta B(q-1,h',j',k')$.

The ζ are numbers equal to ± 1 or 0; they depend on the seven indices q, h, j, k, h', j', k', so that they would be written, if this were necessary, in the form

$$\zeta(q,h,j,k,h',j',k').$$

The indices h', j', k' can run through all values under the $\sum$ sign. We note, meanwhile, that the $B(q-1)$ which bound $B(q,h,j,k)$ must, like $B(q,h,j,k)$ be part of a_j^h; but they can be part of other manifolds $a_{j'}^{k'}$, of a smaller number of dimensions, which are part of a_j^h. We then have

$$h' \leq h, \quad h' \geq q-1.$$

Moreover, if $h' = h$ we have $j' = j$.

If the relations (1), (2), (3) are to define a true substitution they must satisfy certain conditions.

The relations (1, q,i), (3, q,i) give

$$(\alpha) \qquad \sum_k B(q,q,i,k) \equiv \sum \varepsilon_{i,j}^q a_j^{q-1}.$$

If, on the left-hand side, I replace $B(q,q,i,k)$ by its value derived from $2(q,q,i,k)$, and a_j^{q-1} by its value derived from $(1, q-1, j)$, then the two sides must become identical; this is a first condition, but it is evidently not sufficient.

§V. Influence of subdivision on reduced Betti numbers

Let $\sum \alpha B(q,h,j,k)$ be a combination of manifolds b_i^q which represents a closed manifold of q dimensions, so that we have, according to our notations,

$$(1) \qquad \sum \alpha B(1,h,j,k) \equiv 0 \qquad (h \geq q).$$

Among the manifolds b_i^q which occur on the left-hand side of (1), we rewrite those which belong to the same class. Let

$$\mathbf{S} \alpha B(q,h,j,k)$$

be the set of all those which belong to the class a_j^h; the summation sign $\mathbf{S}$ signifies that we only combine manifolds in the same class, whereas $\sum$ signifies that we combine them all.

We shall then have

$$(2) \qquad \mathbf{S} \alpha B(q,h,j,k) = \sum \beta B(q-1,h',j',k'),$$

i.e., the manifold of $q-1$ dimensions

$$\sum \beta B(q-1, h', j', k')$$

is the boundary of the q-dimensional manifold

$$\mathbf{S}\alpha B(q, h, j, k).$$

The manifolds $a_{j'}^{k'}$ should belong to the boundary of a_j^h or they will be confused with a_j^h; in fact $B(q-1, h', j', k')$ belongs to $a_{j'}^{h'}$ and, on the other hand, to one of the $B(q, h, j, k)$ which forms the same part of a_j^h , if then $a_{j'}^{h'}$ does not form part of a_j^h , $B(q-1, h', j', k')$ will form part of a manifold a_k^m common to a_j^h and $a_{j'}^{h'}$, and which will have less than h' dimensions. This is contrary to the definition we have given for these classes.

On the other hand, $a_{j'}^{h'}$ cannot be confused with a_j^h .

In fact, let

$$\mathbf{S}\alpha_1 B(q, h_1, j_1, k_1) = \sum \alpha B(q, h, j, k) - \mathbf{S}\alpha B(q, h, j, k)$$

be the set of manifolds which occur in the left-hand side of (1) and which do not belong to the class a_j^h; we evidently have

$$\mathbf{S}_1 \alpha B(q, h_1, j_1, k_1) \equiv \sum \beta B(q-1, h', j', k').$$

Then $B(q-1, h', j', k')$ must simultaneously be part of $a_j^{h'}$ and one of the $B(q, h_1, j_1, k_1)$ and consequently, of one of the $a_{j_1}^{h_1}$ *different from* a_j^h . Then if $a_{j'}^{h'}$ is confused with a_j^h,

$$B(q-1, h', j', k') = B(q-1, h, j, k)$$

must belong to a manifold a_k^m common to a_j^h and $a_{j_1}^{h_1}$. Then either a_j^h does not form part of $a_{j_1}^{h_1}$ and we shall again have $m < h$, which will again be contrary to the definition of the classes; or else a_j^h will be part of $a_{j_1}^{h_1}$ and we shall have $h_1 > h$. Suppose that I have chosen the class a_j^h which corresponds to the greatest number h. Then we cannot have $h_1 > h$ and $a_{j'}^{h'}$ must belong to the boundary of a_j^h .

The congruence (2) entails the homology

$$\sum \beta B(q-1, h', j', k'). \sim 0 \tag{3}$$

This is because, on the other hand, a_j^h *is simply connected* and since all the manifolds $B(q-1, h', j', k')$ are on the boundary of a_j^h, the left-hand side of (3), representing a closed $(q-1)$-dimensional manifold on that boundary, must form the boundary of a q-dimensional manifold

$$\sum \gamma B(1, h'', j'', k''),$$

likewise situated on the boundary of a_j^h. (There is an exception if we have $h = q$.) Thus we have the congruence

$$\sum \beta B(q-1, h', j', k') \equiv \sum \gamma B(1, h'', j'', k''). \tag{4}$$

Moreover, since $B(q, h'', j'', k'')$ is on the boundary of a_j^h, it will be the same for $a_{j''}^{h''}$; since if $B(q, h'', j'', k'')$ is simultaneously part of $a_{j''}^{h''}$ and a manifold $a_{j'}^{h'}$ on the boundary of a_j^h; either $a_{j''}^{h''}$ is not part of $a_{j'}^{h'}$, and then B must be part of a_k^m, or $m < h''$, and we have seen that this is impossible.

We then have

$$h'' < h.$$

The congruences (2) and (4) give

$$\mathbf{S}\alpha B(q,h,j,k) = \sum \gamma B(q,h'',j'',k'')$$

and, since all the manifolds which occur in this congruence form part of a_j^h or its boundary, and since on the other hand a_j^h is simply connected, we have the homology

$$\mathbf{S}\alpha B(q,h,j,k) \sim \sum \gamma B(1,h'',j'',k'').$$

Then we can replace the set of terms $\mathbf{S}\alpha B(1,h,j,k)$ on the left-hand side of (1) by the set of terms

$$\sum \gamma B(q,h'',j'',k'').$$

If we operate in the same way on all the classes corresponding to a single value of h, the largest, we will have replaced the left-hand side of (1) by

$$\sum \alpha_2 B(q,h_2,j_2,k_2),$$

where the greatest value of the h_2 will be smaller than the greatest value of the h. Moreover, we have the homology

$$\sum \alpha B(q,h,j,k) \sim \sum \alpha_2 B(q,h_2,j_2,k_2).$$

Continuing in this way, we can again diminish the greatest value of the h. We stop when we have $h = q$ everywhere.

We can then finally replace the left-hand side of (1) by

$$\sum \alpha_0 B(q,q,j_0,k_0)$$

and we have, in addition,

$$\sum \alpha B(q,h,j,k) \sim \sum \alpha_0 B(q,q,j_0,k_0), \tag{5}$$

$$\sum \alpha_0 B(q,q,j_0,k_0) \equiv 0. \tag{6}$$

That being done, let us take the congruences which belong to a class determined by $a_{j_0}^h$ in the left-hand side of (6); let them be

$$\mathbf{S}\alpha_0 B(q,q,j_0,k_0).$$

We have

$$\mathbf{S}\alpha_0 B(q,q,j_0,k_0) \equiv \sum \beta_0 B(q-1,h'_0,j'_0,k'_0). \tag{7}$$

We see, as above, that $a_{j'_0}^{h'_0}$ must form part of the boundary of $a_{j_0}^q$, whence $h'_0 < q$ (and, since $h'_0 \geq q-1$, we have $h'_0 = q-1$).

Then let

$$a_{j_0}^q = \sum B(q,q,j_0,k_0) \tag{8}$$

be the equation $(1,q,j_0)$ which defines the subdivision of the manifold $a_{j_0}^q$, and let $B(q,q,j_0,1)$ and $B(q,q,j_0,2)$ be two manifolds appearing on the right-hand side of (8); I claim that these must appear in the left-hand side of (6) with the same coefficient α_0.

Suppose first of all that the two manifolds are neighboring; among the $(q-1)$-dimensional manifolds which serve as their common boundary there will be at least one which does not belong to the boundary of $a^q_{j_0}$, and which consequently is not part of the class $a^q_{j_0}$.

Let $B(q-1, q, j_0, 1)$ be that manifold: it does not belong to any other of the manifolds $B(q, q, j_0, k)$.

Then let

$$(9)\qquad \begin{cases} B(q,q,j_0,1) & \equiv & \varepsilon b_i^{q-1} \\ B(q,q,j_0,2) & \equiv & \varepsilon b_i^{q-1} \\ B(q,q,j_0,k) & \equiv & \varepsilon b_i^{q-1} \quad (k>2) \end{cases}$$

be the congruences $(2, q, q, j_0, 1)$, $(2, q, q, j_0, 2)$, $(2, q, q, j_0, k)$ which tell us the boundaries of the manifolds $B(q, q, j_0)$. Let us see what is the coefficient ε of the manifold $B(q-1, q, j_0, 1)$ in these congruences.

From what we have seen, this coefficient will be $+1$ in the first, -1 in the second, and zero in the others.

Then let α_1 and α_2 be the values of the coefficients α_0 corresponding to the two manifolds $B(1, 1, j_0, 1)$ and $B(q, q, j_0, 2)$.

Combining the congruences (9) furnishes us with a congruence

$$(10)\qquad \mathbf{S}\alpha_0 B(q, q, j_0, k_0) \equiv \varepsilon b_i^{q-1}$$

which must be identical to (8), and the coefficient ε with which $B(q-1, q, j_0, 1)$ appears on the right-hand side of (10) will evidently be $\alpha_1 - \alpha_2$. But $B(q-1, q, j_0, 1)$ cannot appear on the right-hand side of (7), because we have seen that on this right-hand side we must have $\alpha_1 - \alpha_2 = 0$.

Thus the two manifolds $B(q, q, j_0, 1)$ and $B(q, q, j_0, 2)$ must have the same coefficient α_0 if they are neighboring. That will again be true if they are not, for since $a^q_{j_0}$ is in one piece, we can pass from one of these manifolds to the other by a sequence of analogous manifolds, each a neighbor of its predecessor.

Thus the coefficient α_0 is the same for all our manifolds. Whence

$$\mathbf{S}\alpha_0 B(q, q, j_0, k_0) = \alpha_0 \sum B(q, q, j_0, k_0) = \alpha_0 a^q_{j_0}.$$

The congruence (6) and the homology (5) can then be written

$$(5')\qquad \sum \alpha B(q, h, j, k) \sim \sum \alpha_0 a^q_{j_0},$$

$$(6')\qquad \sum \alpha_0 a^q_{j_0} \equiv 0.$$

If a certain number of congruences of the form (1) are distinct, i.e., if any linear combination of their left-hand sides is not homologous to zero, then I claim that the congruences (6′) will be equally distinct, and conversely.

In fact, comparison of the relations (1), (5′) and (6′) shows that if we have

$$\sum \alpha B(Q, H, J, K) \sim 0,$$

we will likewise have

$$\sum \alpha_0 a^q_{j_0} \sim 0,$$

and conversely.

It follows that if the a_i^q and the b_i^q are simply connected, the reduced Betti numbers are the same for the two polyhedra V and V'.

Now let W be any closed manifold of q dimensions, situated on V. We can always construct a polyhedron V' derived from V in the sense of p. 62 of *Analysis situs*, and such that W is a combination of the b_i^q.

We must then conclude that, if the a_i^q are simply connected, the reduced Betti numbers relative to the polyhedron V are identical with the genuine Betti numbers, *defined in the second manner.*

§VI. Return to the proofs of paragraph III

Here we have to return to an essential point in the preceding reasoning. I said above that we have no homologies other than the homologies $(9, q)$ obtained in paragraph III. This is not evident, and it will not even be always true unless we assume that the a_i^q are simply connected.

We show this first for a polyhedron P in four-dimensional space. Consider a certain number of manifolds v_2 or a_i^2 belonging to this polyhedron; I call these its faces, just as the a_i^3, the a_i^1 and the a_i^0 may be called its cells, its edges and its vertices.

Suppose that we have a homology

$$\sum a_i^2 \sim 0$$

between the faces a_i^2.

This homology signifies that there exists a three-dimensional manifold V, forming part of V, which has $\sum a_i^2$ as boundary.

I claim that V *is composed of a certain number of cells of* P.

In fact, if a point of some cell belongs to V it will be the same for any other point of that cell, since we can go from the first point to the second without encountering any face, and consequently, without encountering the boundary of V and without leaving V.

Thus the theorem is evident as far as it concerns polyhedra in four-dimensional space and homologies between their faces.

Now let

$$\sum b_1 \sim 0$$

be a homology among the b_1, which are a certain number of the edges a_i^1. This says that there is a two-dimensional manifold V of which $\sum b_1$ is the boundary.

I denote the set of points common to V and a_j^k by $V(a_i^k)$.

The $V(a_i^3)$ will be two-dimensional manifolds with boundaries that consist either of edges b_1 or of $V(a_j^2)$ where the a_j^2 are the faces which bound the cell a_i^3. In fact, we cannot leave $V(a_i^3)$ without crossing the boundary of V, i.e., traversing one of the b_1 or without crossing the boundary of a_i^3, i.e., traversing a face a_j^2, and, since we remain on V, traversing one of the lines $V(a_j^2)$.

The total manifold V is the set of the $V(a_i^3)$.

Now considering $V(a_i^2)$, we have to distinguish two cases:

1° None of the edges b_1 belongs to a_i^2. We cannot then leave $V(a_i^2)$ without leaving a_i^2, i.e., traversing one of the edges a_j^1; the boundary of $V(a_j^2)$ then consists of the $V(a_j^1)$.

2° Or else one (or more) edge b forms part of a_i^2; in that case it will likewise form part of $V(a_i^2)$; but it can happen that $V(a_i^2)$ is composed of lines other than the edge b_1; these lines are bounded by the points $V(a_j^1)$ or points lying on b_1. These points situated on b_1, or where the other lines which compose $V(a_i^2)$ terminate on the edge b_1 are what I shall call *nodal points.*

In each case $V(a_i^2)$ will be a line or a set of lines, if in fact $V(a_i^2)$ is a surface it is because a_i^2, or a portion of that face, forms part of V. But I have the right to deform V as long as I do not change the boundary $\sum b_1$, and with an infinitely small deformation I can always prevent a region of a_i^2 forming part of V.

For the same reason, I may always assume that $V(a_i^1)$ reduces to one or more points, except if a_i^1 is one of the edges b_1, in which case $V(a_i^1)$ will be that edge itself.

That being given, I can deform V:

1° In such a way that all the $V(a_i^1)$ (other than $V(b_1)$) will be vertices. Let a_j^0 be a vertex of a_i^1. Let M be one of the points which constitute $V(a_i^1)$; around the point M and on V we describe a small closed curve C. Let K be the infinitely small area cut out of V by this curve C. We construct a sort of sleeve, infinitely slender, surrounding the edge a_i^1 and passing through C. At the vertex a_j^0 I take any surface S; it is cut by the sleeve in a very small closed curve C'. Let K' be that portion of the surface S bounded by C'; let H be the surface of the sleeve between C and C'. Thus we imagine a sort of drum, with H as lateral surface and K and K' as the two bases.

We then consider the manifold

$$V' = V - K + H + K'.$$

This manifold has the same boundary as V; but it no longer cuts a_i^1 at M, because we have deleted the portion K of V in which the point M occurs. In return, H does not cut the edge a_i^1, and K' cuts that edge in a_j^0.

By operating in the same way on all the points of intersection of V and a_i^1, we induce all these points to coincide with a_j^0.

2° In such a way that all the nodal points will be vertices.

In fact, let a_i^2 be a face passing through the edge b_1; the intersection of V and a_i^2 comprises other lines apart from b_1; let c be one of these lines, terminating on b_1 in a nodal point D. Let a_j^0 and a_k^0 be the two vertices of b_1. I make a surface S pass through b_1, forming part of P and not cutting a_i^2. Since a_j^0 and a_k^0 are on the boundary of V, I join these two points by a line L situated on V and slightly removed from b_1. Since this line is only slightly removed from b_1, I can pass through it another surface S' which does not pass through b_1 but which cuts S in a line L' only slightly removed from b_1. These three lines L, b_1 and L' have the same extremities a_j^0 and a_k^0.

Let V_1 be the portion of V lying between L and b_1; let S_1 be the portion of S between L' and b_1, and let S_1' be the portion of S' between L and L'.

I replace V by

$$V' = V - V_1 + S_1 + S_1'.$$

V' has the same boundary as V, but $V'(a_i^2)$ does not present nodal points outside of a_j^0 and a_k^0; since if a line analogous to c abuts on a nodal point situated between a_j^0 and a_k^0, the portion of that line c near to the nodal point should be found on S_1, which is impossible because S does not cut a_i^2.

To summarize: we can always assume that the $V(a_i^2)$ are the lines whose extremities are the vertices of a_i^2, having extremities two vertices a_j^0 and a_k^0. We can go from a_j^0 to a_k^0 by following the perimeter of a_i^2; let $\sum a_m^1$ be the set of edges of a_i^2 between a_j^0 and a_k^0. Since the face a_i^2 is assumed simply connected, the line L will divide it into two parts. Let Q be one of these parts, between L and $\sum a_m'$.

Let a_p^3 and a_q^3 be the two components separated by a_i^2. I pass a surface S through the edges $\sum a_m^1$, very little different from the face a_i^2 and situated entirely in the cell a_p^3; I pass a second surface S' through the same edges, very little different from a_i^2 and situated in the cell a_q^3; these two surfaces S and S' cut V along two lines L_1 and L_1' little different from L, and having extremities a_j^0 and a_k^0. Let S_1 be the portion of S between $\sum a_m^1$ and L_1; let S_1' be the portion of S' between $\sum a_m^1$ and A_1'; let V_1 be the portion of V between L_1 and L_1'. It is on V_1 that we will find L.

Now let

$$V' = V - S_1 + S_1 + S_1'.$$

V' has the same boundary as V; V' no longer passes through L, but in return it passes through the edges $\sum a_m^1$.

Operating in the same manner on all the lines like L, we see that we can always assume that all the $V(a_i^2)$ are reduced to combinations of edges.

Since the boundaries of $V(a_i^3)$ are either the b_1 or the $V(a_j^2)$, we see that the boundaries of the $V(a_i^3)$ are combinations of edges of P and, of course, all these edges must belong to a_i^3. Thus $V(a_i^3)$ is a simply or multiply connected surface, bounded by one or more closed lines which themselves are combinations of edges of a_i^3.

Since, in the case where a_i^3 is simply connected, these closed lines will subdivide the surface into a certain number of regions, and since the closed lines are combinations of edges of a_i^3, these regions will be combinations of faces of a_i^3.

We can always find a combination of these regions which has the same boundaries as $V(a_i^3)$. Suppose, for example, that the boundary of $V(a_i^3)$ is composed of three closed lines L, L_1, L_2; the line L will divide the surface of a_i^3 into two regions R and R'; the lines L_1 and L_2 likewise divide this surface into two regions R_1 and R_1', or R_2 and R_2'. I suppose that when we traverse L in a certain sense we have $V(a_i^3)$ on its left, and R and R' on its right; likewise I suppose that when L_1 and L_2 are traversed in a suitable sense they have on their left $V(a_i^3)$ and R_1, or $V(a_i^3)$ and R_2.

Then the manifold $R + R_1 + R_2$ has the same boundary as $V(a_i^3)$; we can then replace $V(a_i^3)$ by $R + R_1 + R_2$.

Operating in the same manner on all the $V(a_i^3)$, we can replace V by another manifold which has the same boundary $\sum b_1$, and which will be a combination of the faces of P.

The theorem is then proved as far as it concerns polyhedra in the space of four dimensions and the homologies between their edges.

It may be proved similarly for any polyhedron.

§VII. Reciprocal polyhedra

Let P be a polyhedron in four-dimensional space; this polyhedron is subdivided into a certain number of manifolds ν_3 which I call *cells*, denoted by a_i^3. These cells are separated from each other by manifolds ν_2 or a_i^2 which I call *faces*; these faces are bounded by manifolds ν_1 or a_i^1 which I call *edges*; and the extremities of the edges are the points ν_0 or a_i^0 which I call *vertices*.

I assume of course that the cells and faces are simply connected.

We choose a point $P(a_1^3)$ in the interior of each cell a_i^3, a point $P(a_i^2)$ in the interior of each face a_i^2, a point $P(a_i^0)$ on each edge, so that each edge is divided into two parts by the point $P(a_i^1)$.

We join each point $P(a_i^2)$ by lines to the vertices of the face a_i^2 and to each of the points $P(a_i^1)$ corresponding to different edges of the face a_i^2. Thus each face is divided into triangles, and the number of these triangles will be double the number of edges of a_i^2. We do the same for all the other faces.

Now consider a cell a_i^3; as we have said, all its faces a_j^2 have been decomposed into triangles T. We construct curvilinear triangles with the point $P(a_i^3)$ as common apex and the various sides of the triangles T as bases. The cell a_i^3 will then be decomposed into tetrahedra with $P(a_i^3)$ as common apex and the different triangles T as bases.

We now distinguish six sorts of lines (which are the edges of our tetrahedra):

The first kind join a vertex a_i^1 and a point $P(a_j^1)$; thus each edge will consist of two lines of the first kind.

The second kind join a point $P(a_i^3)$ and a point $P(a_j^2)$.

The third kind join a point $P(a_i^2)$ and a vertex a_j^0.

The fourth kind join a point $P(a_i^1)$ and a point $P(a_j^2)$.

The fifth kind join a point $P(a_i^3)$ and a point $P(a_j^1)$.

The sixth kind join a point $P(a_i^3)$ and a vertex a_j^0.

The lines of the second kind can be combined in pairs in two ways:

1° What I shall call the line b_i^1 will be formed from two lines of the second kind joining the same point $P(a_i^2)$ to two points $P(a_j^3)$ and $P(a_k^3)$ corresponding to two cells a_j^3 and a_k^3 separated by the face a_i^2. Thus there will be as many lines b_i^1 as faces a_i^2.

2° What I shall call a line c will be formed from two lines of the second kind joining the same point $P(a_i^3)$ to two points $P(a_i^2)$ and $P(a_k^2)$ corresponding to two faces a_j^2 and a_k^2 of the cell a_i^3.

Now we have to define what I shall call the surfaces b_i^2.

Take any a_i^1 and the point $P(a_i^1)$. Suppose that the faces which pass through a_i^1 are successively

$$a_1^2,\quad a_2^2,\quad \ldots,\quad a_q^2,$$

and suppose that the cells to which a_i^1 belongs are successively

$$a_1^3,\quad a_2^3,\quad \ldots,\quad a_q^3$$

in such an order that a_1^2 separates a_1^3 from a_2^3, a_2^2 separates a_2^2 from $a_3^3, \dots$ and finally a_q^2 separates a_q^3 from a_1^3. For the sake of symmetry we agree to denote the cell a_1^3 indifferently by a_1^3 or a_{q+1}^3.

We decompose each cell into tetrahedra and consider, in particular, the tetrahedra which have the point $P(a_i^1)$ as apex. Consider the $2q$ curvilinear triangles

$$P(a_i^1)P(a_k^3)P(a_2^2), \quad P(a_i^1)P(a_{k+1}^3)P(a_k^2) \qquad (k = 1, 2, \dots, q).$$

These $2q$ triangles form a certain polygon which I shall call b_i^2, and which has the set of lines

$$b_1^1, \quad b_2^1, \quad \dots, \quad b_q^1$$

as its boundary.

We now define the volumes b_i^3; the volume b_i^3 will be the set of tetrahedra which have the point a_i^1 as apex; this volume will be a three-dimensional polyhedron, simply connected, with boundary equal to the set of surfaces b_k^2 corresponding to the edges a_k^1 which abut the point a_i^0.

Juxtaposition of the volumes b_i^3 gives a new polyhedron P' which I shall call the *reciprocal polyhedron* of P, and which has cells b_i^3, faces b_i^2, edges b_i^1 and vertices the points $b_i^0 = P(a_i^0)$.

To each cell b_i^3 of P' there corresponds a vertex a_i^0 of P.
To each face b_i^2 of P' there corresponds an edge a_i^1 of P.
To each edge b_i^1 of P' there corresponds a face a_i^2 of P.
To each vertex b_i^0 of P' there corresponds a cell a_i^3 of P.

Moreover, in the sense of paragraph II, there is the same relation, for example, between the edge b_i^1 and the face b_j^2 as between the face a_i^2 and the edge a_j^1.

Then if the congruences characteristic of polyhedron P are written

$$a_i^3 \equiv \sum_j \varepsilon_{i,j}^3 a_i^3, \quad a_i^2 \equiv \sum_j \varepsilon_{i,j}^2 a_i^1, \quad a_i^1 \equiv \sum_j \varepsilon_{i,j}^1 a_j^0,$$

those of the polyhedron P' will be written

$$b_i^3 = \sum_k \varepsilon_{i,j}^1 b_j^2, \quad b_i^2 \equiv \sum_j \varepsilon_{i,j}^2 b_i^1, \quad b_i^1 \equiv \sum_j \varepsilon_{i,j}^3 b_j^0.$$

Now consider a line c formed from two lines of the second kind, joining the same point $P(a_i^3)$ to two points $P(a_j^2)$ and $P(a_k^2)$.

Let a_m^0 and a_p^0 be two vertices, both belonging to the cell a_i^3. Let d and d' be the two lines of the third kind which, respectively, join $P(a_j^2)$ to a_m^0 and $P(a_k^2)$ to a_p^0.

Since a_m^0 and a_p^0 belong to the same cell a_i^3, we can go from one of these vertices to the other by following a broken line consisting of edges a_j^1 belonging to a_i^3.

Let $\sum a_q^1$ be that broken line with extremities a_m^0 and a_p^0; the set of lines $c - d - \sum a_q^1 + d'$ will be a closed line, which I shall express by the congruence

$$c \equiv d + \sum a_q^1 - d'.$$

Since a_i^3 is simply connected, this closed line will be the boundary of a two-dimensional manifold inside a_i^3, which I shall express by the homology

$$c \sim d + \sum a_q^1 - d'.$$

Conversely, let $\sum x_q^1$ be a broken line formed from edges all belonging to a_i^3, the extremities of which are the vertices a_m^0 and a_p^0; these two vertices belong respectively to two faces a_j^2 and a_k^2 which are each part of a_i^3. Let the three lines be

$$c = P(a_j^2)P(a_i^3) + P(a_i^3)P(a_k^2), \quad d = P(a_j^2)a_m^0, \quad d' = P(a_k^2)a_p^0.$$

We have again

$$c \sim d + \sum a_q^1 - d'.$$

Now let a_i^0 be a vertex belonging to two faces a_j^2 and a_k^2. Take the two lines of the third kind

$$d_j = P(a_j^2)a_i^0, \quad d_k = P(a_k^2)a_i^0.$$

We can trace a line L, infinitely slightly removed from the vertex a_i^0, and going from a point of a_j^2 to a point of a_k^2.

Suppose, to fix ideas, that this line traverses three cells, encountering successively the face a_j^2, the cell a_j^3, the face a_m^2, the cell a_m^3, the face a_p^2, the cell a_p^3, and finally the face a_k^2.

We construct the three lines c

$$c_j = P(a_j^2)P(a_j^3) + P(a_j^3)P(a_m^2),$$
$$c_m = P(a_m^2)P(a_m^3) + P(a_m^3)P(a_p^2),$$
$$c_p = P(a_p^2)P(a_p^3) + P(a_p^3)P(a_k^2),$$

and the two lines of the third kind

$$d_m = P(a_m^2)a_i^0, d_p = P(a_p^2)a_i^0.$$

We have

$$c_j \equiv d_j - d_m, \quad c_m \equiv d_m - d_p, \quad c_p \equiv d_p - d_k;$$

and since the three cells a_j^3, a_m^3, a_p^3 are simply connected

$$c_j \sim d_j - d_m, \quad c_m \sim d_m - d_p, \quad c_p \sim d_p - d_k$$

and finally

$$c_j + c_m + c_p \sim d_j - d_k.$$

Then we can always find a broken line consisting of lines c and homologous to $d_j - d_k$, where d_j and d_k are lines of the third kind, abutting on the same vertex.

That being given, let

$$\sum b_i^1 \equiv 0 \tag{1}$$

be a congruence between the edges b_i^1 of the polyhedron P'.

The broken line $\sum b_i^1$ is evidently formed from an even number of lines of the second kind, and in traversing that broken line we successively encounter the q faces

$$a_1^2, \quad a_2^2, \quad \ldots, \quad a_q^2$$

before returning to the face a_1^2, which I shall also denote by a_{q+1}^2; and we encounter q cells

$$a_1^3, \quad a_2^3, \quad \ldots, \quad a_q^3$$

before returning to the cell a_1^3, which I shall also denote by a_{q+1}^3, so that the face a_k^2 will separate the cell a_k^3 from the cell a_{k+1}^3.

Our congruence is then written

$$\sum[P(a_k^3)P(a_k^2)+P(a_k^2)P(a_{k+1}^3)]\equiv 0$$

or, what comes to the same thing

$$\sum[P(a_{k-1}^2)P(a_k^3)+P(a_k^3)P(a_k^2)]=0.$$

Then let a_k^0 be a vertex of the face a_k^2, belonging, consequently, to both the cells a_k^3 and a_{k+1}^3.

Let d_k be the line $P(a_k^2)a_k^0$ of the third kind; we have just seen that there exists a broken line A_k, formed from lines belonging to the cell a_k^3, such that we have the homology

$$P(a_{k-1}^2)P(a_k^3)+P(a_k^3)P(a_k^2)\sim d_{k-1}+A_k-d_k.$$

Adding all these homologies, the left-hand side reduces to

$$\sum[P(a_{k-1}^2)P(a_k^3)+P(a_k^3)P(a_k^2)]=\sum b_i^1,$$

the lines d_k of the third kind disappear and

$$\sum b_i^1\sim\sum A_k$$

remains, so that consequently

$$\sum b_i^1\equiv\sum A_k\equiv 0.$$

So to each congruence $\sum b_i^1\equiv 0$ between the edges of P' there corresponds a congruence $\sum A_k\equiv 0$ between the edges of P, and such that we have

$$\sum b_i^1\sim\sum A_k.$$

Thus if $\sum b_i^1\sim 0$ we have $\sum A_k\sim 0$ and conversely.

Now let

(2) $$\sum A_k\equiv 0$$

be a congruence between the edges of P; suppose that A_k is a broken line formed from edges belonging to the cell a_k^3.

The left-hand side of the congruence (2) is composed of q similar broken lines

$$A_1,\quad A_2,\quad \ldots,\quad A_q$$

among which I shall denote A_1 by A_1 or A_{q+1} and A_q by A_0 or A_q.

Let a_{k-1}^0 and a_k^0 be the two extremities of the line A_k; the vertex a_k^0 will belong simultaneously to the cells a_k^3 and a_{k+1}^3; let $a_k'^2$ be the face of a_k^3, and a_{k+1}^2 be the face of a_{k+1}^3 to which a_k^0 belongs.

Take the lines

$$d_k'=P(a_k^2)a_k^0,\quad d_{k+1}=P(a_{k+1}^2)a_k^0$$

of the third kind, and on the other hand

$$c_k=P(a_k^2)P(a_k^3)+P(a_k^3)P(a_k^2).$$

We have seen that

$$A_k\sim -d_k+c_k+d_k'.$$

On the other hand, the lines d_{k+1} and d_k' abut on the same vertex a_k^0; we have likewise seen that we can find a combination C_k of lines c such that we have

$$C_k\sim d_k'-d_{k-1}.$$

Adding all these homologies, I find

$$\sum A_k \sim \sum c_k + \sum C_k$$

and consequently

$$\sum c_k + \sum C_k \equiv 0.$$

The left-hand side of the latter congruence is a combination of lines c, or, what comes to the same thing, a combination of edges b_i^1 of the polyhedron P', so that I can set

$$\sum c_k + \sum C_k = \sum b_i^1$$

whence

$$\sum A_k \sim \sum b_i^1.$$

In summary, to each congruence between the edges of P there corresponds a congruence between those of P', and conversely, and the necessary and sufficient condition for the left-hand side of one of the conditions to be homologous to zero is that the other shall be.

In other words, the number of distinct congruences between the edges is the same for P and P', if we do not consider congruences to be distinct when a linear combination of their left-hand side is homologous to zero.

In other terms again, the reduced Betti number relative to the edges of P is equal to the reduced Betti number relative to the edges of P'.

We could arrive at the same result by remarking that we can construct a polyhedron which will be simultaneously derived from the polyhedron P and derived from the reciprocal polyhedron P', and applying the theorem of paragraph V.

We shall see later, in paragraph X, that this proposition can be presented in another form.

On the other hand, this permits us to show that the reduced Betti numbers are equal to the genuine Betti numbers more simply than in paragraph V.

In fact, the definition of the polyhedron P' involves a certain arbitrariness: its vertices b_i^0 are not required to lie in the interiors of the cells a_i^3 of P. Under these conditions, we can evidently always choose the polyhedron P' in such a fashion that *any* closed line is a combination of the b_i^1.

§VIII. Proof of the fundamental theorem

Let N_1 be the number of edges of our polyhedron P, N_2 the number of faces, N_3 the number of cells. We form a table by the following rules.

The table has $N_2 + N_3$ columns, N_2 are called the first kind and N_3 the second; it has $N_2 + N_1$ rows, N_2 of the first kind and N_1 of the second. Here are the elements of the table:

1° For an element of the ith row of the first kind I shall write 1 if $i = j$ and 0 if $i \neq j$.

2° The elements belonging to a row of the second kind and a column of the second kind are all zero.

3° The element of the ith column of the first kind and the jth row of the second kind will be $\varepsilon_{i,j}^2$, where $\varepsilon_{i,j}^2$ is the number that gives the relation between the face a_i^2 and the edge a_j^1.

4° The element of the ith row of the first kind and the jth column of the second kind will be $\varepsilon^3_{i,j}$, i.e., the number that gives the relation between the cell a^3_j and the face a^2_i.

If we take an example with two cells, four faces, and three edges, our table will have the following appearance.

$$
(1) \qquad \left\{ \begin{array}{l}
1\,0\,0\,0\,\varepsilon\,\varepsilon \\
0\,1\,0\,0\,\varepsilon\,\varepsilon \\
0\,0\,1\,0\,\varepsilon\,\varepsilon \\
0\,0\,0\,1\,\varepsilon\,\varepsilon \\
\varepsilon\,\varepsilon\,\varepsilon\,\varepsilon\,0\,0 \\
\varepsilon\,\varepsilon\,\varepsilon\,\varepsilon\,0\,0 \\
\varepsilon\,\varepsilon\,\varepsilon\,\varepsilon\,0\,0 \\
\varepsilon\,\varepsilon\,\varepsilon\,\varepsilon\,0\,0
\end{array} \right.
$$

For simplicity, I have not written the indices of the numbers ε.

Now here are the operations that I permit to be performed on the table:

1° Add one column to another *of the same kind*, or subtract.
2° Add one row to another *of the same kind*, or subtract.
3° Permute two columns of the same kind, changing all the signs of one of them.
4° Permute two rows of the same kind, changing all the signs of one of them.

All these transformations, under which the elements of the table remain integers, will be called *arithmetic transformations* of the table.

We can use them to simplify the parts of the table corresponding to columns of the first kind and rows of the second kind, or columns of the second kind and rows of the first kind.

Here is how far we can push the simplification, according to well-known arithmetic theorems. When the reduction terminates, the element of the ith column of the first kind and the jth line of the second kind:

1° will be zero if $i > j$.
2° will equal an integer H_i, which may be zero, if $i = j$.
3° will again be zero if $i < j$ and H_i is prime to H_j.
4° finally will be zero if $j > N_2$.

It will be the same for the element of the ith row of the first kind and jth of the second kind.

The reduction can be pushed further again if we authorize a fifth operation: multiply all the elements of a line or column by the same nonzero number (integer or not).

The corresponding transformations will be called *algebraic transformations* of the table.

We can then assume that the element in the ith column of the first kind and the jth row of the second kind (likewise the element in the ith row of the first kind and the jth column of the second kind) is zero if $i \neq j$. If $i = j$ it can be equal to 0 or 1. If this is the case I call the table *reduced.*

After the fifth operation, the elements belonging to rows and columns of the first kind may no longer be integers; furthermore, the determinant formed from these lines and columns may no longer be 1, but it must remain nonzero.

The table (1) concerns the faces of the polyhedron P and their relations to cells and edges. In the same way we could set one up for the edges of the polyhedron P and their relations to cells and vertices.

Likewise, we could consider the polyhedron P' defined above and construct two tables for the faces and edges of P'.

Compare the table (1) for the faces of P with the table (1′) for the edges of P'.

It follows from the preceding that these tables can be derived from each other by replacing rows by columns.

That being given, we imagine the table (1) for the faces of P and examine how we can derive the Betti number P_2 of P from that table.

First of all, how can we derive the congruences between the faces and the edges?

Consider any column of the first kind; for example, the ith column. We multiply the elements of that column and those of the kth row of the first kind by a_k^2 and add; then equating to the sum obtained by multiplying the elements of that same column and those of the jth row of the second kind by a_j^1, we obtain the congruence

$$a_i^2 \equiv \sum \varepsilon_{i,j}^2 a_j^1,$$

which is indeed one of the congruences (3) of paragraph II. All the other congruences are just combinations.

Now how can we find the homologies between the faces? To do this, imagine for example the ith column of the second kind; we multiply the elements of the kth row of that column by a_k^2, add and equate to zero; we find

$$\sum \varepsilon_{i,k}^3 a_k^2 \sim 0,$$

which is indeed one of the homologies (5) of paragraph II, of which all the others are combinations.

Now what happens to our table (1) *if we apply any algebraic transformation?*

Before the transformation, each column of the first kind corresponds to a face, each column of the second kind to a cell, each row of the first kind to a face, each row of the second kind to an edge.

As we have seen, we obtain as many congruences and homologies as there are columns by multiplying each row of the first kind by the corresponding face, and each row of the second kind by the corresponding edge, and adding.

Suppose now that we apply the second operation, i.e., we add a row of the first kind, which corresponds to a_i^2, to one of the first kind corresponding to a_k^2. We agree to say that the new kth row (that with the ith row added) always corresponds to the manifold a_k^2; but that the new ith row (which has not changed) corresponds to the manifold $a_i^2 - a_k^2$.

If we apply the fifth operation to the kth row of the first kind, multiplying the elements by a constant m, we agree to say that the new kth row corresponds to the manifold $\frac{1}{m}a_k^2$ (purely a symbolic notation, at least when the $\frac{1}{m}$ is not an integer).

As regards the fourth operation, it is only a combination of several operations analogous to the second.

We have thus defined the manifold which corresponds to any of the rows of the first kind of table, after application of any combination of the 2nd, 4th and 5th operations to its rows.

Thanks to the conventions, it will again suffice to obtain the congruences and homologies if we multiply the elements of each line by the corresponding manifold, add and equate to zero, changing the sign of the products so obtained when rows of the second kind are concerned.

Now if we apply the 1st, 3rd and 5th operations to columns of the table we either combine the congruences between them, and the homologies between them, or multiply one congruence and one homology by a constant factor.

Whence the following result:

To derive the congruences from the transformed table, it is necessary to do the following: multiply each row of the first sort by the corresponding manifold (according to our convention) and add; do the same for rows of the second kind; equate the results; thus we have as many congruences as columns of the first kind; all the other possible congruences are simply combinations.

Likewise, to derive the homologies it is necessary to: multiply each line of the first kind by the corresponding manifold, add and equate to zero; thus we have as many homologies as columns of the second kind; all the other possible homologies are simply combinations.

It is important to remark that the congruences and homologies thus obtained are only of symbolic value, since their coefficients can be fractional.

In fact, on the one hand the elements of the transformed table may no longer be integers, while on the other hand the manifold corresponding to a row may be, as I have said, only symbolic.

However, since the coefficients, integral or not, are always commensurable, it will suffice to multiply our congruence or homology by a suitable integer in order to derive a congruence or homology with integral coefficients, which always has a meaning.

Suppose now that we have reduced the table as I have described above. How many distinct homologies are there?

Among the N_3 columns of the second kind we have $N_3 - N_2'$ which are zero, and N_2 which have one element equal to 1 and zeros elsewhere. The first $N_3 - N_2'$ do not yield any homology; each of the N_2' others gives us one of N_2' homologies which are evidently all distinct.

There are then N_2' distinct homologies.

How many distinct congruences are there between faces and their edges?

There are evidently N_2, corresponding to the N_2 columns of the first kind, and these congruences are distinct, since the determinant formed from the lines and columns of the first kind is nonzero.

We now consider the N_1 rows of the second kind in our reduced table; among them there are $N_1 - N_2''$ which are zero, and N_2'' with one element 1 and zeros elsewhere. Among our N_2 congruences we then have N_2'' which contain an edge and $N_2 - N_2''$ which do not. There are then $N_2 - N_2''$ congruences *between faces alone*, and these congruences are all distinct.

We then have, *between faces alone*, $N_2 - N_2' - N_2''$ congruences which remain distinct when we omit those which can be derived from the others by means of the homologies.

The Betti number relative to the faces of P is then

$$N_2 - N_2' - N_2'' + 1.$$

Now we shall look for the Betti number relative to the edges of P.

We shall evidently find it by operating as we have just done on the table (1′) relative to the edges of P'.

But we pass from one table to the other by replacing rows by columns. The numbers which play the same role relative to (1′) as N_2, N_2', N_2'' play relative to (1) are then N_2, N_2'', N_2', respectively.

Then the Betti number relative to the edges of P' is again

$$N_2 - N_2' - N_2'' + 1.$$

Thus the Betti numbers relative to faces of P and edges of P' are equal.

But we have seen above that the Betti numbers relative to edges of P and those of P' are equal, and likewise the Betti numbers for faces of P and those of P'.

Thus the Betti number relative to faces of P is equal to the Betti number relative to edges of P.

Our fundamental theorem is proved then, as far as it concerns the polyhedron P, i.e., for polyhedra of four dimensions.

Without any doubt the proof can be extended to any polyhedron.

§IX. Various remarks

The fundamental theorem is now established, by a proof which differs essentially from that on p. 28 of *Analysis situs*.

However, we are not satisfied with this. We must try to recover the intermediate propositions, in particular that according to which:

The necessary and sufficient condition for the existence of a manifold V such that $\sum N(V, V_i) \neq 0$ is that there is no homology $\sum V_i \sim 0$.

Consider two manifolds, the first V_1 of one dimension, composed of edges of P', the second V_2 of two dimensions, composed of faces of P, such that we have

$$V_1 = \sum \alpha_i b_i^1, \quad V_2 = \sum \alpha_i' a_i^2,$$

the edge b_i^1 being that which corresponds to the face a_i^2, according to the conventions of paragraph VII.

The edge b_i^1 cuts the face a_i^2, and no other face, and it cuts in a way which is represented by the notation of p. 25 of *Analysis situs* as

$$N(V_1, V_2) = \sum \alpha_i \alpha_i'.$$

In what follows we shall suppose that the manifolds V_1 and V_2 are closed, which we express by the congruences

$$(1) \qquad \sum \alpha_i b_i^1 \equiv 0, \quad \sum a_i' \alpha_i^2 \equiv 0.$$

We verify first of all that we have

$$\sum \alpha_i \alpha_i' = 0$$

provided that we have one of the two homologies[19]

$$(2) \qquad \sum \alpha_i b_i^1 \sim 0, \quad \sum \alpha_i' a_i^2 \sim 0.$$

If we have the second homology, for example, this means we have

$$\alpha_i' = \sum_{j=1}^{N_3} \zeta_j \varepsilon_{i,j}^3,$$

where ζ_j is a coefficient not dependent on j.

[19] Cf. *Analysis situs*, pp. 28 and 29.

On another side, the first of the congruences (1) can be derived from one of the following

$$b_i^1 \equiv \sum b_j^0 \varepsilon_{j,i}^3, \tag{3}$$

whence

$$\sum a_i b_i^1 \equiv \sum\sum a_i b_j^0 \varepsilon_{j,i}^3.$$

Equating the coefficient of b_j^0 to zero, we obtain successively

$$\sum \alpha_i \varepsilon_{j,i}^3 = 0, \quad \sum\sum \alpha_i \zeta_j \varepsilon_{j,i}^3 = 0, \quad \sum \alpha_i \alpha_i' = 0.$$

Q.E.D.

We reason the same way if we have the first of the homologies (2).

I now claim that if the second homology (2) does not hold, we can choose the α_i in such a fashion that V_1 remains closed and nevertheless $\sum \alpha_i \alpha_i'$ is not zero.

In fact, to say that the second homology of (2) is not zero is to say that we cannot find numbers ζ_j such that

$$\alpha_i' = \sum \zeta_j \varepsilon_{j,i}^3. \tag{4}$$

To say that V_1 remains closed is to say that the α_i are subject to the conditions

$$\sum \alpha_i \varepsilon_{j,i}^3 = 0. \tag{5}$$

But it is clear that, if the a_i' do not satisfy equations of the form (4), the linear equation $\sum \alpha_i \alpha_i'$ will be distinct from the equations (5); then we can always find numbers α_i which satisfy the equations (5) without satisfying $\sum \alpha_i \alpha_i' = 0$.

We remark furthermore that we have not lost generality in assuming that our manifolds V_1 and V_2 are combinations of the b_i^1 and the a_i^2 from the subdivision of the manifold V into the polyhedra P and P'. Whatever the manifolds V_1 and V_2, we can always subdivide V to form two reciprocal polyhedra P and P' such that V_1 is a combination of edges of the latter, and V_2 is a combination of faces of the former.

It will be necessary to see how the table (1) of paragraph VIII and analogous tables permit us to determine the Betti numbers as Betti himself defined them, and not only the Betti numbers defined in the second manner, i.e., those we have considered up to the present.

Consider for example a table analogous to (1) but relative to edges of the polyhedron P and their relations with the faces and vertices. In particular, consider the columns of the second kind and the rows of the first kind, where the numbers $\varepsilon_{i,j}^2$ appear. Let T be the partial table obtained in this way. With the aid of this table we can form the congruences

$$a_i^2 \equiv \sum \varepsilon_{i,j}^2 a_j^1,$$

whence we deduce the homologies

$$\sum \varepsilon_{i,j}^2 a_j^1 \sim 0. \tag{6}$$

Then we can recognize whether several closed lines formed by combinations of the edges a_j^1 are distinct, *in the sense of the first definition*, i.e., *in the sense of Betti*, which lets us know if they are connected by a homology obtained by combination of the homologies (6) under addition, subtraction or multiplication, but *without division.*

Suppose that we have subjected our table to a series of those transformations which I have called arithmetic in paragraph VIII.

Let $\zeta^2_{i,j}$ be the number which, in the transformed table, appears in the jth row of the first kind and the ith column of the second kind. Let c_j be the manifold which corresponds to the jth row of the first kind in our table, transformed according to the conventions of paragraph VIII. From what we have seen in paragraph VIII, that manifold is none other than a combination of edges a^1_j.

We then have the homologies

$$\sum \zeta^2_{i,j} c_j \sim 0. \tag{6'}$$

These homologies are just the combinations of the homologies (6) that we can obtain *without division*, and conversely we can derive the homologies (6) from the homologies (6′) *without division*, this is a consequence of the arithmetic character of these transformations.

Thus when we want to decide whether two closed lines are distinct in the sense of Betti we can use the homologies (6′) in place of (6).

We can assume that we reduce the table by arithmetic transformations, as I explained in paragraph VIII and, consequently, that $\zeta^2_{i,j}$ is zero: 1° if $i > j$; 2° if $j > N_2$.

When the table is reduced to columns of the second kind and rows of the first kind it will assume, for example, the following form.

$$\begin{matrix} a & 0 & 0 & 0 & 0 \\ e & b & 0 & 0 & 0 \\ f & g & c & 0 & 0 \\ h & k & l & d & 0 \\ 0 & 0 & 0 & 0 & 0 \\ 0 & 0 & 0 & 0 & 0 \end{matrix}$$

I have summed six rows and five columns; I have assumed that one of the numbers $\zeta^2_{i,j}$ is equal to zero so that one of the columns of the table transforms entirely into zeros. I add that if d is equal to 1, the numbers h, k, l which appear in the same row will be zero.

That being given, if d is not equal to zero the two definitions of Betti numbers do not coincide, because we have a homology $dc_4 \sim 0$, whence we can deduce the homology $c_4 \sim 0$ by division. If $d = 1$ we have $h = k = l = 0$, and if c is not equal to 1 we have the homology $cs_3 \sim 0$, and the two definitions do not agree; and so on.

In summary, for the two definitions to agree it is necessary and sufficient that the product $abcd$ equal 1.

To interpret this result, we return to the untransformed table. The product $abcd$ will be the greatest common divisor of all the determinants obtained by suppressing $N_2 - N_3$ lines of the table T, provided these determinants are not all zero (in which case we shall have a column exclusively composed of zeros in the transformed table). If the determinants are all zero, we form others by suppressing a column and $N_2 - N_3 + 1$ rows of the table T; the product $abcd$ will be the greatest common divisor of all these determinants, provided they are not all zero; and so on.

Thus we arrive at the following rule:

Let Δ_p be the greatest common divisor of the determinants obtained by suppressing p rows and $N_2 - N_3 + p$ columns of the table T. The necessary and sufficient condition for the two definitions of Betti numbers to coincide is that the first non-vanishing Δ_p shall equal 1 (the greatest common divisor of several numbers equal to zero being equal to zero by definition).

Suppose that the manifold $V_1 = \sum \alpha_i b_i^1$ considered at the beginning of this paragraph is not the boundary of a two-dimensional manifold, but satisfies the homology $V_1 \sim 0$. In other words, the homology $V_1 \sim 0$ can be deduced from the homologies (6) using division, but not otherwise. In that case we have nevertheless

$$N(V_1, V_2) = \sum \alpha_i \alpha_i' = 0.$$

§X. Arithmetic proof of a theorem of paragraph VII

Here is a way of forming homologies which can be useful to know.

Let b_i^0 be a vertex of the polyhedron P' situated in the interior of a cell a_i^3 of the polyhedron P. On the other hand, let a_k^0 be a vertex of P belonging to the cell a_i^3.

Now let b_i^1 be an edge of P', the extremities of which are b_j^0 and b_h^0, so that one of the congruences (3) (cf. §II) relative to P' will be

$$b_i^1 = b_j^0 - b_h^0.$$

On the other hand, let a_i^2 be the face of P which corresponds to the edge b_i^1 of P' and let a_k^0 be one of the vertices of a_i^2; we have the homology

$$b_i^1 \sim a_k^0 b_j^0 - a_k^0 b_h^0. \tag{1}$$

Let a_i^1 be an edge of P, the extremities of which are a_j^0 and a_h^0, so that one of the congruences (3) relative to P is

$$a_i^1 \equiv a_j^0 - a_h^0.$$

Let a_k^3 be one of the cells of P to which a_i^1 belongs, and b_k^0 the corresponding vertex of P'; we have the homology

$$a_i^1 \sim b_k^0 a_j^0 - b_k^0 a_h^0. \tag{2}$$

I claim now that all the homologies between the a_i^1 can be derived from the homologies (2).

In fact, let a_i^2 be any face of P and let

$$a_i^2 \equiv \sum \varepsilon_{i,j}^2 a_j^1$$

be the congruence of the form (3) to which it corresponds; we deduce the homology

$$\sum \varepsilon_{i,j}^3 a_j^1 \sim 0, \tag{3}$$

and we have seen in paragraph VI that all the homologies between the edges are combinations of the kind just obtained.

Then let a_j^1 be one of the edges of P which appears in the homology (3), and let

$$a_i^1 \equiv a_h^0 - a_l^0.$$

Moreover, let a_k^3 be one of the cells bounded by a_i^2. We have the homology

$$a_j^1 \sim b_k^0 a_h^0 - b_k^0 a_l^0. \tag{2'}$$

If we add the homologies (2′) which are of the form (2), after having multiplied by $\varepsilon^2_{i,j}$, all terms on the right-hand side disappear by virtue of the relations (5) of paragraph II; we then recover the homology (3). Q.E.D.

We prove similarly that all the homologies between the b^1_i can be deduced from the homologies (1).

We have seen above, in paragraph VII, that if we have a congruence

$$\sum a^1_i \equiv 0$$

we can find another congruence between the edges of P',

$$\sum b^1_j \equiv 0,$$

in such a fashion that we have the homology

$$\sum a^1_i \sim \sum b^1_j. \tag{4}$$

I now claim that this homology (4) can be deduced from the homologies (1) and (2).

In fact we decompose the left-hand side of our congruence $\sum a^1_i \equiv 0$ into a certain number of groups, in such a way that edges in the same group belong to the same cell a^3_k. Let $\sum a^1_j$ be one of these groups; we have the congruence

$$\sum a^1_j = a^0_m - a^0_p, \tag{5}$$

where a^0_m and a^0_p are the two extremities of the line formed by the set of edges in this group. I assume that all these edges belong to the cell a^3_k. Let

$$a^1_j \equiv a^0_h - a^0_l$$

be one of the edges; we have the homology

$$a^1_j \equiv b^0_k a^0_h - b^0_k a^0_p, \tag{2''}$$

and in adding all these homologies we find

$$\sum a^1_j \sim b^0_k a^0_m - b^0_k a^0_p. \tag{6}$$

We now add all the homologies (6), as well as the congruences (5) which correspond to different groups. Addition of the congruences (5) must give us the congruence $\sum a^1_i \equiv 0$; it follows that if a vertex a^0_m appears in one of the congruences (5) with the sign $+$, it must appear in another with the sign $-$. Addition of the homologies (6) then gives us

$$\sum a^1_j \sim \sum (b^0_k a^0_m - b^0_g a^0_m). \tag{7}$$

In writing this relation I assume that a^0_m appears in two of the congruences (5), once with the sign $+$ in the congruence which corresponds to the cell a^3_k, and once with the sign $-$ in the congruence which corresponds to the cell a^3_g.

Observe now that b^0_k and b^0_q are two vertices of P', and both these vertices belong to the cell b^3_m. We can then find a line consisting of edges of P', contained in that cell b^3_m, and going from b^0_k to b^0_q. Let $\sum b^1_s$ be that line; we have

$$\sum b^1_s \equiv b^0_k - b^0_g. \tag{5'}$$

Just as we have deduced the homology (6) from the congruence (5) and the homologies (2″), which are of the form (2); so we can deduce the homology

$$\sum b_s^1 \sim a_m^0 b_k^0 - a_m^0 b_q^0 \tag{6'}$$

from the congruence (5′) and homologies of the form (1).

For each term of the right-hand side of (7) there is a homology (6′). Adding these, we find

$$\sum\sum b_s^1 \sim \sum (a_m^0 b_k^0 - a_m^0 b_q^0), \tag{7'}$$

whence

$$\sum a_j^1 + \sum\sum b_s^1 \sim 0 \tag{8}$$

—a homology of the form (4) which has been deduced, as we wanted, from the homologies (1) and (2). Q.E.D.

One may ask why I have deemed it necessary to return to a theorem previously proved in paragraph VII. It may be considered to give an exposé, so to speak, of the geometric nature of the proof in paragraph VII. The present proof, on the contrary, has an arithmetic character; it invokes only properties of the schemas defined in paragraph II and the tables constructed in paragraph VIII; and it remains valid when these schemas and tables correspond to any polyhedron.

What have we assumed, in fact? Simply that if $\alpha_0^p, \alpha_1^p, \alpha_2^p$ are the numbers of vertices, edges and faces belonging to the same cell, and if $\beta_0^p, \beta_1^p, \beta_2^p$ are the number of cells, faces and edges which belong to the same vertex, then we have

$$\alpha_0^p - \alpha_1^p + \alpha_n^p = \beta_0^p - \beta_1^p + \beta_2^p = 2$$

and, in addition, any two vertices a_i^0 and a_k^0 are connected by a homology

$$a_i^0 \sim a_k^0. \tag{9}$$

But we can recognize when a vertex belongs to a face, for example, by applying purely arithmetic rules to the table of paragraph VIII, and in the same manner we can recognize whether a homology such as (9) holds.

§XI. The possibility of subdivision

All of the preceding assumes that any manifold can be subdivided into simply connected manifolds, so as to form a polyhedron P of p dimensions for which the manifolds $a_i^p, a_i^{p-1}, \ldots, a_i^2, a_i^1, a_i^0$ are all simply connected. For example, each manifold of three dimensions is supposed to be divisible into simply connected cells, separated from each other by simply connected faces.

This is what remains to be proved, and it is what I am going to prove. More precisely, I am going to show that every manifold of p dimensions can be subdivided as a polyhedron P for which the manifolds $a_i^p, a_i^{p-1}, \ldots, a_i^2, a_i^1, a_i^0$ are all generalized tetrahedra.

I suppose that the theorem has been proved for manifolds of $p-1$ dimensions, and propose to extend the proof to a manifold of p dimensions.

We present the definition of manifold in the following form, which combines the two definitions given in *Analysis situs*.

We have equations and inequalities:

$$(1)\qquad \begin{cases} x_i = \theta_i(y_1, y_2, \dots, y_q) & (i = 1, 2, \dots, n), \\ f_k(y_1, y_2, \dots, y_q) = 0 & (k = 1, 2, \dots, q - p), \\ \varphi_h(y_1, y_2, \dots, y_q) > 0. \end{cases}$$

These equations and inequalities define a manifold v that is bounded and, in general, not closed. Other, analogous, systems of equations and inequalities define partial manifolds that I call $v_1, v_2, \dots, v_m$.

Two of these manifolds will be called contiguous if they have a common part, and I can suppose that one can pass from any point of one of these manifolds to any point of another without leaving the set of these manifolds. The set makes up the manifold that I call V, and serves to define it.

I assume that this manifold V is orientable.

This is evidently the most general way to define a manifold.

We now consider the partial manifold v_1 defined by the equations (1).

By the implicit function theorem, one can satisfy the equations

$$f_k = 0,$$

by setting

$$y_j = \psi_j(z_1, z_2, \dots, z_p),$$

where the ψ are holomorphic functions of the z; but the series ψ may not converge for all points of the variety v_1.

The conditions for convergence are certain inequalities

$$\eta_k(z_1, z_2, \dots, z_p) > 0.$$

When we replace the y by functions of z, the relations

$$x_i = \theta_i, \quad \varphi_h > 0$$

become

$$x_i = \theta_i'(z_1, z_2, \dots, z_p),$$
$$\varphi_h'(z_1, z_2, \dots, z_p) > 0.$$

Then the set of relations

$$(1')\qquad x_i = \theta_i', \quad \varphi_h' > 0, \quad \eta_k > 0$$

define a certain manifold v_1', in such a way that the set of manifolds analogous to v_1' reproduces the manifold v_1.

In this way we are reduced to the second definition from *Analysis situs.*

This being so, let v_1'' be a manifold satisfying the following conditions: it is contained in v_1'; it consists of all points of v_1' not in common with contiguous manifolds; and consequently, the boundary of v_1'' lies entirely in the intersection of v_1' with its contiguous manifolds.

Thus, for each of the manifolds $v_1', v_2', \dots$ making up V, there corresponds a manifold $v_1'', v_2'', \dots$ satisfying the conditions just described. It is clear that one can arrange things in such a way that each point of V belongs to exactly one of the manifolds v'', or at most to the boundary of another manifold v''.

When the manifold is thus divided into manifolds v'' it constitutes a polyhedron P in the sense of paragraph II. But this polyhedron is not yet suitable, since we do not know that that manifolds v'' are generalized tetrahedra, or even whether they are simply connected.

We consider the manifold v_1'' and let

$$z_1 = 0, \quad z_2 = 0, \quad \ldots, \quad z_p = 0$$

be an *interior* point *of this manifold.* Also consider the one-dimensional manifold

$$z_1 = \alpha_1 t, \quad z_2 = \alpha_2 t, \quad \ldots, \quad z_p = \alpha_p t,$$

where the α are constants, and let t vary from 0 to $+\infty$. This is what I will call a *radius vector.*

Each radius vector meets the boundary of v_1'' in an odd number of points; in fact, when we follow this radius by letting t vary from 0 to $+\infty$ we will leave the manifold v_1''. We may later re-enter and leave several times, but we will ultimately leave once more than we enter.

It may happen that a radius vector meets the boundary of v_1'' at two coincident points.[20] Radius vectors that satisfy this condition will be called *remarkable rays.*

The set of remarkable rays forms one or more manifolds of $p-1$ dimensions, which I will call *remarkable cones.*

The intersections of the remarkable cones with the boundary of v_1'' form one or more manifolds of $p-2$ dimensions that I call U, and these manifolds U divide the boundary of v_1'' into regions that I call R.

A region R cannot be met by a radius vector in more than one point, but by what we have seen there are two cases: when we follow the radius vector by letting t increase from 0 to ∞ we can, at the moment when we encounter R, leave v_1'' or enter it. If the first case, for example, occurs for one of the vectors that encounters R, then it will occur for all the vectors that encounter R.

This leads to a distinction between regions R of the first kind, which are met by vectors leaving v_1'', and regions of the second kind, which are met by vectors entering v_1''.

Since the regions R are of $p-1$ dimensions, it follows from our initial assumption that they can be subdivided into generalized tetrahedra.

We suppose, to fix ideas, that a radius vector meets the boundary of v_1'' three times, in the successive regions R_1, R_2, R_3, so R_1 and R_3 are of the first kind and R_2 is of the second kind.

We subdivide R_1 and R_3 into generalized tetrahedra of $p-1$ dimensions.

If T_1 is one of these subdivisions of R_1, we take all the rays passing through points of T_1 and keep the part between the point $z_i = 0$ and the region R_1 (the part inside v_1''). The set of all these rays forms a generalized tetrahedron of p dimensions, with vertex at the point $z_i = 0$ and base the generalized tetrahedron of dimension $p-1$.

Now let T_3 be one of the subdivisions of R_3. Again we take all the rays passing through the different points of T_3 and keep the part between R_2 and R_3 (the part inside v_1''). This set forms a manifold of $p-1$ dimensions that we could call a *truncated generalized tetrahedron*, whose two bases are T_3 and a generalized tetrahedron of $p-1$ dimensions that I call T_2 and which is part of R_2. It is, in other words, the difference between two generalized tetrahedra, with common vertex $z_i = 0$ and bases T_3 and T_2, respectively.

[20]That is, tangentially. (Translator's note.)

This truncated generalized tetrahedron may be subdivided in its turn, into p generalized tetrahedra, just as a truncated triangular pyramid is divided into three triangular pyramids in the classical theorem.

Thus v_1'' is finally subdivided into generalized tetrahedra.

However, one difficulty remains: we can subdivide the manifolds $v_2'', v_3'', \dots$ as we have subdivided v_1'', and consider the subdivision of v_1'' into the generalized tetrahedra T_1 and that of v_2'' into generalized tetrahedra T_2. The common boundary of v_1'' and v_2'' is thereby subdivided—on the one hand into generalized tetrahedra τ_1 of $p-1$ dimensions that are faces of the T_1, on the other hand into generalized tetrahedra τ_2 of $p-1$ dimensions that are faces of the T_2. *But it is by no means evident that these two subdivisions coincide.*

Consider then the common part of one of the τ_1 and one of the τ_2. By the induction hypothesis, I can subdivide it into generalized tetrahedra σ of $p-1$ dimensions. Thus each of the tetrahedra τ_1 and each of the tetrahedra τ_2 will be subdivided into tetrahedra σ.

Now let τ_1' be one of the manifolds of q dimensions *belonging to* τ_1. (Here I use the word "belonging" in the same sense as when I say that the faces, edges and vertices of an ordinary tetrahedron belong to that tetrahedron, or when I said in paragraph II that the manifolds a_i^q belong to a polyhedron P.) Likewise, let τ_2' be one of the manifolds of q dimensions belonging to τ_2. These two manifolds τ_1' and τ_2' are generalized tetrahedra because, by the definition of generalized tetrahedron, every manifold that *belongs to* a generalized tetrahedron is itself a generalized tetrahedron. Then τ_1' and τ_2' will be found subdivided into generalized tetrahedra σ', of q dimensions, which *belong to* the tetrahedra σ of $p-1$ dimensions.

This will suffice in a pinch. Our manifolds $v_1'', \dots$ are partitioned into generalized tetrahedra T^p of p dimensions, their boundaries into tetrahedra T^{p-1} of $p-1$ dimensions, and so on. However, the tetrahedra T^{p-1} are not those belonging to the tetrahedra T^p, but only subdivisions of them.

But we can go further.

Consider one of the p-dimensional tetrahedra T^p into which v_1'' is subdivided. Recall that we obtained them by subdividing the truncated generalized tetrahedra arrived at above. Consequently, all vertices of T^p are on the boundary of v_1'' (with the exception of tetrahedra with the point $z_i = 0$ as a vertex, but these cause no difficulty).

Suppose, for example, that the points common to T^p and the region that I called R_3 above form a q-dimensional generalized tetrahedron T^q *belonging to* T^p, and that the points common to T^p and the region R_2 form a tetrahedron of $p-q-1$ dimensions, T^{p-q-1}, belonging to T^p.

The tetrahedra T^p and T^{p-q-1} are analogous to the tetrahedra τ_1' treated above; they can therefore be subdivided into tetrahedra analogous to those I called σ'. Let $S_1^q, S_2^q, \dots$ be the tetrahedra, analogous to the σ', which are the subdivisions of T^q, and let $S_1^{p-q-1}, S_2^{p-q-1}, \dots$ be the tetrahedra analogous to σ' which are the subdivisions of T^{p-q-1}. I claim that we can subdivide T_k^p into tetrahedra of p dimensions in such a way that the manifolds $S_1^q, S_2^q, \dots, S_1^{p-q-1}, S_2^{p-q-1}, \dots$ belong to T^p.

To show this, suppose first that T^p is a *rectilinear tetrahedron* (cf. end of §II). We know that a rectilinear tetrahedron is completely defined when we know its

$p+1$ vertices. Then T^p is the rectilinear tetrahedron whose vertices are those of T^q and T^{p-q-1}.

Suppose that T^q is decomposed into g subtetrahedra

$$S_1^q, \quad S_2^q, \quad \ldots, \quad S_g^q$$

and that T^{p-q-1} is decomposed into h subtetrahedra

$$S_1^{p-q-1}, \quad S_2^{p-q-1}, \quad \ldots, \quad S_h^{p-q-1}.$$

One can then check that T^p is decomposed into gh subtetrahedra whose vertices are those of

$$S_i^q \text{ and } S_k^{p-q-1} \quad (i=1,2,\ldots,g; k=1,2,\ldots,h).$$

If the tetrahedron T^p is not rectilinear, the result still holds because any tetrahedron is homeomorphic to a rectilinear tetrahedron.

Thus our manifold is decomposed into tetrahedra of p dimensions, so as to form a polyhedron whose members each belong to one of the tetrahedra.

One is thus freed of all doubts on the subject of subdividing a manifold V in the form of a polyhedron P for which all the a_i^q are simply connected.

Second Supplement to Analysis Situs

Proceedings of the London Mathematical Society 32 (1900), pp. 277–308.

Introduction.

I published a work entitled "*Analysis situs*" in the *Journal de l'École Polytechnique*; I dealt with the same problem a second time in a memoir bearing the title "*Complément à l'Analysis situs*" which was published in the *Rendeconti del Circolo Matematico di Palermo.*

Nevertheless, the question is far from exhausted, and I shall doubtless be forced to return to it several times. This time, I confine myself to certain considerations in the way of simplifying, clarifying and completing results previously acquired.

References carrying simply an indication of paragraph or page apply to the first memoir, that in the *Journal de l'École Polytechnique*; references where these indications are preceded by the letter c are associated with the memoir in the *Rendeconti.*

When references refer to paragraphs of the present memoir, I shall preface them by the letters *2c*.

§1. Review of the principal definitions

Consider a closed manifold of p dimensions. We assume that this manifold is subdivided so as to form a polyhedron P of p dimensions. The elements of that polyhedron are called the a_i^p; they are separated from each other by $(p-1)$-dimensional manifolds called the a_i^{p-1}; these are separated from each other by $(p-2)$-dimensional manifolds called the a_i^{p-2}; and so on until we arrive at the vertices of the polyhedron which are called the a_i^0.

All these manifolds are simply connected, i.e., homeomorphic to hyperspheres.[21]

If a manifold a_i^q is bounded by the a_j^{q-1} I shall write the congruence

$$a_i^q \equiv \sum \varepsilon_{ij}^q a_j^{q-1}, \tag{1}$$

where the ε are equal to $+1, -1$ or 0 (c, §II, p. 78).

We write, as well, the homology

$$\sum \varepsilon_{ij}^q a_j^{q-1} \sim 0. \tag{2}$$

We combine the congruences (1) and the homologies (2) by addition, subtraction, multiplication and sometimes division.

[21]Poincaré seems to be unconsciously assuming the Poincaré conjecture here, or else forgetting his definition of simply connected from p. 52. (Translator's note.)

Among the congruences between a_i^q and a_i^{q-1} obtained by combination of the congruences (1) we distinguish those which contain only the a_i^q, i.e., those from which the a_i^{q-1} have disappeared.

We sometimes denote the a_i^0 by the name *vertices*, the a_i^1 by the name *edges*, the a_i^2 by the name *faces*, the a_i^3 by the name *cells*, the a_i^4 by the name *hypercells*.

To a polyhedron P there corresponds a dual polyhedron P' (c, §VII) which has elements b_i^p in place of a_i^b; b_i^{p-1} in place of $a_i^{p-1}, \ldots$, and finally b_i^0 in place of a_i^0.

There is a correspondence between the two polyhedra such that b_i^{p-q} corresponds to a_i^q. The two polyhedra result from subdivision of the same manifold V.

Between the elements of P' we have congruences

$$b_i^q = \sum \varepsilon_{ji}^{p-q+1} b_j^{q-1} \tag{1'}$$

analogous to the congruences (1); we can also write

$$b_i^q = \sum \varepsilon_{ij}'^q b_j^{q-1}$$

by setting

$$\varepsilon_{ji}^{p-q+1} = \varepsilon_{ij}'^q.$$

We have another relation between the elements of P and those of P'.

Recall the notation $N(V, V')$ (§9, p. 25). We then have

$$N(a_k^q, b_i^{p-q}) = 0$$

if i is not equal to k, and

$$N(a_i^q, b_i^{p-q}) = \pm 1.$$

It remains to see whether we must take the + sign or − sign.

To find out, consider two corresponding elements of P and P' which I shall call a_i^q and b_i^{p-q}; also consider two corresponding elements a_j^{q-1} and b_j^{p-q+1} such that a_i^{q-1} belongs to a_i^q and b_i^{p-q} belongs to b_j^{p-q+1}.

I can always choose my coordinates in such a way that the equations of the a_i^q are

$$F_1 = F_2 = \cdots = F_{p-q} = 0, \tag{3}$$

where the F are the functions of the coordinates $y_1, y_2, \ldots, y_p$ which define the position of a point on the manifold V.

Likewise let

$$\Phi_1 = \Phi_2 = \cdots = \Phi_{q-1} = 0 \tag{4}$$

be the equations of b_j^{p-q+1}; I can then assume that the equations of the a_j^{q-1} are obtained by adjoining the equation $\psi = 0$ to the equations (3), and those of b_i^{p-q} are obtained by adjoining the equation $\psi = 1$ to the equations (4). I can arrange for the *same* function ψ to be on the left-hand side of these two equations.

Then among the inequalities which, along with the equations (3), complete the definition of a_i^p, we must have

$$\psi > 0.$$

Likewise, among the inequalities which, along with the equations (4) complete the definition of b_j^{p-q+1}, we must have

$$\psi < 1.$$

If we want ε_{ij}^q to be equal to $+1$ it is necessary by our conventions that the equations of a_j^{q-1} be placed in the following order:

$$F_1 = F_2 = \cdots = F_{p-1} = \psi = 0$$

and if we want $a_{ji}^{\prime p-q+1}$ to be $+1$ at the same time, it is necessary for the equations of b_i^{p-q} to be placed in the following order:

$$\Phi_1 = \Phi_2 = \cdots = \Phi_{q-1} = 1 - \psi = 0.$$

The number $N(a_i^q, b_i^{p-q})$ depends on the sign of the Jacobian of

$$F_1, \quad F_2, \quad \ldots, \quad F_{p-q}, \quad \Phi_1, \quad \Phi_2, \quad \ldots, \quad \Phi_{q-1}, \quad 1-\psi.$$

Likewise, the number $N(a_j^{q-1}, b_j^{p-q+1})$ depends on the sign of the Jacobian of

$$F_1, \quad F_2, \quad \ldots, \quad F_{p-q}, \quad \psi, \quad \Phi_1, \quad \Phi_2, \quad \ldots, \quad \Phi_{q-1}.$$

We can always assume that the functions F, Φ and ψ have been chosen so that these determinants do not vanish in the domain considered.

We then see that the two determinants are of the same sign if q is even and of opposite sign when q is odd.

In the first case we have

$$N(a_i^q, b_i^{p-q}) = N(a_j^{q-1}, b_j^{p-q+1})$$

and in the second case

$$N(a_i^q, b_i^{p-q}) = -N(a_j^{q-1}, b_j^{p-q+1}).$$

Since we can always assume

$$N(a_i^0, b_i^p) = +1,$$

we find successively

$$N(a_i^1, b_i^{p-1}) = -1, \; N(a_i^2, b_i^{p-2}) = 1, \; N(a_i^3, b_i^{p-3}) = +1, \; N(a_i^4, b_j^{p-4}) = +1, \; \ldots.$$

Apart from this, the number $N(a_i^q, b_i^{p-q})$ does not depend on g.

That being given, we can use the numbers ε_{ij}^q to form a table that I call T_q, where the number ε_{ij}^q occupies the ith row and jth column. Then in the table T_q there are as many rows as the a_i^q and as many columns as the a_j^{q-1}.

I let α_p be the number of a_i^q, so that the table T_q has α_q rows and α_{q-1} columns. In particular, the table T_1 gives us the relation between the edges and the vertices, the table T_2 gives that between the faces and the edges, etc.

I call the table for P', corresponding to T_q for P, the table T'_q. We see that the table T'_q is obtained from the table T_{p-q+1} by interchanging rows and columns.

We have denoted (c, §III, p. 82) by $\alpha_q - \alpha'_q$ the number of distinct homologies between the a_i^q and by $\alpha_q - \alpha''_q$ the number of distinct congruences between the a_i^q (the a_j^{q-1} being eliminated); and by

$$P_q = \alpha'_q - \alpha''_q + 1$$

the Betti number corresponding to the a_i^q.

We have let β_q, β'_q and β''_q be the numbers analogous to α_p, α'_q and α''_q in connection with the polyhedron P', so that

$$\beta_q = \alpha_{p-q}.$$

§2. Reduction of tables

Consider a table T consisting of integers arranged in a certain number of lines and columns. Our tables T_q are examples.

Suppose that we allow the following operations to be applied to this table:

1° Add one column to another, or subtract;

2° Permute two columns and change the sign of one of them;

3° Make the same operations on rows.

By combination of these operations we can subject the columns to any linear substitution with integral coefficients and determinant 1. The same is true for the rows.

What is the greatest degree of simplicity we can achieve in the table by means of these operations? This is what we are going to examine.

Suppose first of all, to fix ideas, that the table T does not have more rows than columns.

LEMMA I. *Let*

$$\begin{vmatrix} a_1 & a_2 & a_3 & a_4 & a_5 \\ b_1 & b_2 & b_3 & b_4 & b_5 \\ c_1 & c_2 & c_3 & c_4 & c_5 \end{vmatrix}$$

be a table T, assumed to be of three rows and five columns to fix ideas.

I assume that the fifteen numbers a, b, c are relatively prime to each other; I claim then that we can always find three numbers $\alpha_1, \beta_1, \gamma_1$ such that the five numbers

$$h_{1i} = \alpha_1 a_i + \beta_1 b_i + \gamma_1 c_i \qquad (i = 1, 2, 3, 4, 5)$$

are relatively prime.

For this purpose the numbers $\alpha_1, \beta_1, \gamma_1$ must first of all be relatively prime to each other. If this condition is satisfied, we can find six other numbers $\alpha_2, \beta_2, \gamma_2$; $\alpha_3, \beta_3, \gamma_3$ such that the determinant

$$\begin{vmatrix} \alpha_1 & \beta_1 & \gamma_1 \\ \alpha_2 & \beta_2 & \gamma_2 \\ \alpha_3 & \beta_3 & \gamma_3 \end{vmatrix} = 1.$$

We then set

$$h_{ki} = \alpha_k a_i + \beta_k \beta_i + \gamma_k c_i \qquad (i = 1, 2, 3, 4, 5;\ k = 1, 2, 3).$$

The rule for multiplication of determinants gives us

$$\begin{vmatrix} h_{11} & h_{12} & h_{13} \\ h_{21} & h_{22} & h_{23} \\ h_{31} & h_{32} & h_{33} \end{vmatrix} = \begin{vmatrix} \alpha_1 & \beta_1 & \gamma_1 \\ \alpha_2 & \beta_2 & \gamma_2 \\ \alpha_3 & \beta_3 & \gamma_3 \end{vmatrix} \cdot \begin{vmatrix} a_1 & b_1 & c_1 \\ a_2 & b_2 & c_2 \\ a_3 & b_3 & c_3 \end{vmatrix}$$

$$= \begin{vmatrix} a_1 & b_1 & c_1 \\ a_2 & b_2 & c_2 \\ a_3 & b_3 & c_3 \end{vmatrix} = \Delta \quad \text{say.}$$

This shows that the greatest common divisor of the three numbers h_{11}, h_{12}, h_{13}, and consequently that of the five numbers h_{1i}, divides Δ. It must likewise divide all determinants obtained by suppressing two columns in the table and consequently the greatest common divisor, M, of all these determinants.

Let p be any prime factor of M. Since our fifteen numbers a, b, c are relatively prime to each other, at least one of them, c for example, will not be divisible by p.

If we then put

$$\alpha_1 \equiv 0, \quad \beta_1 \equiv 0, \quad \gamma_1 \equiv c_5^{p-2} \pmod p, \tag{1}$$

it follows that

$$h_{15} \equiv c_5^{p-1} \equiv 1 \pmod p$$

so that the greatest common divisor of the five numbers h_{1i} will not be divisible by p.

We obtain a system of congruences analogous to (1) for each of the prime factors of M.

Then the greatest common divisor of the five numbers h_{1i} will not be divisible by any of the prime factors of M; and since it divides M it will be equal to 1.

1ST COROLLARY: If we subject the rows of the table to the linear substitution

$$\begin{matrix} \alpha_1 & \beta_1 & \gamma_1 \\ \alpha_2 & \beta_2 & \gamma_2 \\ \alpha_3 & \beta_3 & \gamma_3 \end{matrix}$$

it is clear that the elements of the ith column

$$a_i, \quad b_i, \quad c_i$$

become

$$h_{1i}, \quad h_{2i}, \quad h_{3i}$$

whence the consequence:

If the elements of the table are relatively prime to each other we can reduce the table to one where the elements of the first row are relatively prime to each other.

2ND COROLLARY. *If the elements of the table have greatest common divisor δ we can reduce the table to one where the elements of the first row have greatest common divisor δ.*

THEOREM: *Let m be the number of columns and n the number of rows ($m \geq n$); let M_0 be the greatest common divisor of all the determinants obtained by suppressing any $m - n$ columns of the table; let M be the greatest common divisor of all the determinants obtained by suppressing $m - n + 1$ columns and one row; let M_2 be that of the determinants obtained by suppressing $m - n + 2$ columns and two rows, etc.; finally let M_{n-1} be that of the determinants obtained by suppressing $m - 1$ columns and $n - 1$ rows, in other words the greatest common divisor of all the elements.*

These numbers $M_0, M_1, \ldots, M_{n-1}$ are not altered by the operations made on the rows, or on the columns.

It goes without saying that the number M_k will be considered to be zero if all the corresponding determinants are zero.

We can then enunciate our corollary in the following form:

3RD COROLLARY: *We can reduce the table to one where the greatest common divisor of the elements in the first row is M_{n-1}.*

LEMMA II. *By a transformation of columns we can reduce the table to one where the first element of the first row becomes M_{n-1}, and all other elements of the first row become zero.*

In fact we are going to subject the columns (assumed to be five in number, as above) to the linear substitution

$$(2) \qquad \begin{vmatrix} \alpha_1 & \alpha_2 & \alpha_3 & \alpha_4 & \alpha_5 \\ \beta_1 & \cdots & \cdots & \cdots & \beta_5 \\ \gamma_1 & \cdots & \cdots & \cdots & \gamma_5 \\ \delta_1 & \cdots & \cdots & \cdots & \delta_5 \\ \zeta_1 & \cdots & \cdots & \cdots & \zeta_5 \end{vmatrix}$$

with determinant 1. Let

$$a_1, \quad a_2, \quad a_3, \quad a_4, \quad a_5$$

be the elements of the first row. After the reductions the table has experienced above, the greatest common divisor of these five numbers has become M_{n-1}. We can then choose the substitution (2) in such a way that we have

$$\sum \alpha_i a_i = M_{n-1}, \quad \sum \beta_i a_i = \sum \gamma_i a_i = \sum \delta_i a_i = \sum \zeta_i a_i = 0,$$

then, after the transformation, the elements of the first row will be

$$M_{n-1}, \quad 0, \quad 0, \quad 0, \quad 0.$$

LEMMA III. *I claim that we can reduce to zero all elements of the first column, except the first, which remains equal to M_{n-1}, by a transformation of rows.*

In fact, after the reductions made previously, the elements of the first column (supposing three rows) are

$$M_{n-1}, \quad q_2 M_{n-1}, \quad q_3 M_{n-1},$$

where q_2 and q_3 are integers, since by our hypothesis all elements are divisible by M_{n-1}.

If we then subtract the first row q_2 times from the second, and q_3 times from the third, the first column becomes

$$M_{n-1}, \quad 0, \quad 0.$$

Meanwhile, the first row has not changed.

If we suppress the first row and first column of the table now, there remains a table T' of $m-1$ columns and $n-1$ rows in which the numbers

$$\frac{M_0}{M_{n-1}}, \quad \frac{M_1}{M_{n-1}}, \quad \ldots$$

play the same role as the numbers $M_0, M_1, \ldots$ do in connection with the table T.

In particular, the greatest common divisor of the elements of T' is $\frac{M_{n-2}}{M_{n-1}}$.

We can continue the reduction, now operating only on the $m-1$ remaining columns and the $n-1$ remaining rows. The first row does not change because its last $m-1$ elements are zero, nor does the first column, because its last $n-1$ elements are zero.

Thus we can operate on the table T' as we have on the table T. After this new reduction

1° All the elements of the first row and first column are zero, except the first in each, which remains equal to M_{n-1}.

2° All the elements of the second row and second column have become zero, except the second in each, which has become $\frac{M_{n-2}}{M_{n-1}}$.

3° If we suppress the first two rows and the first two columns, we obtain a table T'' of $m-2$ columns and $n-2$ rows, in which the numbers

$$\frac{M_0}{M_{n-2}}, \quad \frac{M_1}{M_{n-2}}, \quad \ldots, \quad \frac{M_{n-3}}{M_{n-2}}$$

play the same role as $M_0, M_1, \ldots, M_{n-1}$ do in connection with the table T. And so on.

At the end of the reduction the element in the ith row and jth column is zero if i is unequal to j; the element in the ith row and the ith column is equal to $\frac{M_{n-i}}{M_{n-i+1}}$.

The n numbers

$$M_{n-1}, \quad \frac{M_{n-2}}{M_{n-1}}, \quad \frac{M_{n-3}}{M_{n-2}}, \quad \ldots, \quad \frac{M_1}{M_2}, \quad \frac{M_0}{M_1} \tag{3}$$

may be called the *invariants* of the table T.

We may remark:

1° Each of these invariants divides its successor;

2° Any of these invariants can be zero, but if one of them is, so are all its successors.

If the table T has more rows than columns the reduction is made in the same manner, except that the roles of rows and columns are interchanged.

We then have $m < n$; the number M_0 will be the greatest common divisor of the determinants obtained by suppressing $n-m$ rows; in general, M_i will be the greatest common divisor of the determinants obtained by suppressing any $n-m+i$ rows and i columns. Finally, the greatest common divisor of the elements of the table will be M_{m-1}.

In general, the number of invariants will be the smaller of the two numbers m and n.

§3. Comparison of the tables T_q and T'_q

The table T_q tells us the relations between the a_i^q and the a_j^{q-1} in the polyhedron P. Each row of the table corresponds to an a_i^q and each column to an a_j^{q-1}. Likewise each row of this table corresponds to a congruence

$$a_i^q \equiv \sum \varepsilon_{ij}^q a_j^{q-1} \tag{1}$$

between the a_i^q and the a_j^{q-1}, and a homology

$$\sum \varepsilon_{ij}^q a_j^{q-1} \sim 0 \tag{2}$$

between the a_j^{q-1}.

Now what happens if we reduce the table T_q by the operations of the preceding paragraph? Each row of the reduced table corresponds to a linear combination of the a_i^q, and each column to a linear combination of the a_j^{q-1}. I have explained (c, §VIII, p. 99) the rules by which these linear combinations must be formed. Here is how these rules may be summarized.

Suppose that we pass from the table T_q to the reduced table by application of a certain linear substitution S to the rows of T_q and another linear substitution σ to the columns of T_q. Let σ' be the substitution contragredient to σ (by that I mean that if we have two series of α_{q-1} variables x_i and y_i and we apply the substitution σ to the first series and σ' to the second, then the form $\sum x_i y_i$ will not be altered).

Now suppose that S changes a_i^q into

$$c_i^q = \sum_{j=1}^{\alpha_q} \lambda_{ij} a_i^q$$

and that σ' changes a_i^{q-1} into

$$d_i^{q-1} = \sum_{i=1}^{a_{q-1}} \mu_{ij} a_j^{q-1}.$$

We shall make the ith row of the reduced table correspond to the linear combination c_i^q, and the ith column to the linear combination d_i^{q-1}.

In our reduced table all these elements are zero except those in the ith row and ith column which, by the preceding paragraph, are given by the formula

$$\frac{M_{n-1}}{M_{n-i-1}}.$$

For brevity I denote by ω_i^q this element in the ith row and column; and I agree to regard ω_i^q as zero if i is greater than the smaller of the numbers α_q and α_{q-1} (the numbers of rows and columns).

Then, corresponding to the ith row of the reduced table, we have the congruence

$$c_i^q \equiv \omega_i^q d_i^{q-1} \tag{1'}$$

and the homology

$$\omega_i^q d_i^{q-1} \sim 0. \tag{2'}$$

The congruences and the homologies (1′) and (2′) result from the congruences and homologies (1) and (2) by addition, subtraction, multiplication, *but without division*, and conversely.

If $\alpha_{q-1} > \alpha_q$ and if $i > \alpha_q$, ω_i^q is zero, so that the congruence and homology (1′) and (2′) reduce to

$$c_i^q \equiv 0 \text{ and } 0 \sim 0.$$

The numbers ω_i^q are those which I called the *invariants* of the table T_q in the preceding paragraph. Suppose that *among these invariants we have γ_q which are nonzero*; we shall have, of course,

$$\gamma_q \leq \alpha_q, \quad \gamma_q \leq \alpha_{q-1}.$$

Among the congruences (1′), the first γ_q will include both c_i^q and d_i^{q-1} because ω_i^q will be nonzero. On the other hand, the last $\alpha_q - \gamma_q$ will be written

$$c_i^q \equiv 0,$$

and will not contain the a_i^{q-1}; it is clear that all these congruences will be distinct, and that we obtain in this way all the congruences between the a_i^q from which the a_j^{q-1} have been eliminated. We then have

$$\alpha_q - \alpha_q'' = \alpha_q - \gamma_q, \quad \alpha_q'' = \gamma_q.$$

Now among the homologies (2′), the last $\alpha_q - \gamma_q$ reduce to identities, but the first γ_q are distinct; we then have

$$\alpha_{q-1} - \alpha_{q-1}' = \gamma_q,$$

whence the Betti number

$$P_q = \alpha_q - \gamma_{q+1} - \gamma_q + 1.$$

We now compare the table T_q with the table T_{p-q+1} relative to the dual polyhedron P'. This table, which may be derived from T_q by interchanging lines and columns has $\beta_{p-q+1} = \alpha_{q-1}$ rows and $\beta_{p-q} = \alpha_q$ columns. The number γ_q is the same for both tables so that we get

$$\beta''_{p-q+1} = \gamma_q = \alpha''_q, \quad \beta_{p-q} - \beta'_{p-q} = \gamma_q,$$
$$\beta'_{p-q} = \beta_{p-q} - \gamma_q = \alpha_q - \alpha''_q,$$

whence

$$\beta'_{p-q+1} = \alpha_{q-1} - \gamma_{q-1},$$

and for the Betti number P'_{p-q+1} relative to the polyhedron P'

$$P'_{p-q+1} = \beta'_{p-q+1} - \beta''_{p-q+1} + 1 = \alpha_{q-1} - \gamma_{q-1} - \gamma_q + 1.$$

We deduce from this that

$$P'_{p-q} = P_q$$

which, when we recall that the Betti numbers relative to the dual polyhedra P and P' are the same, shows that the Betti numbers equally distant from the extremes are equal.

Now let us return to the homologies (2′). If we concede the right to divide homologies by nonzero integers, the first γ_q homologies will give us

$$d_i^{q-1} \sim 0 \quad (i = 1, 2, \ldots, \gamma_q),$$

and the most general of the homologies between the a_j^{q-1} will be written

$$\sum_{i=1}^{\gamma_q} \lambda_i d_i^{q-1} \sim 0, \tag{3}$$

where the λ_i are any integers. If, on the other hand, division of homologies is not permitted, the most general homology will be written

$$\sum_{i=1}^{\gamma_q} \lambda_i \omega_i^q d_i^{q-1} \sim 0, \tag{4}$$

where the λ_i are integers. For the two definitions of Betti number (*c*, §I, p. 75) to coincide it is necessary and sufficient that the two formulas (3) and (4) agree, i.e., that all the nonzero invariants ω_i^q be equal to 1 (§IX, p. 103).

We now consider the linear combinations of the a_j^{q-1} which will be homologous to zero by virtue of the homologies (3), and ask ourselves which among these combinations will remain distinct if, abandoning the homologies (3), we confine ourselves to the homologies (4) without admitting the right to divide homologies.

We see immediately that the number of these expressions which are distinct in this way is precisely the product

$$\omega_1^q \omega_2^q \cdots \omega_{\gamma_q}^q.$$

But, referring to the notations of the preceding paragraph, we see that this product is none other than one of the numbers of the sequence

$$M_0, \quad M_1, \quad M_2, \quad \ldots$$

and precisely the first member of this sequence which is nonzero (*c*. §IX, p. 103).

The preceding shows how it is important to distinguish two kinds of manifolds.

Those of the first kind, which I call *manifolds without torsion*, are those for which all the invariants of the tables T_q are equal to 0 or 1; for which, consequently, the two formulas (3) and (4) agree and the two definitions of Betti numbers are in accord.

Those of the second kind, which I call *manifolds with torsion*, are those for which certain of these invariants are not equal to either 0 or 1, and for which, consequently, the two definitions of Betti numbers are not in accord. In this case we always adopt the second definition (c, §I), except where the contrary is stated.

The justification of this nomenclature for the presence of invariants greater than 1 is due, as we shall see later, to a circumstance truly comparable to a twisting of the manifold on itself.

§4. Application to some examples

Desirous of applying the preceding to the examples given in *Analysis situs* (p. 32 ff.), I must first of all make a distinction between several kinds of polyhedra.

Ordinary polyhedra of the first kind are those for which all the a_i^q are simply connected (homeomorphic to hyperspheres) and such that all the elements of these a_i^q are distinct; for example, in ordinary space the tetrahedron will be a polyhedron of the first kind since it has four faces which are triangles and consequently simply connected (homeomorphic to circles), and all of these triangles are distinct and likewise their sides and vertices.

Polyhedra of the second kind are those for which all the a_i^q are simply connected but such that not all the elements of these a_i^q are distinct. Take for example a torus in ordinary space; from a point A on the surface of the torus draw a meridian and a parallel. These two cuts do not separate the surface of the torus into two regions; however, they render it simply connected. The surface is homeomorphic to a rectangle when rendered simply connected in this way, with two opposite sides corresponding to the two sides of the meridian cut and the other two corresponding to the two sides of the parallel cut. Thus the torus is a species of polyhedron with a single face; that face is a quadrilateral, and hence simply connected, but the four sides of that quadrilateral are not distinct—two are identified with the meridian cut and two with the parallel cut; likewise the four vertices are not distinct but all four are identified with the point A. The polyhedron so defined is then a polyhedron of the second kind.

Finally, polyhedra of the third kind are those for which not all the a_i^q are simply connected.

Properties of polyhedra of the first kind extend for the most part to those of the second kind. Nevertheless, we note one difference. In a polyhedron of the first kind, each a_j^{p-1} separates two a_i^q and does not belong to any other a_i^p. Consequently, in each column of the table T_p we shall have one of the numbers ε_{ij}^p equal to $+1$, another equal to -1, and all the others 0.

It is not the same with polyhedra of the second kind. It can happen that two of the a_j^{p-1} for the same a_i^p are not distinct. In that case, after crossing that a_j^{p-1} we shall find ourselves in the same a_i^p as before. Thus we consider our torus, as always, to be a polyhedron with a single face; after having crossed the meridian cut, for example, we find ourselves on the same face again. It happens then that this a_j^{p-1} is not in the relation to this a_j^p; actually, it is simultaneously in the direct and inverse relation, so that the two compensate, and the corresponding number ε_{ij}^p is equal to zero. In that case, all the numbers ε_{ij}^p which appear in the corresponding column of the table T_p are zero.

In the examples in question (p. 32 ff.) the manifolds are closed, three-dimensional ones which we see can be regarded as polyhedra of the second kind. Each of these polyhedra has a single cell (which in the first, third and fourth examples is a cube, in the fifth an octahedron), but the faces of this cell are identified in pairs.

1st example:

1st face $ABDC = A'B'D'C'$, 1st edge $AB = CD = A'B' = C'D'$;
2nd face $ACC'A = BDD'A'$, 2nd edge $AC = BD = A'C' = B'D'$;
3rd face $CDD'C' = ABB'A'$, 3rd edge $AA' = BB' = CC' = DD'$;
One unique cell, one unique vertex.

The three tables T_1, T_2, T_3 are composed entirely of zeros. All their invariants are therefore zero.

3rd example:

1st face $ABDC = B'D'C'A'$, 1st edge $AB = B'D' = C'C$;
2nd face $ABB'A' = C'CDD'$, 2nd edge $AC = DD' = B'A'$;
3rd face $ACC'A' = DD'B'B$, 3rd edge $AA' = C'D' = DB$;
4th edge $CD = BB' = A'C'$;
1st vertex $A = B' = C' = D$, 2nd vertex $B = D' = A' = C$.

Table T_3

$$\left|\begin{matrix} 0 & 0 & 0 \end{matrix}\right|$$

Table T_2

$$\left|\begin{matrix} +1 & -1 & -1 & -1 \\ +1 & +1 & -1 & +1 \\ -1 & +1 & -1 & -1 \end{matrix}\right|$$

Table T_1

$$\left|\begin{matrix} +1 & -1 \\ +1 & -1 \\ +1 & -1 \\ -1 & +1 \end{matrix}\right|$$

Table T_3 has no nonzero invariant; table T_1 has two which are 0 and 1; table T_2 has three which are 1, 2 and 2.

4th example:

1st face $ABDC = B'D'C'A'$, 1st edge $AA' = CC' = BB' = DD'$;
2nd face $ABB'A' = CDD'C'$, 2nd edge $AB = CD = B'D' = A'C'$;
3rd face $ACC'A' = BDD'B'$, 3rd edge $AC = BD = D'C' = B'A'$;
One unique cell, one unique vertex.

Table T_3

$$\left|\begin{matrix} 0 & 0 & 0 \end{matrix}\right|$$

Table T_2

$$\left|\begin{matrix} 0 & 0 & 0 \\ 0 & +1 & +1 \\ 0 & +1 & -1 \end{matrix}\right|$$

Table T_1

$$\left|\begin{matrix} 0 \\ 0 \\ 0 \end{matrix}\right|$$

Tables T_1 and T_3 have no nonzero invariants; table T_2 has three which are 1, 2 and 0.

5th example:

1st face $ABC = FED$,	1st edge $AB = FE$,	1st vertex $A = F$;
2nd face $ACE = FDB$,	2nd edge $AC = FD$,	2nd vertex $B = E$;
3rd face $AED = FBC$,	3rd edge $AE = FB$,	3rd vertex $C = D$;
4th face $ADB = FCE$,	4th edge $AD = FC$;	
	5th edge $BC = ED$;	
	6th edge $CE = DB$.	

Table T_3

$$\begin{vmatrix} 0 & 0 & 0 & 0 \end{vmatrix}$$

Table T_2

$$\begin{vmatrix} +1 & -1 & 0 & 0 & +1 & 0 \\ 0 & +1 & -1 & 0 & 0 & +1 \\ 0 & 0 & +1 & -1 & +1 & 0 \\ -1 & 0 & 0 & +1 & 0 & +1 \end{vmatrix}$$

Table T_1

$$\begin{vmatrix} +1 & -1 & 0 \\ +1 & 0 & -1 \\ +1 & -1 & 0 \\ +1 & -1 & 0 \\ 0 & +1 & -1 \\ 0 & -1 & +1 \end{vmatrix}$$

The invariants are

$$0 \text{ for } T_3; \quad 2,\ 1,\ 1 \text{ and } 1 \text{ for } T_2; \quad 0,\ 1 \text{ and } 1 \text{ for } T_1.$$

We now pass to the sixth example (p. 36). We have seen (§14, p. 45) that the fundamental equivalences may be written

$$\begin{aligned} C_1 + C_2 &\equiv C_2 + C_1, \\ C_1 + C_3 &\equiv C_3 + \alpha C_1 + \gamma C_2, \\ C_2 + C_3 &\equiv C_3 + \beta C_1 + \delta C_2. \end{aligned}$$

To write the homologies which yield the fundamental homologies for addition and multiplication, *but without division*, it suffices to permit permutation of terms in the fundamental equivalences; thus we find

$$0 \sim 0; \quad (\alpha - 1)C_1 + \gamma C_2 \sim 0; \quad \beta C_1 + (\delta - 1)C_2 \sim 0.$$

The determinant

$$(\alpha - 1)(\delta - 1) - \beta\gamma$$

is equal to

$$2 - \alpha - \delta.$$

Now let μ be the greatest common divisor of the four numbers

$$\alpha - 1, \quad \delta - 1, \quad \beta, \quad \gamma;$$

examination of the homologies that we have written shows that the two invariants of the table T_2 which are not equal to 0 or 1 are equal to

$$\mu \quad \text{and} \quad \frac{2 - \alpha - \delta}{\mu}.$$

(The number μ can also be equal to 1.)

As far as the invariants of the tables T_1 and T_3 are concerned, we shall see later that they are always equal to 0 or 1.

For example, let

$$\alpha = -1, \quad \beta = 1, \quad \gamma = -1, \quad \delta = 0.$$

We have

$$\mu = 1, \quad 2 - \alpha - \delta = 3$$

so that one of the invariants is 3 and the other 1.

This can also be verified by forming the table T_2. Let

$$(x+1, y, z), \quad (x, y+1, z), \quad (-x+y, -x, z+1)$$

be the three substitutions of the group G, which I shall call S_1, S_2 and S_3, and which correspond to the three fundamental contours C_1, C_2, C_3 (§13, p. 43).

The manifold being studied is generated by the cube $ABCDA'B'C'D'$ (§10, p. 32). But the face $ABCD$ must be considered to be decomposed into two triangles ABD and ACD, and likewise the face $A'B'C'D'$ into two triangles $D'A'B'$ and $C'D'A'$.

It is easy to see that the face $ABB'A'$ is changed into $CDD'C'$ by the substitution S_2, the face $ACC'A'$ into $BDD'B'$ by the substitution S_1, the face ABD into $D'A'B'$ by the substitution $S_3S_1S_2$, the face ACD into $C'D'A'$ by the substitution S_3S_2.

Our polyhedron then has:

1° A single cell;

2° Four faces, namely,

$$\begin{aligned}
&\text{1st face } ABB'A' = CDD'C',\\
&\text{2nd face } ACC'A' = BDD'B',\\
&\text{3rd face } ABD = D'A'B',\\
&\text{4th face } ACD = C'D'A';
\end{aligned}$$

3° Four edges, namely,

$$\begin{aligned}
&\text{1st edge } AA' = BB' = CC' = DD',\\
&\text{2nd edge } AB = CD = D'A',\\
&\text{3rd edge } AC = BD = C'D' = A'B',\\
&\text{4th edge } AD = C'A' = D'B';
\end{aligned}$$

4° A single vertex.

The tables T_1 and T_3 are composed entirely of zeros.

We now pass to the *example of Heegaard.* Let $x_1, x_2, y_1, y_2, z_1, z_2$ be the coordinates of a point in the space of six dimensions; let

$$\begin{aligned}
x &= x_1 + x_2\sqrt{-1} = |x|e^{\xi\sqrt{-1}},\\
y &= y_1 + y_2\sqrt{-1} = |y|e^{\eta\sqrt{-1}},\\
z &= z_1 + z_2\sqrt{-1} = |z|e^{\zeta\sqrt{-1}}.
\end{aligned}$$

Our manifold has the equations

$$z^2 = xy, \quad x_1^2 + x_2^2 + y_1^2 + y_2^2 = 1,$$

whence

$$|z|^2 = |xy|, \quad \zeta = \frac{\xi + \eta}{2}, \quad |x^2| + |y^2| = 1.$$

To obtain the whole manifold we need the following ranges of the variables:

1° $|x|$ from 0 to 1, with $|y|$ varying at the same time from 1 to 0.

2° y from 0 to 2π.

3° $\xi + \eta$ from 0 to 4π.

The polyhedron obtained has a single cell defined by the inequalities

$$0 < |x| < 1, \quad 0 < \eta < 2\pi, \quad 0 < \xi + \eta < 4\pi.$$

It has two faces defined by the following relations:

1st face:

$$\eta = 0, \quad 0 < |x| < 1, \quad 0 < \xi < 4\pi;$$

this face is identical with

$$\eta = 2\pi, \quad 0 < |x| < 1, \quad -2\pi < \xi < 2\pi.$$

2nd face:

$$\xi + \eta = 0, \quad 0 < |x| < 1, \quad 0 < \eta < 2\pi;$$

this face is identical with

$$\xi + \eta = 4\pi, \quad 0 < |x| < 1, \quad 0 < \eta < 2\pi.$$

It has three edges defined by the following relations:

1st edge:

$$\xi = \eta = 0, \quad 0 < |x| < 1;$$

this edge is identical with the following three:

$$\begin{array}{lll} \xi = 0, & \eta = 2\pi, & 0 < |x| < 1; \\ \xi = -2\pi, & \eta = 2\pi, & 0 < |x| < 1; \\ \xi = \eta = 2\pi, & & 0 < |x| < 1. \end{array}$$

2nd edge:

$$x_1 = x_2 = 0, \quad 0 < \eta < 2\pi.$$

3rd edge:

$$y_1 = y_2 = 0, \quad -2\pi < \xi < 0,$$

identical with the following two:

$$\begin{array}{ll} y_1 = y_2 = 0, & 0 < \xi < 2\pi; \\ y_1 = y_2 = 0, & 2\pi < \xi < 4\pi. \end{array}$$

Finally, there are two vertices, to wit:

1st vertex:

$$x_1 = x_2 = 0, \quad \eta = 0,$$

identical with

$$x_1 = x_2 = 0, \quad \eta = 2\pi.$$

2nd vertex:

$$y_1 = y_2 = 0, \quad \xi = -2\pi,$$

identical with the following three:

$$\begin{array}{ll} y_1 = y_2 = 0, & \xi = 0; \\ y_1 = y_2 = 0, & \xi = 2\pi; \\ y_1 = y_2 = 0, & \xi = 4\pi. \end{array}$$

The table T_3 is composed entirely of zeros; while T_1 and T_2 are written

$$T_2 = \begin{vmatrix} 0 & 0 & +2 \\ 0 & 1 & 1 \end{vmatrix}; \quad T_1 = \begin{vmatrix} 0 & -1 \\ 0 & 0 \\ 0 & 0 \end{vmatrix}.$$

We see that the invariants are 0, 2 and 1 for T_2, 0 and 1 for T_1.

§5. Generalization of a theorem in the first supplement

I want to return to one of the questions treated in the previous memoir (c, §X). In that place I envisaged the case $p = 3$, and I want to explain how we can extend the same reasoning to the general case. Here is what is involved:

Let P and P' be dual polyhedra: we consider on the one hand the elements a_i^p of P, on the other hand the elements b_i^q of P'. I assume that we have found a congruence

$$\sum \lambda_i a_i^q \equiv 0 \tag{1}$$

between the a_i^q; I claim that we can associate this congruence with a congruence between the b_i^q

$$\sum \mu_i b_i^q \equiv 0 \tag{2}$$

such that we have the homology

$$\sum \lambda_i a_i^q \sim \sum \mu_i b_i^q. \tag{3}$$

Conversely, to each congruence of the form (2) there is a congruence of the form (1) such that the left-hand sides of the two congruences are connected by the homology (3).

Such is the theorem which is to be proved. I have given a simple proof in the case $p = 3$, and it is a question of extending that proof to the general case. I shall first of all make a preliminary remark.

Consider the congruences

$$a_i^q \equiv \sum \varepsilon_{ij}^q a_j^{q-1}. \tag{4}$$

We know that by forming linear combinations of them we can eliminate the a_j^{q-1} and obtain congruences of the form

$$\sum \zeta_i a_i^q \equiv 0. \tag{5}$$

The number of distinct congruences of the form (5) is what we have called $\alpha_q - \alpha_q''$.

Suppose now that we consider the different elements a_i^h of the polyhedron p, where the number h of dimensions must be greater than p, but can equal $q+1$, $q+2, \ldots, p-1$ or p. We fix the value of the number h once and for all.

We then divide the congruences (4) into groups, putting two congruences in the same group if the two corresponding a_i^q belong to the same a_i^h; it is clear that we shall have as many groups as there are a_i^h, and the same congruence may be found in several groups, because an a_i^q forms part of more than one a_i^h.

By linearly combining congruences (4) *of the same group* we can eliminate the a_j^{q-1} and obtain congruences of the form

$$\sum \zeta_i' a_i^q \equiv 0. \tag{5'}$$

The congruences (5′) evidently form part of the system (5), because the latter system consists of *all* distinct congruences of this form that we can obtain by combining the congruences (4). On the other hand, the system (5) may contain congruences that are not in the system (5′). In fact we obtained the latter system by imposing restrictions on the right to combine the congruences (4), because we have combined only congruences in the same group.

I now claim that the congruences (5′) entail the homology

$$\sum \zeta_i' a_i^q \sim 0. \tag{6}$$

In fact, the congruence (5′) is a congruence between the elements of the polyhedron a_i^h and, *since by hypothesis this polyhedron is simply connected*, this congruence must entail the corresponding homology.

Conversely, if the homology (6) holds, the corresponding congruence will form part of the system (5′). In fact, the validity of the homology (6) between the elements of the polyhedron a_i^h must entail the corresponding congruence, and that congruence must be capable of derivation from the fundamental congruences of the form (4) *relative to the polyhedron* a_i^h, i.e., belonging to the same group.

It follows that the number of distinct congruences of the system (5′) is $\alpha_q - \alpha_q'$.

The system (5′) thus always remains the same, whatever value is attributed to the number h.

We see at the same time that this consideration allows us to find the Betti number P_q by considering only the table T_q, provided we know in addition whether two congruences (4) belong to the same group or not.

Now we introduce a notion which can be considered as a generalization of the notion of pyramid. Let a_q be a domain contained in a hyperplane P_q of q dimensions; let b_m be a domain contained in another hyperplane P_m' of m dimensions. I assume that these two hyperplanes have no common point. I can then make exactly one hyperplane Π of $q+m+1$ dimensions pass through these two hyperplanes.

That being given, we connect each of the points of the domain a_q to each of the points of the domain b_m by lines. The set of all these lines generates a certain domain contained in the hyperplane Π, having $q+m+1$ dimensions, which I designate by the notation $a_q b_m$, and which I can call a generalized rectilinear pyramid.

If in fact the domain a_q reduces to a plane polygon ($q = 2$) and the domain b_m to a point ($m = 0$), then the domain $a_q b_m$ reduces to an ordinary pyramid with a_q as base and b_m as apex.

All figures homeomorphic to a generalized rectilinear pyramid may be called generalized pyramids.

That being given, consider an element a_i^q of the polyhedron P and an element b_j^m of the dual polyhedron P'; this element b_j^m corresponds to an element a_j^{p-m} of the polyhedron P. I assume that the element a_i^q forms part of the element a_j^{p-m}; we then have

$$q < p - m; \quad p \geq q + m + 1.$$

I remark also that each point of b_j^m forms part of the a_k^p which contains a_j^{p-m}, and consequently of the a_k^p which contains a_i^q. It suffices to show this for the vertices of b_j^m; but if b_k^o is one of these vertices it will be in the interior of a_k^p, and since b_k^o belongs to b_j^m, a_j^{p-m} will belong to a_k^p by virtue of the definition of the dual polyhedron itself.

Given that, we can define a system of lines L in the interior of each of the a_k^p, such that exactly one line passes through any two points in the interior of that a_k^p. The system of lines L then has the same qualitative properties of a system of straight lines. This holds since a_k^p is assumed simply connected.

We now join each of the points b_j^m to each of the points of a_i^q by a line L situated in the a_k^p to which a_i^q and the point considered in b_j^m belong.

The set of these lines L generates a figure I shall call $a_i^q b_j^m$, which will be homeomorphic to a generalized rectilinear pyramid and which will have $q+m+1 \leq p$ dimensions.

What is the boundary of this manifold $a_i^q b_j^m$? Suppose that we have the congruences

$$a_i^q \equiv \sum \varepsilon_{ik}^q a_h^{q-1}, \quad b_j^m \equiv \sum \varepsilon_{jk}^{\prime m} b_k^{m-1}.$$

The boundary is composed of generalized pyramids $a_h^{q-1} b_i^m$ and $a_i^q b_k^{m-1}$, and we have

$$(7) \qquad a_i^q b_j^m \equiv \sum \varepsilon_{ik}^q a_h^{q-1} b_j^m + \sum \varepsilon_{jk}^{\prime m} a_i^q b_k^{m-1}.$$

This will no longer be true if we have $m = 0$. In that case, in fact, the manifold a_i^q has $q = (q+m+1) - 1$ dimensions; it then must form part of the boundary of $a_i^q b_j^m$, and the congruence (7) will become

$$(7') \qquad a_i^q b_j^o \equiv \sum \varepsilon_{ih}^q a_h^{q-2} b_j^o - a_i^q$$

(the terms in ε' disappearing); likewise for $q = 0$ we have

$$(7'') \qquad a_i^o b_j^m \equiv \sum \varepsilon_{jk}^{\prime m} a_i^o b_k^{m-1} + b_j^m.$$

From the congruences (7), (7′) and (7″) we deduce the homologies

$$(8) \qquad \sum \varepsilon_{ih}^q a_h^{q-1} b_j^m \sim - \sum \varepsilon' m_{jk} a_i^q b_k^{m-1},$$

$$(8') \qquad a_i^q \sim \sum \varepsilon_{ik}^q a_h^{q-1} b_j^o,$$

$$(8'') \qquad b_j^m \sim - \sum \varepsilon_{jk}^{\prime m} a_i^o b_k^{m-1}.$$

The congruence (8′) assumes that a_i^q forms part of a_j^p; this is what we have envisaged elsewhere [c, §X, p. 103 eq. (2)].

Suppose now that we have found a congruence of the form

$$(9) \qquad \sum \lambda_{ij} a_i^q b_j^m \equiv 0.$$

I claim that we can find a congruence of the same form where the number q is increased by one and the number m decreased by one, and of such a kind that the left-hand sides of the two congruences are homologous.

In fact, by virtue of (7) we have identically

$$\sum \lambda_{ij} a_i^q b_j^m \equiv \sum \lambda_{ij} \varepsilon_{ih}^q a_h^{q-1} b_j^m + \sum \lambda_{ij} \varepsilon_{jk}^{\prime m} a_i^q b_k^{m-1}.$$

We must then have (annulling the coefficient of $a_i^{q-1} b_j^m$ on the right-hand side)

$$\sum_i \lambda_{ij} \varepsilon_{ih}^q = 0.$$

We deduce the congruence

$$(10) \qquad \sum_i \lambda_{ij} a_i^q \equiv \sum_i \lambda_{ij} \varepsilon_{ik}^q a_h^{q-1} \equiv 0.$$

All the elements a_i^q which appear on the left-hand side of (10) belong to a_j^{p-m}; but by hypothesis a_j^{p-m} *is simply connected*; all congruences between its elements then entail the corresponding homology, so that we have

$$\sum_i \lambda_{ij} a_i^q \sim 0,$$

whence

$$\sum_i \lambda_{ij} a_i^q \equiv \sum_\rho \mu_{\rho j} a_\rho^{q+1},$$

where the μ are integer coefficients and the a_p^{q+1} are elements belonging to a_j^{p-m}.

But

$$\sum_\rho \mu_{\rho j} a_\rho^{q+1} \equiv \sum_{\rho i} \mu_{\rho j} \varepsilon_{\rho i}^{q+1} a_i^q.$$

We then have

$$\lambda_i = \sum_\rho \mu_{\rho j} \varepsilon_{\rho i}^{q+1}.$$

The congruence (9) can then be written

$$\sum \mu_{\rho i} \varepsilon_i^{q+1} a_i^q b_j^m \equiv 0$$

(the summation extends over the three indices ρ, i, j).

But we can form the following homology which is none other than one of the homologies (8)

(11) $$\sum \varepsilon_i^{q+1} a_i^q b_j^m \sim -\sum \varepsilon_{jk}^{\prime m} a_\rho^{q+1} b_k^{m-1}.$$

We then have

$$\sum \lambda_{ij} a_i^q b_j^m \sim -\sum \mu_{\rho j} \varepsilon_{kj}^{\prime m} a_p^{q+1} b_k^{m-1},$$

which proves the theorem claimed.

The case $m = 0$ is of course left to one side and must be treated separately. In this case the homology (11) must be replaced by the following which is one of the homologies (8′)

(11′) $$\sum \varepsilon_{\rho i}^{q+1} a_i^q b_j^o \sim a_\rho^{q+1},$$

whence

$$\sum \lambda_{ij} a_i^q b_j^o \sim \sum \mu_{\rho j} a_\rho^{q+1}.$$

Then, corresponding to the congruence

(9′) $$\sum \lambda_{ij} a_i^q b_j^o \equiv 0,$$

we have the congruence

$$\sum \mu_{\rho j} a_\rho^{q+1} \equiv 0,$$

which is of the form (1), and the left-hand sides of the two congruences will be homologous.

Now let

(2) $$\sum \lambda_j b_j^q \equiv 0$$

be a congruence of the form (2); by a homology analogous to (8″) we have

$$b_j^q \sim \sum \varepsilon_{jk}^{\prime q} a_i^o b_k^{q-1}$$

if b_j^{p-1} is one of the elements of P' which belongs to b_i^q.

We then have the homology

$$\sum \lambda_j b_j^q \sim -\sum \lambda_j \varepsilon_{jk}^{\prime q} a_i^o b_k^{q-1}$$

so that our congruence (2) will correspond to a congruence

$$-\sum \lambda_j \varepsilon_k^{\prime q} a_i^o b_k^{q-1} \equiv 0, \tag{12}$$

the left-hand side of which is homologous to that of (2).

Thus if we have a congruence of the form (2) we deduce the congruence (12), which is a congruence of the form (9), where the numbers we have called q and m above have the values 0 and $q-1$ respectively. We next deduce another congruence of the form (9) but where the two numbers have values 1 and $q-2$, and so on. We finally arrive at a congruence of the form (9′), i.e., a congruence where the two numbers have values $q-1$ and 0, and we then deduce a congruence of the form (1).

The left-hand sides of all these congruences are homologous to each other.

Thus the theorem enunciated at the beginning of this paragraph is proved.

In order to draw all the admissible consequences, we must realize that it is important to distinguish several kinds of homologies. Let v_q be any q-dimensional manifold forming part of our manifold v, and let v_{q-1} be its boundary, expressed by the congruence

$$v_q \equiv v_{q-1}.$$

We deduce the homology

$$v_q \sim 0.$$

The homologies obtained in this way are the fundamental homologies.

In combining fundamental homologies by addition, subtraction, and multiplication, we obtain others which are *homologies without division.* Finally, by combining addition, multiplication, and division we obtain others again, which are *homologies with division.*

Well, *all the homologies we have encountered in this paragraph are homologies without division.*

That said, we return to our tables T_q and T_q' and their invariants, and in particular to those invariants which are not equal to 0 or 1, which we call the *torsion coefficients.*

Suppose that we have the following homology:

$$\sum k\lambda_i a_i^q \sim 0, \tag{13}$$

where the λ_i are relatively prime integers, and that (13) is a homology without division, while the homology

$$\sum \lambda_i a_i^q \sim 0 \tag{14}$$

cannot be obtained without division. From what we have seen in one of the preceding paragraphs, this says that k is one of the torsion coefficients of the table T_q.

We have the congruence

$$\sum \lambda_i a_i^q \equiv 0. \tag{14′}$$

Using the procedure of this paragraph, we can deduce from (14′) a congruence between the b_i^q which I shall write as

$$\sum \mu_i b_i^q \equiv 0. \tag{14''}$$

Moreover, by the theorem we have just established, we will have

$$\sum \lambda_i a_i^q \sim \sum \mu_i b_i^q.$$

This is a homology without division, and we deduce immediately, also without division, that

$$\sum k\lambda_i a_i^q \sim \sum k\mu_i b_i^q,$$

from which we have, without division,

$$\sum k\mu_i b_i^q \sim 0,$$

whereas we do not have

$$\sum \mu_i b_i^q \sim 0$$

without division, as this would imply

$$\sum \lambda_i a_i^q \sim 0,$$

contrary to hypothesis.

This says that k is a torsion coefficient of the table T'_q.

Thus the torsion coefficients of the two tables T_q and T'_q are equal (the proof is easy to complete) and, when we observe that the two tables T'_q and T_{p-q} have the same invariants, we conclude that *the tables equally distant from the extremes have the same torsion coefficients.*[22]

We can arrive at the same result by another method.

We have seen in one of the previous memoirs (§16) how to define the operation we have called annexation; I suppose that two elements of a polyhedron, a_i^q and a_j^q for example, are separated from each other by an element a_k^{q-1}, and this is the only element of q dimensions common to a_i^q and a_j^q, and lastly that a_k^{q-1} does not belong to any q-dimensional elements other than a_i^q and a_j^q. We then have $\varepsilon_{ik}^q = 1, \varepsilon_{jk}^q = -1$; all the other ε_{hk}^q will be zero, and likewise all the products $\varepsilon_{ih}^q \varepsilon_{jh}^q$.

Under these conditions, we can annex the two elements a_i^q and a_j^q to each other by suppressing the element a_k^{q-1}. What is the effect of this operation on our tables T_k? The table T_q loses a row and a column; the table T_{q-1} loses a row. One of the invariants, equal to 1, of T_q disappears; as far as the table T_{q-1} is concerned, it loses an invariant if there are no longer more rows than columns. In that case the invariant lost is equal to zero. All the other invariants of the two tables are unchanged; so the two tables retain their torsion coefficients.

Now it is easy to form a polyhedron derived simultaneously from P and P'. We can then return from this polyhedron to either P or P' by regular annexations. Since these annexations do not alter the torsion coefficients, they must be the same for the tables T_q and T'_q.

[22]It is T'_q and T_{p-q+1} which have the same invariants (2*c*, §3): see the classical treatises on topology for a correct enunciation of the Poincaré duality theorem.

§6. Internal torsion of manifolds

Consider one of our tables T_q. We say that a sequence of distinct elements of that table, arranged in a certain order, forms a *chain* if each element of odd rank belongs to the same row as the following element and the same column as the preceding element. The chain will be *closed* if the last element is the same as the first. It is clear that a closed chain always contains an odd number of elements and an even number of *distinct* elements. For example, the elements

$$\varepsilon_{11}^q, \quad \varepsilon_{12}^q, \quad \varepsilon_{22}^q, \quad \varepsilon_{23}^q, \quad \varepsilon_{33}^q, \quad \varepsilon_{31}^q, \quad \varepsilon_{11}^q \tag{1}$$

form a closed chain.

Since all the elements of the table T_q are equal to 0, +1 or −1, the product of the distinct elements of a closed chain will always be 0, +1 or −1.

Suppose that the elements of the chain (1) have the following values:

$$\varepsilon_{12}^q = \varepsilon_{23}^q = \varepsilon_{31}^q = 1, \quad \varepsilon_{11}^q = \varepsilon_{22}^q = \varepsilon_{33}^q = -1;$$

the product of the elements of the chain will be −1. Consider then the three manifolds a_1^q, a_2^q, a_3^q and the three manifolds $a_1^{q-1}, a_2^{q-1}, a_3^{q-1}$. By suppressing the manifolds a_1^{q-1}, a_2^{q-1} and a_3^{q-1} we annex the other three manifolds a_1^q, a_2^q and a_3^q to each other, and the manifold obtained,

$$a_1^q + a_2^q + a_3^q,$$

is an orientable manifold.

If on the contrary we have

$$\varepsilon_{12}^q = \varepsilon_{23}^q = \varepsilon_{31}^q = 1, \quad \varepsilon_{22}^q = \varepsilon_{33}^q = -1, \quad \varepsilon_{11}^q = 1,$$

we can again suppress a_1^{q-1}, a_2^{q-1} and a_3^{q-1} and obtain a manifold $a_1^q + a_2^q + a_3^q$ by annexation; but *this manifold will be nonorientable.*

More generally, if all the elements of the chain (1) are equal to +1 or −1 then, if we first suppress a_2^{q-1} and a_3^{q-1}, we obtain the manifold

$$a_1^q - \varepsilon_{12}^q\varepsilon_{22}^q a_2^q + \varepsilon_{12}^q\varepsilon_{22}^q\varepsilon_{13}^q\varepsilon_{23}^q a_3^q. \tag{2}$$

Suppressing a_1^{q-1} next, we see that the manifold (2) is henceforth formed from a closed chain of a_i^q in the sense of paragraph 8 (p. 19) of *Analysis situs*, and that chain is orientable or nonorientable according as the product of the distinct elements of the chain (1) equals −1 or +1.

In the first case we say that the chain (1) is orientable, and in the second case, that it is nonorientable.

We are then led to distinguish three categories among the closed chains formed with the aid of elements of the tables T_q:

1° *Null* chains, i.e., those for which the product of the elements is zero.

2° *Orientable* chains.

It is easy to see that these are those for which the product of the elements is +1 if the number of elements is a multiple of 4, or those where this product is −1 if the number of elements is a multiple of 4, plus 2.

3° *Nonorientable* chains.

These are those where the product is −1 if the number of elements is a multiple of 4, or +1 if the number is a multiple of 4, plus 2.

That being given, we say that the table T_q (or more generally, any table or determinant, the elements of which are all 0, +1 or −1) is *orientable* if it does not contain any nonorientable chains.

It follows from this definition that:

An orientable table remains orientable if we change all the signs of a column, or all the signs of a row; or again, if we permute two colums or two rows.

THEOREM: *An orientable determinant cannot equal* 0, +1 *or* −1.

In fact, by changing the signs of columns where necessary, we can always arrange for the elements of the first row to be 0 or +1.

Suppose for example that the first two elements of the first row are +1, then if I subtract the first column from the second the value of the determinant will not change. I claim that the determinant remains orientable.

Indeed, consider a chain in the original determinant with first and last element belonging to the second column and all the other elements in other columns. Let a and c be the first and last elements; let ξ be the product of all the other elements in the chain; let b and d be the elements in the first column next to a and c, respectively.

The product of the elements of our chain which I call the chain (1) will be $ac\xi$ and we have

$$ac\xi = 0 \text{ or } 1 \text{ if the number of elements} \equiv 0 \pmod 4,$$

$$ac\xi = 0 \text{ or } -1 \text{ if the number of elements} \equiv 2 \pmod 4.$$

The product of the elements of the chain which I call (2), and which consists of the corresponding elements in the new determinant will be

$$(a-b)(c-d)\xi,$$

and in fact the elements of our chain do not change, except the elements a and c which become $a-b$ and $c-d$.

The chain formed in the original determinant by the two elements of the first row and the elements a and b will be orientable or null, so that we must have

$$a-b=0 \quad \text{or} \quad a=0 \quad \text{or} \quad b=0.$$

We must likewise have

$$c-d=0 \quad \text{or} \quad c=0 \quad \text{or} \quad d=0.$$

If $(a-b)$ or $(c-d)$ is zero then the theorem is proved because the product $(a-b)(c-d)\xi = 0$.

If $b = d = 0$ we have

$$(a-b)(c-d)\xi = ac\xi,$$

and the theorem is proved because the products of the chains (1) and (2) are the same, the number of elements is the same, and (1) is orientable or null.

If $a = c = 0$ we have

$$(a-b)(c-d)\xi = bd\xi.$$

The chain (3) which belongs to the original determinant and has the same elements as the chain (1), except that a and c are replaced by b and d is, I claim, orientable or null; it has the same number of elements as (2) and its product is $bd\xi$, equal in this case to the product of (2). Then the chain (2) is orientable or null in this case also.

If $a = d = 0$ we have

$$(a-b)(c-d)\xi = -bc\xi.$$

This time it is necessary to consider a chain (4) in the original determinant, the elements of which are the two elements of the first row, the elements b and c, and the elements of the chain (1), save a and c. This chain (4) must be orientable or null.

It contains two elements more than the chain (2).

Its product is equal to $bc\xi$, and consequently, equal but of opposite sign to the product of (2).

Then (2) is orientable or null.

Lastly, if $b = c = 0$ we have

$$(a-b)(c-d)\xi = -ad\xi,$$

and we show just as in the preceding case that the chain (2) is orientable or null.

We have just treated the case of chains where two elements belong to the second column. The result is the same whatever the number of elements in the second column, a number which must, however, always be even.

If this number is zero, the theorem is evident, since the chain in the new determinant is no different from that in the old.

Suppose that the number is 4, to fix ideas. Let a, c, e, g be the four elements in the second column, and imagine that we encounter in succession the element a, various elements ξ belonging to other columns, the elements c and e, various elements η belonging to other columns, and finally 9. Our chain will be closed.

$a\xi ce\eta g$ may be decomposed into two closed chains $a\xi ca$, $e\eta ge$ and it will be orientable provided these two are. Thus we are reduced to the case of chains having two elements in the second column.

I should add that all the elements of the new determinant are 0, +1 or −1. In fact, since we have

$$a-b=0 \quad \text{or} \quad a=0 \quad \text{or} \quad b=0,$$

we have

$$a-b=0,\ a \text{ or } -b,$$

whence

$$a-b=0,\ 1 \text{ or } -1.$$

That being given, we subtract the first column from all the columns with +1 as the first element. The determinant retains its values, it remains orientable, but all the elements of the first row are zero except the first, which is +1.

This reasoning is applicable in all cases except that where the first row is zero; but then the determinant is zero and the theorem is evident.

Now if we suppress the first row and the first column we obtain a new determinant equal to the first and, like it, orientable. We operate in the same fashion on this new determinant, which has one row and column less than its predecessor, and we finally arrive at a determinant which has a single element, 0, +1 or −1.

Our determinant is then equal to 0, +1 or −1.

1ST COROLLARY: *If a table T_q is orientable, its invariants are all 0 or 1.*

2ND COROLLARY: *If a polyhedron and all its tables T_q are orientable, i.e., if we cannot compose a nonorientable manifold from its elements a_i^q, then this polyhedron has no torsion coefficients.*

We see that the existence of torsion coefficients (which is a necessity because of the distinction between the definitions of the Betti numbers, or between homologies with and without division) is due to the fact that the elements of a polyhedron can engender nonorientable manifolds. i.e., the polyhedron is, so to speak, twisted onto itself.

This is what justifies the expression "torsion coefficients" or that of manifolds with or without torsion.

If the manifold V formed by the set of elements a_i^p of the polyhedron P is not itself nonorientable, the two tables T_1 and T_p are orientable.

In fact, each line of one, and each column of the other has all elements zero, except one equal to $+1$ and one equal to -1. Then if a chain is not null its elements are in equal pairs with opposite signs. It is then orientable.

It follows that the two extreme tables T_1 and T_p have all their invariants equal to 0 or 1. This explains why we do not encounter torsion coefficients with polyhedra in ordinary space; these polyhedra do not involve more than the two tables T_1 and T_2.

This is no longer true if the manifold V is nonorientable. Thus the manifold considered in the seventh example (§15, p. 54) can be subdivided into polyhedra and, depending on the manner of subdivision given, we find for the table T_2

$$\left| \, 2 \, \right|, \quad \begin{vmatrix} +1 & +1 \\ +1 & -1 \end{vmatrix}, \quad \ldots .$$

In order to avoid making this work too prolonged, I confine myself to stating the following theorem, the proof of which will require further developments:

Each polyhedron which has all its Betti numbers equal to 1 *and all its tables* T_q *orientable is simply connected, i.e., homeomorphic to a hypersphere.*[23]

[23]This is Poincaré's first (erroneous) step towards the Poincaré conjecture. He disproved the present conjecture with the construction of the so-called *Poincaré homology sphere* in the Fifth Supplement, and replaced it by the (correct) conjecture that any simply connected and finite 3-manifold is homeomorphic to a hypersphere. (Translator's note.)

On Certain Algebraic Surfaces; Third Supplement to Analysis Situs

Bulletin de la Sociétè Mathématique de France, 30 (1902), pp. 49–70.

We propose to study, from the viewpoint of *analysis situs*, the surface

$$z = \sqrt{F(x, y)}, \tag{1}$$

where F is a polynomial. We suppose that the curve

$$F(x, y) = 0$$

has no ordinary points where $\frac{\partial F}{\partial x}$ is nonzero, or even where $\frac{\partial F}{\partial x} = 0$, without $\frac{\partial^2 F}{\partial x^2}$ or $\frac{\partial F}{\partial y}$ vanishing; nor ordinary double points where

$$\frac{\partial F}{\partial x} = \frac{\partial F}{\partial y} = 0$$

without $\frac{\partial^2 F}{\partial x^2}$ or $\frac{\partial^2 F}{\partial x^2}\frac{\partial^2 F}{\partial y^2} - \left(\frac{\partial^2 F}{\partial x \partial y}\right)^2$ vanishing.

First we treat y as a constant. Then equation (1) represents an algebraic curve, and we know that the coordinates x and y of a point on that curve can be expressed as Fuchsian functions of the same auxiliary variable u. We consider the corresponding Fuchsian group and its generating Fuchsian polygon. In general, the Fuchsian polygon that corresponds to a curve of genus p is a polygon of $4p$ sides, and we can suppose that the conjugate pairs of sides are those of ranks $4q+1, 4q+3$ and $4q+1, 4q+4$ (corresponding to the so-called *normal* periods of abelian functions), or else that opposite sides are conjugate.

We adopt the latter hypothesis.

We recall that the angle sum of a polygon is equal to 2π. However, in the particular case of the curve (1) one has the case called *hyperelliptic*, where the corresponding abelian functions are hyperelliptic functions.

In that case we know that our Fuchsian polygon admits a center of symmetry and it may be decomposed into two polygons of $2p+1$ sides, each symmetric to the other with respect to the center.

To go further, I have to clarify what I mean by the word *symmetry*. At this point we take the viewpoint of non-Euclidean geometry:

- Non-Euclidean lines are circles orthogonal to the fundamental circle.
- Two figures are symmetric with respect to a non-Euclidean line when we can map one onto the other by an inversion (transformation by reciprocal radii) that leaves that line fixed.

- Two figures are congruent[24] when they are symmetric to a third with respect to non-Euclidean lines, or else when they are both congruent to a third figure.
- Finally, we say that two figures are symmetric with respect to a center when they are each symmetric to a third figure in non-Euclidean lines passing through that center.

That being given, our polygon R with $4p$ sides may be decomposed into two polygons R' and R'' with $2p+1$ sides, each of which is symmetric to the other with respect to a center.

We may assume that the polynomial $F(x, y)$ is not divisible by any square. Under these conditions, the equation in x

$$F(x, y) = 0$$

does not have a double root, except for certain singular values of y. There are $2p+2$ simple roots that I call

$$x_0, \quad x_1, \quad x_2, \quad \ldots, \quad x_{2p+1}.$$

The root x_0 corresponds to the $2p + 1$ vertices of the polygon R', while the roots $x_1, x_2, \ldots, x_{2p+1}$ correspond to the midpoints (as always, from the non-Euclidean point of view) of the $2p + 1$ sides.

In the plane of x we can make $2p + 1$ cuts

$$C_1, \quad C_2, \quad \ldots, \quad C_{2p+1},$$

from the point x_0 to the points $x_1, x_2, \ldots, x_{2p+1}$, in such a way that the two sides of the cut C_i correspond on the polygon R' to the two halves of the ith edge.

In the case where $p = 1$ (that is, when the equation $F = 0$ is of fourth degree in x) the polygon reduces to a parallelogram, the polygons R' and R'' to rectilinear triangles, and the Fuchsian functions to elliptic functions.

Now, if we let y vary continuously, under what conditions does this variable return to its initial value?

Our Fuchsian group will vary in a continuous manner, along with $x_0, x_1, x_2, \ldots, x_{2p+1}$ and the Fuchsian polygon R. When y makes a return trip the Fuchsian group will become itself again. The points x_i will in general be permuted among themselves and the polygon R will become another polygon R_1, equivalent to R, which I will say is also able to generate the same Fuchsian group.

Take, for example, the case $p = 1$. The polygon R is a parallelogram whose sides ω and ω' represent the magnitude and direction of the two periods of an elliptic function. When y makes a return trip, our parallelogram becomes R_1, whose sides again represent in length and direction two periods of the same elliptic function. However, the two periods will not in general be the same as ω and ω', but two equivalent periods

$$\alpha\omega + \beta\omega', \quad \gamma\omega + \delta\omega',$$

where $\alpha, \beta, \gamma, \delta$ are four integers such that $\alpha\delta - \beta\gamma = 1$.

This brings us into contact with *analysis situs*. Suppose that $p = 1$ and suppose that we agree to give y any one of the values on a certain closed contour K, and x any complex value, and let z be defined by equation (1). The set of these triples

[24]Poincaré calls them "equal," but later (p. 189) he switches to the more appropriate word "congruent." Unfortunately, this skates very close to conflict with Poincaré's use of the word "congruence" for the boundary relation in homology theory. (Translator's note.)

x, y, z form a certain closed manifold of three dimensions. What are the properties of this manifold from the viewpoint of *analysis situs*?

At each point of the manifold I assign three real variables ξ, η, ζ defined as follows. ζ is a function of y that increases by 1 whenever y describes its complete contour. As for ξ and η, they are linear functions of the real and imaginary parts of the elliptic integral u defined by equation (1). The linear functions are such that ξ and η change to $\xi + 1, \eta$ when the elliptic integral is augmented by ω and to $\xi, \eta + 1$ when it is augmented by ω' (in such a way that $u = \xi\omega + \eta\omega'$). Under these conditions, we return to the same point on the manifold V when ξ, η, ζ change to

$$\xi + 1, \quad \eta, \quad \zeta,$$

or to

$$\xi, \quad \eta + 1, \quad \zeta,$$

or to

$$\delta\xi - \gamma\eta, \quad \beta\xi + \alpha\eta, \quad \zeta + 1.$$

Because, if ξ_1 and η_1 are what ξ and η become when ζ changes to $\zeta + 1$, we have

$$u = \xi\omega + \eta\omega',$$

and, on the other hand,

$$u = \xi_1(\alpha\omega + \beta\omega') + \eta_1(\gamma\omega + \delta\omega').$$

More generally, ξ, η, ζ are subject to any transformation in the group G generated by these three transformations.

Here we recognize the group considered in *Analysis situs*, p. 36, example 6°. The manifold V is homeomorphic to the manifold in example 6° and it has G as its *fundamental group* (cf. *Analysis situs* p. 38).

The general definition of the manifold V is the same, except that we do not assume that $p = 1$. Then R is a curvilinear Fuchsian polygon. Again we introduce the three variables ξ, η, ζ, with ζ being defined as above. As for ξ and η, they are bi-uniform functions of ζ and are the real and imaginary parts of the variable u, in such a way that each complex value of u corresponds to a unique pair of values ξ and η, and conversely. To each value of ζ there corresponds a Fuchsian group and the Fuchsian polygon R for this group. This group will be generated by $2p$ substitutions

$$S_1, \quad S_2, \quad \ldots, \quad S_{2p}.$$

The substitution S_k changes ξ and η to

$$\varphi_k(\xi, \eta, \zeta), \quad \psi_k(\xi, \eta, \zeta)$$

in such a way that we return to the same point of the manifold V when we change ξ, η, ζ to

$$\varphi_k(\xi, \eta, \zeta), \quad \psi_k(\xi, \eta, \zeta), \quad \zeta.$$

Moreover, we can define the functions ξ and η in such a way that φ_k and ψ_k do not depend on ζ.

We consider the figure formed by the Fuchsian polygon $R = R' + R''$ and its transforms under the substitutions in the Fuchsian group. This polygon and its transforms cover the interior of the fundamental circle. The figure formed in this way is deformed continuously as ζ varies continuously, but it remains homeomorphic to itself. We can then set up a correspondence between each point M_0 of the initial figure and a unique point M in each of its consecutive positions in such a way that:

1° The point M varies continuously when ζ varies continuously.
2° When M_0 is a vertex of R, M remains a vertex of R; when M_0 is on an edge of R, M remains on an edge of R.
3° If two points M_0 and M_0' are congruent (that is, transforms of each other under substitutions in the Fuchsian group), then the points M and M' are likewise congruent.

I can then assume that there are two auxiliary variables ξ and η which are the same for the the point M_0 and the point M; I can suppose, for example, that they are the coordinates of the point M_0.

Under these conditions, φ_k and ψ_k are independent of ζ.

When y makes a round trip, so that ζ increases by 1, the polygon R becomes, through successive deformation, a polygon R_1 equivalent to R.

The point M becomes a point M_1 whose coordinates will be

$$\theta(\xi, \eta), \quad \theta_1(\xi, \eta).$$

We see that we return to the same point of V when ξ, η, ζ are changed to

$$\theta(\xi, \eta), \quad \theta_1(\xi, \eta), \quad \zeta + 1,$$

or, more generally, when ξ, η, ζ are subjected to one of the substitutions in the group G generated by the $2p + 1$ substitutions that change ζ and η to

$$\varphi_k, \quad \psi_k, \quad \zeta \quad (k = 1, 2, \dots, 2p),$$

or to

$$\theta, \quad \theta_1, \quad \zeta + 1.$$

This group G will then be the fundamental group of the manifold V.

We first remark that this group is not simple. Let G' be the group generated by the $2p$ substitutions $(\varphi_k, \psi_k, \zeta)$ and let Σ be the substitution $(\theta, \theta_1, \zeta+1)$. Then I claim that G' is an invariant subgroup of G. It suffices to show that G' commutes with Σ. In fact, Σ changes the polygon R to an equivalent polygon without changing the Fuchsian group. But the Fuchsian group is none other than the group generated by the $2p$ substitutions $(\xi, \eta; \varphi_k, \psi_k)$. We see that this group commutes with the substitution $(\xi, \eta; \theta, \theta_1)$ and the theorem follows immediately.

Now consider a manifold V defined as follows:

We represent the variable y on a sphere. On this sphere we distinguish the ordinary points, for which the equation $F(x, y)$ has multiple roots, and the singular points, for which this equation has multiple roots.

Let O be an ordinary point and let $A_1, A_2, \dots, A_q$ be the singular points. We join O to each of the latter points by disjoint cuts $OA_1, OA_2, \dots, OA_k$.

In addition, we draw around each of the singular points a circle of very small radius that we call its *protective circle.*

To construct the manifold V we give y any value not inside one of the protective circles, x any complex value, and z one of the two values defined by equation (1).

To each value of y there corresponds a Fuchsian polygon R, and this polygon is completely determined as long as y varies without crossing the cuts OA, since it is only when y makes a complete circuit around one of the singular points A that the polygon R can change.

The polygon R and its transforms under the Fuchsian group form a figure which, when y varies, is deformed in a continuous manner while remaining homeomorphic to itself. We let y_0 be the initial value of y, let R_0 the corresponding polygon, and let M_0 be a point in the plane of R_0. We can set up a correspondence between M_0 and a point M in the plane of R, in such a way that the coordinates of M are continuous and bi-uniform functions of those of M_0, M is a vertex or on an edge of R if M_0 is a vertex or on a side of R_0, and M and M' are congruent when M_0 and M_0' are congruent.

It follows that we can assign two auxiliary variables ξ and η to the point M which are none other than the coordinates of M_0. Under these conditions, the Fuchsian group is generated by the $2p$ substitutions,

$$S_1, \quad S_2, \quad \dots, \quad S_{2p},$$

such that S_k changes ξ and η to $\varphi_k(\xi, \eta)$, $\psi_k(\xi, \eta)$, and *the functions* φ_k *and* ψ_k *are independent of* y.

When y makes a circuit around the singular point A_i, R changes to an equivalent polygon R_1. It follows that ξ and η change to

$$\theta_i(\xi, \eta), \quad \theta_i'(\xi, \eta),$$

where θ_i and θ_i' are bi-uniform and continuous functions of ξ and η such that, when the point ζ, η is a vertex or on a side of R_0, the point θ_i, θ_i' is a vertex or on a side of the polygon R_1^0 analogous to R_1 and equivalent to R_0.

Now consider a second Fuchsian group that I call Γ, such that y is a Fuchsian function, with group Γ, of the auxiliary variable $\zeta + i\zeta'$. The corresponding Fuchsian polygon P will be of the second family (that is, all its vertices will be on the circle at infinity and all its angles zero) and of genus 0. Its vertices correspond to the values $A_1, A_2, \dots, A_q$ of the variable y.

Corresponding to the q singular points $A_1, A_2, \dots, A_q$ there are q substitutions

$$\Sigma_1, \quad \Sigma_2, \quad \dots, \quad \Sigma_q$$

which generate the group Γ. And the substitution Σ_i will change ζ and ζ' to

$$\chi_i(\zeta, \zeta'), \quad \chi_i'(\zeta, \zeta').$$

It follows that we return to the same point of the manifold V when the four variables ξ, η, ζ, ζ' are subject to one of the substitutions in the group G generated by the $2p + q$ substitutions that change these variables to

$$\begin{array}{lllll} \varphi_k(\xi, \eta), & \psi_k(\xi, \eta), & \zeta, & \zeta' & (k = 1, 2, \dots, 2p); \\ \theta_i(\xi, \eta), & \theta_i'(\xi, \eta), & \chi_i(\zeta, \zeta'), & \chi_i'(\zeta, \zeta') & (i = 1, 2, \dots, q). \end{array}$$

The first $2p$ of these substitutions generate a group G' (none other than the Fuchsian group applied to ξ and η, while leaving the two variables ζ and ζ' constant). Since this Fuchsian group commutes with the substitutions Σ_i, we conclude, as above, that G' is an invariant subgroup of G.

The group G can be regarded as the fundamental group of the manifold V, provided we assume, as above, that y is not permitted to penetrate the protective circles.

In fact, let N be a point in the four-dimensional space whose coordinates are ξ, η, ζ, ζ'. To each point N there corresponds a point of V, but to each point of V there corresponds an infinity of points of N; I say that the latter points are congruent to each other.

Following the definitions given in *Analysis situs* (cf. p. 39), to each substitution in the fundamental group of V there corresponds a closed contour K on V, with initial point a certain fixed point of V chosen once and for all (this is the point I called M_0 in *Analysis situs*). Let N_0 be one of the points N corresponding to this fixed point of V. Then, corresponding to our closed contour K, we have a line N_0BN_0' in the space $(\xi, \eta, \zeta, \zeta')$, going from N_0 to a congruent point N_0'.

It is clear that two lines N_0BN_0' and N_0CN_0' with the same extremities give the same substitution in the fundamental group. It suffices to show that the closed curve $N_0BN_0'CN_0$ bounds a surface, since the corresponding closed contour on V then bounds a surface that can be shrunk to a point by continuous deformation.

Thus it remains to show that the region of four-dimensional space where the point N moves is simply connected. What is this region? First of all, the point ξ, η can move throughout the interior of the fundamental circle, which is simply connected. Likewise, the point ζ, ζ' can cover the polygon P and its transforms under the Fuchsian group Γ. It, therefore, can also move throughout the whole interior of the fundamental circle, if we ignore the existence of the protective circles. But since y cannot penetrate the protective circles, it is necessary to remove small regions from P in the neighborhood of each vertex, and likewise for the transforms of P. It is therefore necessary to remove infinitely many small disks from the interior of the fundamental circle, but centered on the fundamental circle. Since the remaining area is no less simply connected, the region in which the point $N = (\xi, \eta, \zeta, \zeta')$ moves is also simply connected. Q.E.D.

Thus, to a point N_0' (or, if you prefer, to the substitution in the group G that changes N_0 to N_0') there corresponds a unique substitution in the fundamental group. It follows that the fundamental group is isomorphic to G, and we know, moreover, that the isomorphism is not meriedric.[25]

This is how it happens that there are no points N_0' (other than N_0) for which the corresponding substitution is the identity, that is, such that the closed contour on V corresponding to the line N_0BN_0' can be reduced to a point by continuous deformation.

We then need to find out whether, by describing an infinitely small contour on V, it can happen that the point N undergoes a nonidentity substitution in G. Now, when we describe an infinitely small contour on V, the variable y also describes an infinitely small contour on its plane. This contour cannot enclose any of the singular points A_i, because each of them has a protective circle which y cannot penetrate. We can therefore arrange, by subjecting our contour to an infinitely small deformation, that y remains constant and the variable x describes an infinitely small contour on its plane. If the latter contour does not enclose any of the singular points x_i, the point N whose coordinates are ξ, η, ζ, ζ' returns to its original value, and it is subject only to the identity substitution. If the contour encloses just one singular point, the variable z will change sign and the cycle is not closed on V. Finally, it cannot happen that the contour encloses two singular points, because it is infinitely small, and two singular points cannot be infinitely close together except when y approaches one of the points A_i, and we cannot approach the points A_i because of the protective circles.

In summary, when we describe a closed cycle on V the substitution experienced by N is always the identity. Thus the isomorphism between G and the fundamental

[25]As mentioned on p. 40, Poincaré is using the obsolete terminology of "holoedric isomorphism" and "meriedric isomorphism" for what we now call "isomorphism" and "homomorphism." (Translator's note.)

group is holoedric. In other words, since the fundamental group is defined only up to isomorphism, the group is none other than G.

One sees the role played by the protective circles in the preceding reasoning. We now suppress the protective circles and suppose that x and y can take any complex values, so that V is the manifold defined by equation (1).

First, the fundamental group will always be isomorphic to G; I do not have to change anything in that part of the argument. But it remains to find out whether this isomorphism is meriedric. To find out, I am going, as above, to study what happens when one describes an infinitely small cycle on V.

If, when this cycle is described, y does not turn around a singular point A_i, or come infinitely close to A_i, then the preceding reasoning applies and the substitution undergone by N is the identity. Now suppose that y describes a very small closed circle around A_i. Then ζ and ζ' change to χ_i and χ_i' and N undergoes either the substitution $(\theta_i, \theta_i', \chi_i, \chi_i')$, which I call T_i for short, or a substitution in the group G'.

Let us be more precise. When we describe a circle, the point x describes an infinitely small closed contour in its plane. At the same time, the points

$$x_0, \quad x_1, \quad x_2, \quad \dots, \quad x_{2p+1},$$

because of the variations in y, describe very small curves. Two of these points, which I call x_a and x_b, are very close to each other when y is in the neighborhood of A_i. The other points x_k describe closed contours when y turns around A_i. As for x_a and x_b, it can happen that they change places while each describes a very small arc, so that the union of these two arcs is a small closed curve. Otherwise, they do not change places, so each of them describes a closed curve.

If x does not turn around any of the singular points x_k, then the point N undergoes the substitution T_i, corresponding to which we must have the identity substitution in the fundamental group.

If x turns around a point x_k other than x_a and x_b, then z changes sign and the cycle is not closed. This case must therefore be excluded.

If x turns around the points x_a and x_b, then the sum of the arguments of $x - x_a$ and $x - x_b$ increases by 2π or 4π. The first case must be excluded because z changes sign; we examine the second.

We go back to the Fuchsian polygon R and the two partial polygons R' and R''. Corresponding to the point x_a there is a certain point u_a on R' in the middle of one of the sides (from the non-Euclidean viewpoint). Let s_a be the substitution that changes a point u in the plane of R to the symmetric point with respect to u_a (from the non-Euclidean viewpoint).

Let u_a' be a point congruent to u_a, the transform of u_a by a substitution S in the Fuchsian group, and let s_a' be a substitution that changes a point to its symmetric point with respect to u_a'. We evidently have

$$s_a' = S^{-1} s_a S.$$

We now consider the different points of the plane of R that correspond to x_b. Among these points, I distinguish those that tend to u_a when y tends to A_i without crossing the cuts OA; I call them u_b. I denote by u_b' the transforms of the u_b by S. I define s_b and s_b' relative to u_b and u_b' just as s_a and s_a' were defined relative to u_a and u_a'.

It is clear that

$$s_a^2 = s_b^2 = s_a'^2 = s_b'^2 = I,$$

that $s_a s_b$ and $s_b s_a$ belong to the Fuchsian group, and that

$$s_b' = S^{-1} s_b S.$$

Moreover, $s_a s_b$ and $s_b s_a$ are inverses of each other.

This being so, when x describes its contour around x_a and x_b the point ξ, η undergoes the substitution $s_a s_b$ (or the substitution $s_b s_a$, depending on the sense in which the contour is described) or, more generally, one of the substitutions in the Fuchsian group.

The point N then undergoes the substitution T_i followed by one of the substitutions S' in G' or, what amounts to the same thing, a substitution S'' in G' followed by T_i.

Thus, the substitution $T_i S' = S'' T_i$ in the group G again corresponds to the identity substitution in the fundamental group.

Since we have already seen that T_i corresponds to the identity substitution, we conclude that S' and S'' likewise correspond to the identity substitution.

Once again it can happen that, when we describe a small closed cycle on V, y does not turn around A_i, but it remains very close to A_i. In that case, ζ and ζ' return to their initial values and, at the same time, x describes a closed contour in its plane. We may suppose that the latter contour encloses the two singular points x_a and x_b because, when y is in the neighborhood of A_i, these two points x_a and x_b are close to each other. Then the point u undergoes a substitution in the Fuchsian group and the point N undergoes a substitution S''' in G' for which the corresponding substitution in the fundamental group is again the identity.

If we then recall that our group G is generated by the subgroup G' and the substitutions T_i, we see that all the substitutions T_i and certain of the substitutions in G' correspond to the identity substitution. Thus the fundamental group will be isomorphic to G' (and, consequently, to a Fuchsian group), because all the T_i correspond to the identity substitution, and this isomorphism will in general be meriedric, because certain substitutions in G' will correspond to the identity.

Before going further, a distinction is necessary. The singular points x_a and x_b may change places when y turns around A_i, or they may not. In the first case there is no difficulty: the part of the manifold (1) in the neighborhood of the point $y = A_i$, $x = x_a = x_b$ may be replaced by the part of the manifold

$$z^2 = y - x^2$$

in the neighborhood of the origin, and every closed cycle that stays close to this point is reducible to a point in such a way that the corresponding substitution in the fundamental group cannot be the identity substitution. In that case, T_i, S' and S'' correspond to the identity substitution, as we have explained.

In the second case the surface (1) presents a cone point and the portion of the manifold (1) in the neighborhood of $y = A_i$, $x = x_a = x_b$ may be replaced by the portion of the manifold

$$z^2 = y^2 - x^2$$

in the neighborhood of the origin.

Then we can make two equally legitimate conventions. Suppose that a closed contour on V may be shrunk to a point, *but only by crossing the cone point.* We can admit that the corresponding substitution in the fundamental group is again the identity or, in other words, we can treat the cone point as an ordinary point of the manifold. In that case, T_i, S', and S'' again correspond to the identity.

Or else we can make the opposite convention of treating the cone point as a singular point which it is forbidden to cross. In that case T_i again corresponds to the identity substitution, but this is not the case for S'''. Then we always have $S'' = S'''$.

A question arises here. Picard has shown (*Théorie des fonctions algébriques de deux variables*, t. I, pp. 85 ff.) that, if an algebraic surface is the most general for its degree, then the linear cycles can be reduced to points in such a way that the Betti number P_1 equals 1.

It does not follow immediately that the fundamental group reduces to the identity substitution. In fact, Picard shows that each linear cycle is *homologous to zero*, and to show that the fundamental group reduces to the identity substitution it is necessary to show that each linear cycle is *equivalent* to zero. For the difference between homologies and equivalences, see *Analysis situs*, p. 40.[26]

It is therefore necessary to return to this question from a new point of view. We begin with the case where $p = 1$, that is, where our Fuchsian polygon R is a parallelogram. Then all the substitutions in G', and hence all those in the fundamental group, commute with each other.

The group G' is generated by two substitutions that I call s and s_1. Suppose that we are given any cycle; to this cycle there corresponds a substitution in G' that we can write as, for example,

$$s^\alpha s_1^{\alpha_1} s^\beta s_1^{\beta_1} s^\gamma s_1^{\gamma_1}.$$

Since the substitutions in G' commute, we can write it equally well as

$$s^{\alpha+\beta+\gamma} s_1^{\alpha_1+\beta_1+\gamma_1}.$$

If, as shown by Picard, each cycle is homologous to zero, this means that, among the possible cycles, there are two corresponding to substitutions in G'

$$s^a s_1^b, \quad s^c s_1^d$$

(where a, b, c, d are four integers whose determinant is nonzero) and which are reducible to a point. It then follows that $s^a s_1^b$ and $s^c s_1^d$ correspond in the fundamental group to the identity substitution, which I write

$$s^a s_1^b \equiv I, \quad s^c s_1^d \equiv I.$$

We conclude, recalling that s and S_1 commute, that

$$s^\varepsilon \equiv I, \quad s_1^\varepsilon \equiv I, \quad \text{where} \quad \varepsilon = ad - bc.$$

We see that, in this case, the fundamental group consists of a finite number of substitutions—at most ε^2.

But we can go further, to the case where $p > 1$.

Suppose, to fix ideas, that $p = 2$, and let

$$a, \quad b, \quad c, \quad d, \quad e, \quad f$$

abbreviate the six singular points of the plane that we previously called $x_0, \ldots, x_5$.

[26]It seems here that Poincaré has finally overcome his confusion between null-homologous and null-homotopic curves, previously shown on p. 39 and p. 77. However, he confuses them yet again on p. 156. (Translator's note.)

First suppose $y = 0$. We join the point $x = 0$ to the points a, b, c, d, e, f by rectilinear cuts, in such a way that we encounter the cuts in the order

$$Oa, \quad Ob, \quad Oc, \quad Od, \quad Oe, \quad Of$$

when turning around the point O.

We now let y vary continuously, but without crossing any of the cuts OA_i. At the same time the points $a, b, \dots$ move continuously, but without changing order or turning around each other. The cuts $Oa, \dots$ may cease to be rectilinear, but they maintain the same order in a circuit around O.

When we cross the cut Oa, the variable u (the argument of the Fuchsian function) undergoes a transformation that I call a and which is a sort of symmetry analogous to the transformation s_a defined above, p. 141 (symmetry with respect to u_a).

I define the transformations b, c, d, e, f likewise. It is clear that we have

$$a^2 = b^2 = c^2 = d^2 = e^2 = f^2 = I, \quad abcdef = I,$$

and that these are the only relations that hold among them. The Fuchsian group consists of all possible products of these transformations *up to an even number.*

When the point y is near A_i (but without having crossed the cut OA_i), the four cuts Oa, Ob, Oc, Od look like the solid lines in Figure 1.

After the the point y has described a contour around the point A_i, the cuts Oa and Od are deformed and take the form shown by the dotted lines in the figure. Moreover, they are permuted in such a way that the solidly drawn cut Oma becomes the dotted cut $Om'd$, while the cut Ond becomes the cut $On'a$.

We now draw a closed contour in the plane, from any fixed point M_0. If this contour successively crosses the cuts Oa, Oc, Oa, Ob, Of, Oe, for example, then it is equivalent to the substitution $acabfe$.

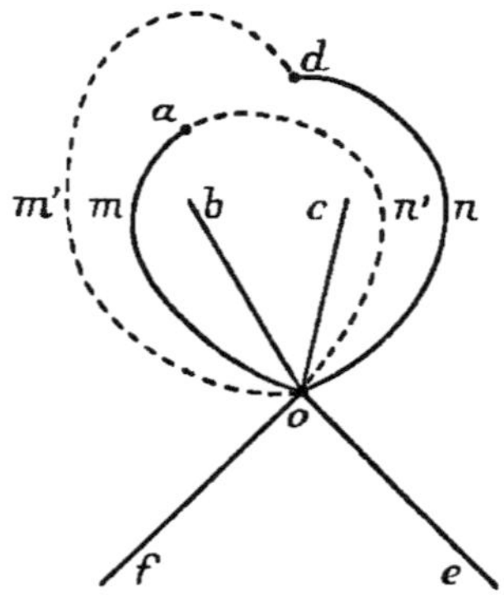

Fig. 1.

That being so, we consider a closed contour that crosses exactly one of the cuts

(1) $$Oma, \quad Ob, \quad Oc, \quad Ond, \quad Oe, \quad Of.$$

As y loops around A_i it is transformed into a contour crossing exactly one of the cuts

$$Om'd, \quad Ob, \quad Oc, \quad On'a, \quad Oe, \quad Of. \tag{2}$$

It is easy to see from the figure that a contour crossing one of the cuts (2) will cross certain of the cuts (1) in a certain order, and conversely, thus producing a certain combination of the substitutions a, b and c.

Suppose, for example, that M_0 is outside the contour $Om'dnO$. The cycle that crosses

$Om'd$	will cross	Ond	and is equivalent to	d,
Ob	will cross	Ond, Oma, Ob, Oma and Ond	and is equivalent to	$dabad$,
Oc	will cross	Ond, Oma, Oc, Oma and Ond	and is equivalent to	$dacad$,
$Om'a$	will cross	Ond, Oma and Ond	and is equivalent to	dad,
Oe	will cross	Oe	and is equivalent to	e,
Of	will cross	Of	and is equivalent to	f.

Consequently, the transformation T_i will change the substitutions

$$a, \quad b, \quad c, \quad d, \quad e, \quad f, \tag{3}$$

respectively to

$$d, \quad dabad, \quad dacad, \quad dad, \quad e, \quad f. \tag{4}$$

We have previously written down the relation

$$T_i S' = S'' T_i$$

and showed that the two substitutions S' and S'' correspond to the same substitution in the fundamental group, so that we can write

$$S' \equiv S''. \tag{5}$$

Since S' is a certain combination of (and even number of) the substitutions (3) and S'' is the corresponding combination of the substitutions (4) it suffices, for the equivalence (5) to hold, that we have

$$a \equiv d, \quad b \equiv dabad, \quad c \equiv dacad, \quad d \equiv dad.$$

But all these equivalences reduce to

$$ad \equiv 1$$

or, which comes to the same thing, $a \equiv d$.

We then saw that the substitution S''' in G' must correspond to the identity substitution in the fundamental group, which I wrote

$$S''' \equiv 1.$$

Here we see that S''' is none other than ad, so that we recover the equivalence

$$ad \equiv 1,$$

which is (with $T_i \equiv 1$) the only one that we can derive by consideration of the singular point A_i.

It remains for us to examine the case where the point

$$x = a, \quad y = A_i$$

is a conical point of the surface $z = F(x, y)$. We begin again with a figure analogous to Figure 1. When the point y turns around A_i, the cuts Oma and Ond drawn as solid lines are changed into the cuts $Om'a$ and $On'd$ drawn as dotted lines (Figure 2).

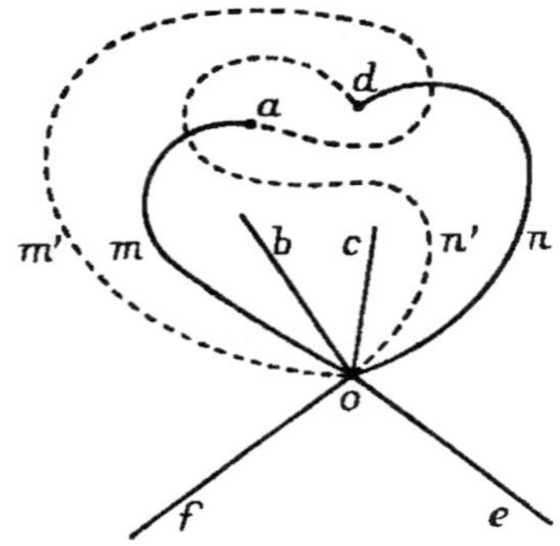

Fig. 2.

By reasoning as usual, and considering the different cycles that originate at M_0 and cross one of the cuts, we see that the cycle that crosses

$Om'a$	will cross	Ond, Oma and Ond,
Ob	will cross	$Ond, Oma, Ond, Oma, Ob, Oma, Ond, Oma, Ond$,
Oc	will cross	$Ond, Oma, Ond, Oma, Oc, Oma, Ond, Oma, Ond$,
$On'd$	will cross	Ond, Oma, Ond, Oma, Ond,
Oe	will cross	Oe,
Of	will cross	Of.

These cycles will therefore be equivalent to the respective combinations

$$dad, \quad dadabadad, \quad dadacadad, \quad dadad, \quad e, \quad f.$$

That is, T_i will transform the substitutions

$$a, \quad b, \quad c, \quad d, \quad e, \quad f$$

into the substitutions

$$dad, \quad dadabadad, \quad dadacadad, \quad dadad, \quad e, \quad f.$$

If we treat the conical point as an ordinary point of the manifold, then we again have

$$S' \equiv S'', \quad S''' \equiv 1.$$

The first condition implies

$$a \equiv dad, \quad b \equiv dadabadad, \quad c \equiv dadacadad, \quad d \equiv dadad. \tag{6}$$

The second gives simply

$$ad \equiv 1,$$

which also implies the conditions (6).

We now consider the conical point as one that it is not permitted to cross. Then we again have $S' \equiv S''$ and, consequently, the conditions (6). But we do not also have $S''' \equiv 1$, that is, $ad \equiv 1$.

The conditions (6) reduce to just one:

$$(ad)^2 \equiv 1.$$

Thus if we forbid the crossing of a conical point we do not have $ad \equiv 1$, but we do have $(ad)^2 \equiv 1$.

This means that the cycle that turns around the two points a and d is not the boundary of a two-dimensional manifold contained in V, but when this cycle is taken twice it does form the boundary of a two-dimensional manifold contained in V, at least if we assume that the manifolds of one or two dimensions are not all pushed away from the conical point.

It is easy to relate this to a well known fact. We have already remarked that the portion of V in the neighborhood of a conical point is homeomorphic to the portion of the manifold $z^2 = x^2 - y^2$, of $z^2 = xy$, in the neighborhood of the origin.

We therefore let W be the four-dimensional manifold $z^2 = xy$ and suppose that *we exclude the origin, which is a conical point.* The preceding teaches us that if C is a closed one-dimensional cycle on W then we do not have the equivalence

$$C \equiv 0,$$

but we do have the equivalence

$$2C \equiv 0.$$

Now consider the three-dimensional manifold

$$z^2 = xy, \quad |x^2| + |y^2| = 1,$$

which I call W'. This is the manifold due to Heegaard (cf. *Premier Supplément* [sic] *à l'Analysis situs* (*Rendiconti del Circulo matematico di Palermo*, vol. XIII, 1899)).

To each point x, y, z of W there corresponds a point of W':

$$\frac{x}{\sqrt{|x^2| + |y^2|}}, \quad \frac{y}{\sqrt{|x^2| + |y^2|}}, \quad \frac{z}{\sqrt{|x^2| + |y^2|}}.$$

If a point of W describes a cycle C, the corresponding point of W' describes a cycle C'. But it is evident that if we have $C \equiv 0$ on W then we have $C' \equiv 0$, and conversely. (Recall that the equivalence $C \equiv 0$ means that there is a two-dimensional manifold in W with boundary C.)

So if we have $2C \equiv 0$ on W without having $C \equiv 0$ then on W' we have a cycle C' with $2C' \equiv 0$ but without having $C' \equiv 0$.

We recall that the existence of such a cycle C' is one of the characteristic properties of Heegaard's manifold.

Nothing is now easier than finding the fundamental group. This group is meriedrically isomorphic to the Fuchsian group generated by all combinations of *an even number* of the substitutions a, b, c, d, e, f, which are subject to the relations

$$a^2 = b^2 = c^2 = d^2 = e^2 = f^2 = 1, \quad abcdef = 1.$$

But if the two points a and d can be exchanged when y turns around A_i we have $ad \equiv 1$, whence

$$a \equiv d.$$

I claim that the same relation holds if a is exchanged with d when y describes any closed cycle, enclosing, for example, not just one singular point but two singular points A_i and A_k. If, for example, a is exchanged with b when y turns around A_i, and b with d when y turns around A_k (so that a is exchanged with d when y describes a contour enclosing both these singular points), then we have

$$a \equiv b, \quad b \equiv d.$$

Consequently,

$$a \equiv d.$$

Q.E.D.

Thus if the singular points a, b, c, d are exchangeable we have

$$a \equiv b \equiv c \equiv d.$$

If the polynomial $F(x, y)$ is indecomposable, the roots of the equation

$$F(x, y) = 0$$

(considered as an equation in x) will be exchangeable when we vary y arbitrarily.

Our $2p+2$ singular points (which are six in number, a, b, c, d, e, f, if $p = 2$) are then exchangeable and we have

$$a \equiv b \equiv c \equiv d \equiv e \equiv f.$$

An arbitrary substitution in the fundamental group, which reduces to a combination of an even number of factors a, b, c, d, e, f, then reduces to an even power of a, that is, to the identity.

Thus, if the polynomial F is indecomposable, the fundamental group consists only of the identity substitution.

If the polynomial F decomposes into two factors, $F = F_1F_2$, we need to distinguish two kinds of singular points: those satisfying the equation $F_1 = 0$ and those satisfying the equation $F_2 = 0$. Suppose, for example, that a, b, c, d satisfy $F_1 = 0$ and e, f satisfy $F_2 = 0$. We then have

$$a \equiv b \equiv c \equiv d, \quad e \equiv f.$$

We do not have $a \equiv e$ (if the conical points are not regarded as ordinary points); but we do have $(ae)^2 \equiv 1$, so that the fundamental group consists only of the two substitutions

$$1, \quad ae.$$

It remains to see whether this number is not further reduced by considering the relation

$$abcdef = 1. \tag{7}$$

We observe that we can always make the degree of F even (up to a homographic transformation), because the number of singular points is $2p + 2$. We must then distinguish between the case where the degrees of F_1 and F_2 are both even—in which case relation (7) is an identity and the fundamental group is not further reduced—and the case where the degrees of F_1 and F_2 are both odd, in which case relation (7) reduces to $ae \equiv 1$ and the fundamental group consists only of the identity substitution.

If F decomposes into three factors, $F = F_1F_2F_3$, so that a is one of the roots of $F_1 = 0$, b is one of those of $F_2 = 0$, and c is one of those of $F_3 = 0$, then we have

$$a^2 \equiv b^2 \equiv c^2 \equiv 1, \quad (ab)^2 \equiv (bc)^2 \equiv (ac)^2 \equiv 1,$$

which shows that the fundamental group reduces to four substitutions

$$1, \quad ab, \quad bc, \quad ac.$$

This number may again be reduced, because of relation (7), if two of the factors are of even degree.

Finally, if F decomposes into n factors, $F = F_1 F_2 \cdots F_n$, and if a_i is one of the roots of $F_i = 0$, then we have

$$a_i^2 \equiv 1, \quad (a_i a_k)^2 \equiv 1 \quad (i, k = 1, 2, \ldots, n). \tag{8}$$

If S is any substitution in the fundamental group, then it will be the product of an even number of the substitutions a_i. But the relations (8) imply

$$a_i a_k \equiv a_k a_i.$$

We can therefore permute the factors of S so as to write it in the form

$$a_1^{\varepsilon_1} a_3^{\varepsilon_2} \cdots a_n^{\varepsilon_n} \quad (\varepsilon_1 + \varepsilon_2 + \cdots + \varepsilon_n \equiv 0, \bmod 2).$$

After that, with the help of the relations (8), we can reduce all the exponents to 0 or 1, so that S reduces to the product of k different factors among $a_1, a_2, \ldots, a_n$, where k is even and the order of the factors does not matter. There are 2^{n-1} such combinations, so the fundamental group consists of 2^{n-1} substitutions. This number may be halved by means of relation (7) if two or more of the factors are of odd degree.

Cycles on Algebraic Surfaces; Fourth Supplement to Analysis Situs

Journal de Mathématiques, 8 (1902), pp. 169–214.

§1. Introduction

The beautiful works of Picard on *Algebraic surfaces* have, for a long time, impressed on me the importance of the notion of cycles of one, two and three dimensions. I have thought that it should be possible to apply, to the notion in question, the principles I have expounded in *Analysis situs* and its first two supplements (centenary volume of the *Journal de l'École Polytechnique*; *Rendiconti del Circolo Matematico di Palermo*, vol. XIII; *Proceedings of the London Mathematical Society*, vol. XXXII), and I have obtained certain partial results that were announced in a note in *Comptes rendus* and which complete those of Picard at certain points.

Given a closed manifold V of p dimensions, I consider other manifolds contained in it, closed or not, of fewer dimensions. I denote by W_q such a manifold of q dimensions.

If $\sum W_q$ is a set of q-dimensional manifolds and $\sum W_{q-1}$ is a set of $(q-1)$-dimensional manifolds, the congruence

$$\sum W_q \equiv \sum W_{q-1}$$

signifies (by definition) that $\sum W_{q-1}$ is the complete boundary of the set of manifolds $\sum W_q$. I express the same fact without showing $\sum W_q$ by writing the relation

$$\sum W_{q-1} \sim 0,$$

which I call a *homology*.

Then the congruence

$$\sum W_q \equiv 0$$

signifies that the manifold $\sum W_q$ is closed.

If we have $\sum W_q \equiv 0$ without having $\sum W_q \sim 0$ (or $n \sum W_q \sim 0$ for some integer n), I say that the manifold $\sum W_q$ is a cycle of q dimensions.

Now let

$$f(x, y, z) = 0 \tag{1}$$

be the equation of an algebraic surface, which defines a manifold of four dimensions. To each value of y there corresponds a Riemann surface which will in general be of genus p. I suppose that the genus does not fall for $y = 0$ or $y = \infty$, but that it falls for q singular points

$$y = A_1, \quad y = A_2, \quad \dots, \quad y = A_q.$$

In the plane of y values I draw q cuts $OA_1, OA_2, \dots, OA_q$. Let S be one of the Riemann surfaces. When y varies *without crossing any of the cuts*, the surface S will vary, but it will remain homeomorphic to itself in the sense that any two of these Riemann surfaces correspond point-to-point in a one-to-one and continuous manner.

Any one of the surfaces S can be decomposed as a polyhedron P; let F be the number of faces, B the number of edges, and C the number of its vertices. Another surface S', corresponding point-to-point with S, has a corresponding subdivision as a polyhedron P', with faces, edges and vertices corresponding to the faces, edges and vertices of the polyhedron P.

Now suppose that y, starting from a point infinitely close to one of the cuts, describes an almost closed contour ending at another point infinitely close to the initial point *but on the other side of the cut.* The surface S will be transformed into a surface that differs from it by an infinitely small amount; but a point of the first surface will correspond, in general, to an entirely different point on the second surface.

The polyhedron P will therefore be transformed into a very different polyhedron P'.

On the other hand, to different points in the plane of y values dissected by our cuts we can make correspond the points of a polygon Q with $2q$ edges $\alpha_i\beta_i$ and $\alpha_i\beta_{i+1}$: the edges $\alpha_i\beta_i$ and $\alpha_i\beta_{i+1}$ correspond to the two sides of the cut OA_i, the point α_i to A_i, and the points β_i and β_{i+1} to O. Needless to say, I write β_1 or β_{q+1} indifferently, likewise β_2 or β_{q+2}, in order to have symmetry of notation.

We are now going to describe a subdivision of the manifold V as a polyhedron of four dimensions.

To each face F_i of P there corresponds a hypercell of H that I also call F_i. To each edge B_i of P there corresponds a cell B_i of H; likewise, to each of the edges $\alpha_i\beta_i$ or $\alpha_i\beta_{i+1}$ of Q, combined with each of the faces F_k of P, there corresponds a cell $\alpha_i\beta_iF_k$ or $\alpha_i\beta_{i+1}F_k$ of H.

To each vertex C_i of P there corresponds a face C_i of H. To each vertex α_i or β_i of Q, combined with each of the faces F_k of P, there corresponds a face α_iF_k or β_iF_k of H. To each edge $\alpha_i\beta_i$ or $\alpha_i\beta_{i+1}$ of Q, combined with each of the edges B_k of P, there corresponds a face $\alpha_i\beta_iB_k$ or $\alpha_i\beta_{i+1}B_k$ of H.

To each vertex C_k of P, combined with each of the edges $\alpha_i\beta_i$ or $\alpha_i\beta_{i+1}$ of Q, there corresponds an edge $\alpha_i\beta_iC_k$ or $\alpha_i\beta_{i+1}C_k$ of H. To each vertex α_i of β_i of Q, combined with each edge B_k of P, there corresponds an edge α_iB_k or β_iB_k of H.

Finally, to each vertex α_i or β_i of Q, combined with each of the vertices C_k of P, there corresponds a vertex α_iC_k or β_iC_k of H.

Several observations are appropriate. First, for $y = A_i$ the polyhedron P degenerates in such a way that certain faces disappear. If, for example, the face F_k disappears for $y = A_i$, the corresponding face α_iF_k of the polyhedron H does not exist.

Likewise, though this can be avoided, one can conceive of an edge B_k disappearing for $y = A_i$. In that case, the edge α_iB_k does not exist.

On the other hand, suppose we fix a value of i and let the index k take all possible values. We then consider the set of cells $\alpha_i\beta_iF_k$ and the set of cells $\alpha_i\beta_{i+1}F_k$.

These two sets are identical [in extent], even though in general the cell $\alpha_i\beta_iF_k$ is not identical to the cell $\alpha_i\beta_{i+1}F_k$, nor to another cell $\alpha_i\beta_{i+1}F_l$.

It can also happen that certain of the faces $\alpha_i\beta_iB$ are identical to certain of the faces $\alpha_i\beta_{i+1}B$, or certain of the edges $\alpha_i\beta_iC$ to certain of the edges $\alpha_i\beta_{i+1}C$.

On the other hand, we compare the different faces β_iF_k. The Riemann surface S_0 corresponding to the point O will be found subdivided into a polyhedron in q different ways, according as we consider the point O to correspond to the vertex β_1, or to β_2, ..., or to β_q. These are the q modes of subdivision that generate the faces βF. Then if m is the number of faces of P, we have the identities

$$(2)\quad \beta_iF_1+\beta_iF_2+\cdots+\beta_iF_m=\beta_jF_1+\beta_jF_2+\cdots+\beta_jF_m \quad (i,j=1,2,\ldots,q).$$

It can happen that certain of the faces are identical, but this is not always the case. It can equally well happen that certain of the edges βF, or certain of the vertices βC, are identical.

Finally, as a result of the degeneration of P for $y=A_i$, it can happen that certain of the faces α_iF, or edges α_iB, or vertices α_iC, are identical.

To sum up, our partial manifolds—hypercells, cells, faces, edges or vertices—can be divided into four categories as shown in the following table.

Nature of manifold	Category: 1	$\alpha\beta$	α	β
Hypercells	F_k	-	-	-
Cells	B_k	$\alpha_i\beta_iF_k, \alpha_i\beta_{i+1}F_k$	-	-
Faces	C_k	$\alpha_i\beta_iB_k, \alpha_i\beta_{i+1}B_k$	α_iF_k	β_iF_k
Edges	-	$\alpha_i\beta_iC_k, \alpha_i\beta_{i+1}C_k$	α_iB_k	β_iB_k
Vertices	-	-	α_iC_k	β_iC_k

One cannot have identity between two manifolds of different categories. Two manifolds of category 1 are always distinct.

There cannot be identity between two manifolds of category $\alpha\beta$ unless the index i of the α is the same for both (without this, the corresponding values of y will be on two different cuts OA_i, OA_j). The indices of the β, however, must be different. For example, there can be identity between $\alpha_i\beta_iF_k$ and $\alpha_i\beta_{i+1}F_h$, but not between $\alpha_i\beta_iF_k$ and $\alpha_i\beta_iF_h$. Two manifolds of category α cannot be identical unless the index of α is the same.

Before going further, we are going to modify our conventions slightly, in order to avoid the inconvenience resulting from identities such as (2) between two sums of faces, despite lack of identity between individual faces in the sums.

For any point M on the cut OA_i the corresponding Riemann surface can be decomposed into a polyhedron in two different ways, according as one views the point M as belonging to one side of the cut or the other. We superimpose the two modes of subdivision by drawing the edges of one subdivision and then those of the other. In this way we obtain a certain polyhedron that I call P', which we can arrange to remain homeomorphic to itself as M traverses the whole cut OA_i (see below, §5).

I let F'_k, B'_k, C'_k denote the faces, edges and vertices of P'. Each of the faces F in the first mode of subdivision is decomposed into a certain number of faces F', and it is the same for each of the faces F in the second mode. Thus each face F' belongs to exactly one face F in the first mode, and to exactly one face F in the second mode.

Each of the edges B in each of the two modes of subdivision is decomposed into a certain number of edges B'. Each edge B' belongs to at least one edge B in each of the two modes, and possibly to an edge in each mode. But, in any case, it does not belong to two different edges in the same mode.

Finally, the vertices C' will be the vertices in the two modes, together with the points where edges in the first mode intersect edges in the second mode.

Likewise, we have seen that the surface S_0 corresponding to the point O may be decomposed into a polyhedron in q different ways. We superimpose these q modes of subdivision, obtaining a polyhedron P'' whose faces, edges and vertices I call F''_k, B''_k, C''_k. Each of the faces F in one of the q modes, and also each of the faces F' corresponding to a polyhedron P'_i obtained by viewing the point O as a member of the cut OA_i, is decomposed into a certain number of faces F''. Each face F'' belongs to exactly one face F' of the polyhedron P'_i, and to exactly one face F in each of the q modes of subdivision.

Each of the edges B in the q modes, and each of the edges B' of the various polyhedra P'_i, is decomposed into a certain number of edges B''. Each edge B'' belongs to one of the edges B in one of the q modes, and to one of the edges B' of one the polyhedra P'_i. It may belong to two different edges in two different modes, or to two edges B' in different polyhedra P', but not to two edges B in the same mode or to two edges B' in the same polyhedron.

The vertices C'' are the vertices in the q modes together with the intersection points of edges in different modes.

None of this changes when we pass to manifolds of category 1; we move on to category $\alpha\beta$. Each of the cells $\alpha_i\beta_i F_k$ or $\alpha_i\beta_{i+1}F_k$ will be decomposed into subcells $\alpha_i\beta_i F'_k$ or $\alpha_i\beta_{i+1}F'_k$; likewise, each of the faces $\alpha\beta B$ will be decomposed into faces $\alpha\beta B'$. To the edges $\alpha\beta C$ we must adjoin, as we have just seen, other edges corresponding to the intersections of edges B belonging to different modes. The latter edges are what I call the edges $\alpha\beta C'$.

One sees that the cells $\alpha_i\beta_i F'_k$ are identical with the cells $\alpha_i\beta_{i+1}F'_k$; likewise for the faces $\alpha_i\beta_i B'_k$ and $\alpha_i\beta_{i+1}B'_k$ and for the edges $\alpha_i\beta_i C'_k$ or $\alpha_i\beta_{i+1}C'_k$. But it is important to remark that the cell $\alpha_i\beta_i F_k$ is decomposed into subcells $\alpha_i\beta_i F'$ which are not the same, in general, as the subcells into which the cell $\alpha_i\beta_{i+1}F_k$ is divided. The same observation applies to the faces $\alpha_i\beta_i B_k$ and $\alpha_i\beta_{i+1}B_k$.

We pass to category α. When the point M comes to A_i, the two modes of decomposition of the surface S into a polyhedron P coincide. On the other hand, this polyhedron degenerates, as I have said, so that certain of the manifolds $\alpha_i F'_k, \alpha_i B'_k, \alpha_i C'_k$ may disappear or coincide.

We pass to category β. Then we have the submanifolds $\beta_i F''_k, \beta_i B''_k, \beta_i C''_k$.

The faces $\beta_1 F''_k, \beta_2 F''_k, \dots, \beta_q F''_k$ are identical, but the face $\beta_i F''_k$ is decomposed into subfaces $\beta F''$ which are not the same, in general, as those into which we decompose the face $\beta_2 F_k$, or the face $\beta_3 F_k$, etc. The same observation holds for the edges.

This being so, I first make the following remark:

A manifold of category α is always homologous to a sum of manifolds belonging to other categories.

Consider, for example, the face $\alpha_i F''_k$. It belongs to the cell $\alpha_i\beta_i F'_k$ which admits, in addition, the face $\beta_i F'_k$ and those of the faces $\alpha_i\beta_i B'_h$ that correspond to the edges B'_h belonging to the face F'_k of the polyhedron P'. Then, if we have the congruence (for the polyhedron P)

$$F'_k \equiv \sum \varepsilon_q B'_q,$$

where the ε_q are equal to $+1$, -1 or 0, we will have the congruence

$$\alpha_i\beta_i F'_k \equiv \alpha_i F'_k - \beta_i F'_k + \sum \varepsilon_q \alpha_i\beta_i B'_q$$

and, consequently, the homology

$$\alpha_i F'_k \sim \beta_i F'_k - \sum \varepsilon_q \alpha_i\beta_i B'_q.$$

The manifolds appearing on the right-hand side of this homology belong to the categories β and $\alpha\beta$, so the theorem is shown, and we can establish the same for $\alpha_i B'_k$ and $\alpha_i C'_k$.

§2. Three-dimensional cycles

I now pass to the study of homologies and congruences between manifolds, beginning with the following remark:

We can always assume that our congruences do not contain a manifold of category α.

Suppose, in fact, that

$$\sum \alpha_i A + H \equiv 0$$

is a congruence in which the $\alpha_i A$ are p-dimensional manifolds of category α (corresponding to a manifold A with polyhedron P or P'), and H is a combination of p-dimensional manifolds of other categories.

We then have, on our polyhedron P or P', the congruence

$$A \equiv \sum \varepsilon a,$$

where the ε are integers and the a are manifolds of dimension one less than that of A. We then have the congruence

$$\alpha_i\beta_i A \equiv \alpha_i A - \beta_i A + \sum \varepsilon \alpha_i\beta_i a,$$

and hence the homology

$$\alpha_i A \sim \beta_i A - \sum \varepsilon \alpha_i\beta_i a.$$

Combining this homology with the congruence

$$\sum \alpha_i A + H \equiv 0,$$

we find the congruence

$$\sum \beta_i A - \sum\sum \varepsilon \alpha_i\beta_i a + H \equiv 0,$$

which does not contain any manifold of category α. Q.E.D.

To obtain homologies between cells, it suffices to consider those that follow from hypercells.

Suppose that, on the polyhedron P, we have the congruence

$$F_k \equiv \sum \varepsilon_q B_q,$$

where the ε are $+1$, -1 or 0. Then we have, for the four-dimensional polyhedron, the congruence

$$F_k \equiv \sum \varepsilon_q B_q + \sum \alpha_i \beta_i F_k - \sum \alpha_i \beta_{i+1} F_k,$$

and, consequently, the homology

$$\sum \varepsilon_q B_q \sim \sum \alpha_i \beta_{i+1} F_k - \sum \alpha_i \beta_i F_k. \tag{1}$$

We also recall that $\alpha_i \beta_i F_k$, and likewise $\alpha_i \beta_{i+1} F_k$, can be replaced by the sum of several subcells $\alpha_i \beta_i F'$.

Here is a first consequence: let $\sum \zeta_q B_q$ be any combination of cells B_q, where the ζ_q are integers. I suppose that, on the polyhedron P, we have the homology

$$\sum \zeta_q B_q \sim 0.$$

That is, the set of these edges (each multiplied by the coefficient ζ_q) forms a cycle on the surface S that is continuously deformable to a point.[27]

Then on the polyhedron P we have the congruence

$$\sum \zeta_q B_q \equiv \sum \theta_k F_k,$$

where the θ are integers. So on the four-dimensional polyhedron we have the congruence

$$\sum \theta_k F_k \equiv \sum \zeta_q B_q + \sum \theta_k \alpha_i \beta_i F_k - \sum \theta_k \alpha_i \beta_{i+1} F_k,$$

and hence the homology

$$\sum \zeta_q B_q \sim \sum \theta_k \alpha_i \beta_{i+1} F_k - \sum \theta_k \alpha_i \beta_i F_k.$$

Thus, if a combination of edges B is homologous to zero on P, the corresponding combination of cells B will be homologous to a combination of cells of the category $\alpha\beta$.

We now seek the congruences between cells.

These congruences are of the form

$$\sum \zeta_q B_q + \sum \theta'_k \alpha_i \beta_i F'_k \equiv 0, \tag{2}$$

where the ζ and θ' are integers. *I claim first that on the polyhedron P we have*

$$\sum \zeta_q B_q \equiv 0,$$

that is, the set of edges (weighted by the coefficients ζ) forms one or more cycles on the surface S.

In fact, suppose we have the congruence

$$B_q \equiv \sum \varepsilon_h C_h$$

on P. Then, on our four-dimensional polyhedron we have the congruence

$$B_q \equiv \sum \varepsilon_h C_h + H,$$

where H denotes a combination of faces not belonging to category 1. On the other hand, we have

$$\alpha_i \beta_i F'_k \equiv H,$$

[27]For the last time, Poincaré forgets that a null-homologous curve is not necessarily null-homotopic. (Translator's note.)

where H has the same meaning. We deduce that

$$\sum \zeta_q B_q + \sum \theta'_k \alpha_i \beta_i F'_k \equiv \sum \zeta_q \varepsilon_h C_h + H,$$

where H again has the same meaning.

Since the right-hand side must be identically zero, we must have, identically,

$$\sum \zeta_q \varepsilon_h C_h = 0.$$

Then, on the polyhedron P, we have

$$\sum \zeta_q B_q \equiv \sum \zeta_q \epsilon_h C_h = 0. \tag{3}$$

Q.E.D.

One then has, by virtue of (2),

$$\sum \theta'_k \alpha_i \beta_i F'_k \equiv \sum \zeta_q \alpha_i \beta_{i+1} B_q - \sum \zeta_q \alpha_i \beta_i B_q. \tag{4}$$

Now let $S(M)$ be the Riemann surface S at the point where y becomes equal to M on the cut OA_i. Let MF'_k be the face of the polyhedron P' corresponding to this position of the point M; let MP be the limit to which the polyhedron P tends when y approaches M from the side $\alpha_i\beta_i$ of the cut. Let (MP) be the limit to which the same polyhedron tends when y approaches M from the side $\alpha_i\beta_{i+1}$ of the cut. Let MB_q and (MB_q) be the edges B_q of the two polyhedra MP and (MP).

If we take the congruence (4) and, in each of the manifolds involving y, retain only the points that also belong to $S(M)$, we obtain a new congruence

$$\sum \theta' MF'_k \equiv \sum \zeta_q (MB_q) - \sum \zeta_q MB_q. \tag{5}$$

The expressions $\sum \zeta_q (MB_q)$ and $\sum \zeta_q MB_q$ represent two cycles on the surface $S(M)$; let Ω'_i and Ω_i be these two cycles. From the congruence (5) one derives the homology

$$\Omega'_i - \Omega_i \sim 0, \tag{6}$$

which must hold on the surface $S(M)$.

Suppose that y, starting from the point M on the side $\alpha_i\beta_i$ of the cut, loops around the singular point A_i and returns to the point M on the other side of the cut. The cycle $\sum \zeta_q B_q$ is thereby deformed continuously, with initial value $\sum \zeta_q MB_q = \Omega_i$ and final value $\sum \zeta_q (MB_q) = \Omega'_i$.

Thus the homology says that Ω_i is homologous to its transform Ω'_i (under the transformation of cycles of S induced when y takes a turn around the singular point A_i). The cycle $\sum \zeta_q B_q$ therefore remains homologous to itself under a sequence of these transformations, and hence also under a sequence of analogous transformations corresponding to other singular points. Cycles which enjoy this property may be called *invariant cycles.* We will come back to this notion later. We see that every congruence of the form (2) corresponds to an invariant cycle. *I claim that, conversely, to every invariant cycle there corresponds a congruence of the form* (2).

Suppose that $\sum \zeta_q B_q$ is an invariant cycle. On $S(M)$ we have

$$\sum \zeta_q (MB_q) \sim \sum \zeta_q MB_q, \tag{6}$$

and, consequently, we can find integers θ' such that

$$\sum \theta'_k MF'_k \equiv \sum \zeta_q (MB_q) - \sum \zeta_q MB_q. \tag{5}$$

This yields the congruence

$$\sum \theta'_k \alpha_i \beta_i F'_k \equiv \sum \zeta_q \alpha_i \beta_{i+1} B_q - \sum \zeta_q \alpha_i \beta_i B_q + \sum \theta'_k \alpha_i F'_k - \sum \theta'_k \beta_i F'_k.$$

Here the sign $\sum$ is with respect to the index k, with the index i held constant. But from it we deduce

$$\sum \zeta_q B_q + \sum\sum \theta'_k \alpha_i \beta_i F'_k \equiv \sum\sum \theta'_k \alpha_i F'_k - \sum\sum \theta'_k \beta_i F'_k,$$

where the double $\sum$ sign is over the two indices i and k, since we evidently have

$$\sum \zeta_q B_q \equiv \sum\sum \zeta_q \alpha_i \beta_i B_q - \sum\sum \zeta_q \alpha_i \beta_{i+1} B_q.$$

It remains to show that

$$\sum \theta'_k \alpha_i F'_k = 0, \quad \sum\sum \theta'_k \beta_i F'_k = 0.$$

We begin with the first of these equations.

We return to the congruence (5) and let the point M tend to A_i, so the limit MF'_k reduces to $\alpha_i F'_k$ and Ω_i coincides with Ω'_i and with $\sum_q \alpha_i B_q$. The congruence (5) then becomes

$$\sum \theta'_k \alpha_i F'_k \equiv 0. \tag{7}$$

We examine the meaning of this congruence. First suppose that, for $y = A_i$, the surface S does not decompose; that is, the curve

$$f(x, A_i, z) = 0$$

is indecomposable. Then the congruence can hold only under two conditions: either we have the identity

$$\sum \theta'_k \alpha_i F'_k = 0,$$

and, on the left-hand side (with a coefficient θ' different from zero) the faces $\alpha_i F'_k$ disappear as a result of the degeneration of the polyhedron mentioned above; or else the combination $\sum \theta'_k \alpha_i F'_k$ represents (one or more times) the Riemann surface in its entirety, so that

$$\sum \theta'_k \alpha_i F'_k = nS.$$

But, for the surface S corresponding to the point M, we can write

$$S = \sum MF'_k,$$

because the set of faces F'_k of the polyhedron P'_i enables the surface S to be recovered in its entirety.

Moreover, we have

$$\sum MP'_k \equiv 0.$$

We therefore have the congruence

$$\sum (\theta'_k - n) MF'_k \equiv \Omega'_i - \Omega_i \tag{5'}$$

and, on the other hand, when M tends to A_i,

$$\sum \alpha_i F'_k = S$$

and consequently

$$\sum (\theta'_k - n) \alpha_i F'_k = 0.$$

The second condition is therefore reduced to the first; it suffices to change θ'_k to $\theta'_k - n$, and this is permissible because the congruence (5) is thereby replaced by a congruence (5′) of the same form.

Now suppose that the curve $F(x, A_i, z)$ decomposes, for example, that the corresponding Riemann surface decomposes into two subsurfaces S_1 and S_2.

Then our congruence (7) can hold provided that we have

$$\sum \theta'_k \alpha_i F'_k = n_1 S_1 + n_2 S_2,$$

where n_1 and n_2 are integers. This at least is what we have to fear, but we can see in several ways that it does not happen.

The simplest is to argue as follows:

We begin with the Riemann surface S_0 that corresponds to the point O. We let y vary continuously from O to A_i along the cut OA_i. The Riemann surface S deforms continuously, remaining homeomorphic to itself. Let $S(M)$ be the surface S corresponding to the point M. To each point of $S(M)$ we can make correspond a point of S_0 by passing from S_0 to $S(M)$ by continuous deformation. On $S(M)$ we consider the two cycles Ω_i and Ω'_i; these correspond to two cycles on S_0 that I call U_i and U'_i.

When the point M traverses OA_i in a continuous motion, the two cycles U_i and U'_i undergo a continuous movement on the surface S_0. When M is very close to A_i the two cycles Ω_i and Ω'_i, and consequently the two cycles U_i and U'_i, are very close to one another. When the point M comes to O the two cycles Ω_i and U_i become identical and reduce to

$$\sum \zeta_q \beta_i B_q = \Omega_i^0;$$

the two cycles Ω'_i and U'_i become identical and reduce to

$$\sum \zeta_q \beta_{i+1} B_q = \Omega_{i+1}^0.$$

Suppose that the point M varies from A_i to a certain point M_0 on the cut OA_i. The two cycles U_i and U'_i, which initially coincide, then separate and sweep out a certain region R on S_0. This region corresponds on $S(M_0)$ to a region formed by a certain number of faces of the corresponding polyhedron P'_i, because it is bounded by the two cycles Ω_i and Ω'_i that are formed from certain edges of this polyhedron. We can write the congruence

$$\sum \theta'_k M F'_k \equiv \Omega'_i - \Omega_i,$$

where $\sum \theta'_k M F'_k$ represents the region we have just defined.

This region shrinks to zero when M approaches A_i, in which case

$$\sum \theta'_k \alpha_i F'_k = 0. \tag{8}$$

We now show that

$$\sum \sum \theta'_k \beta_i F'_k = 0,$$

and for this we have to study the sum $\sum \theta'_k \beta_i F'_k$.

From the preceding, the latter sum is none other than the region swept out on the surface S_0 by the two cycles U_i and U'_i when the point M varies from A_i to O. It is bounded by the two cycles Ω_i^0 and Ω_{i+1}^0.

We see that

$$\sum \theta'_k \beta_1 F'_k \equiv \Omega_2^0 - \Omega_1^0,$$
$$\sum \theta'_k \beta_2 F'_k \equiv \Omega_3^0 - \Omega_2^0,$$
$$\vdots$$
$$\sum \theta'_k \beta_q F'_k \equiv \Omega_1^0 - \Omega_q^0,$$

whence, by addition,

$$\sum\sum \theta'_k \beta_i F'_k \equiv 0.$$

This congruence shows that the combination $\sum\sum \theta'_k \beta_i F'_k$ reduces to zero or to a certain multiple of the surface S_0.

We now explore whether the latter situation can occur. Suppose that y describes a small closed contour infinitely close to the cuts OA_i, first on one side of OA_1, then on the other side, then the two sides of OA_2, and so on, and finally the two sides of OA_q. We consider a cycle on the corresponding Riemann surface; this cycle first coincides with Ω_1^0 when y is at O on the first side of OA_1; while y is on the first side of OA_1 this cycle is continuously deformed, as is our cycle Ω_1 and the corresponding points on S_0 forming the cycle U_1; when the point y returns from A_1 to O from the second side this cycle becomes none other than our cycle Ω'_1, and the corresponding points on S_0 form the cycle U'_1; when y returns to O this cycle coincides with Ω_2^0; when y describes the two sides of OA_2 this cycle coincides first with Ω_2, later with Ω'_2, and the corresponding points on S_0 first form the cycle U_2, later the cycle U'_2, and so on. What we want to know is whether the moving cycle, which coincides successively with $U_1, U'_1, U_2, U'_2, \ldots, U_q, U'_q$ and in its initial and final positions with Ω_1^0, describes the entire surface S_0.

To find out, we have to see how the cycles deform.

Consider the equation

$$f(x, y, z) = 0,$$

where y is held constant. The singular points of z as a function of x are given by the equation

$$\frac{\partial f}{\partial z} = 0.$$

We consider a cycle on the Riemann surface. To this cycle there corresponds, in the plane of x, a certain contour enclosing a certain number of these singular points. As y varies, these singular points move, and if we do not want the cycle to pass through one of the singular points it is necessary to deform it to avoid the moving singular points. Whatever the displacements of these singular points may be, provided that two of them do not merge into one, it will always be possible to deform the cycle continually so as not to pass through any of them. We can likewise choose a certain number of fixed points and deform the cycle in such a way that it does not pass through either the singular points or the fixed points, provided that none of the singular points come into coincidence with each other, or with the fixed points.

As y varies, the singular points are displaced, and the corresponding points on S_0 are likewise displaced. They do not come into coincidence unless y comes to one of the points A_i, but we make y turn around these points, approaching closely but not attaining them. On the other hand, these points describe lines that we can find on S_0, and we can find a region ρ on S_0 not crossed by any of these lines. It is the points of this region ρ that play the role of the fixed points I mentioned above. We can then deform our cycle in such a way that it neither passes through one of the singular points nor enters the region ρ. It therefore cannot generate the entire surface S_0. The second possible situation must therefore be rejected, so we always have the identity

$$\sum\sum \theta'_k\beta_i F'_k = 0. \tag{9}$$

In conclusion we remark that *there are no congruences between cells of category $\alpha\beta$ alone.* Suppose that, indeed,

$$\sum \theta'_k\alpha_i\beta_i F'_k \equiv 0$$

is such a congruence. First we find that

$$\sum \theta'_k\alpha_i\beta_i F'_k \equiv \sum \theta'_k\varepsilon'_h\alpha_i\beta_i B'_h + \sum \theta'_k\alpha_i F'_k - \sum \theta'_k\beta_i F'_k,$$

so that we have

$$\sum \theta'_k\varepsilon'_h\alpha_i\beta_i B'_h = 0, \quad \sum \theta'_k\alpha_i F'_k = 0, \quad \sum \theta'_k\beta_i F'_k = 0,$$

and, consequently, on the polyhedron P'_i (by reasoning like that used to deduce (5) and (4)),

$$\sum \theta'_k M F'_k \equiv 0.$$

The set of faces MF'_k (weighted by numerical coefficients θ'_k) of the polyhedron P'_i must then be congruent to zero; that is, it must form a closed surface which cannot be the entire surface $S(M)$. The sum $\sum \theta'_k\alpha_i F'_k$ then represents the entire surface $S(A_i)$, and we cannot then have

$$\sum \theta'_k\alpha_i F'_k = 0.$$

Thus, *our congruence is impossible.*

We know that, to obtain all the three-dimensional cycles, it suffices to find all combinations of cells that are congruent to zero without being homologous to zero.

First, we have seen that such a combination must contain cells of category 1; let $\sum \zeta_q B_q$ be the set of these cells. The set of the corresponding edges must form a cycle on the surface S (it is the cycle Ω_i). This cycle cannot be homologous to zero. Indeed, if it is, then the combination $\sum \zeta_q B_q$ will be homologous to a combination of cells of category $\alpha\beta$. One can then replace $\sum \zeta_q B_q$ by this combination in the left-hand side of our congruence. The left-hand side then contains only cells of category $\alpha\beta$, which we have seen to be impossible.

Finally, our cycle Ω_i must be *invariant*, that is, it must change into a homologous cycle Ω'_i when y turns around A_i. But when y turns around one of the singular points A_i, the cycles on the Riemann surface undergo one of the substitutions in the *Picard group.* The cycle Ω_i must therefore be invariant under the Picard group.

Thus, every three-dimensional cycle of V corresponds to a cycle on the surface S that is invariant under the Picard group.

Conversely, consider a cycle invariant under the Picard group. If Ω_i is one position of this cycle on the polyhedron P'_i and if Ω'_i is what becomes of Ω_i when y turns around A_i, then we have

$$\Omega_i \sim \Omega'_i.$$

We can find integers θ' so as to satisfy the congruence (5). The congruence (4) likewise holds. But we have seen that, under these conditions, the identities (8) and (9) hold, so that the congruence (4) can be written

$$\sum\sum \theta'_k \alpha_i \beta_i F'_k \equiv \sum \zeta_q \alpha_i \beta_{i+1} B_q - \sum \zeta_q \alpha_I \beta_i B_q,$$

whence

$$\sum \zeta_q B_q + \sum\sum \theta'_k \alpha_i \beta_i F'_k \equiv 0.$$

The left-hand side of this congruence represents a cycle of three dimensions.

To sum up: to the extent that the Picard group admits distinct invariant cycles, the manifold V admits distinct three-dimensional cycles.

A better way to represent three-dimensional cycles is to suppose that we have not only

$$\Omega_i \sim \Omega'_i$$

but identically

$$\Omega_i = \Omega'_i,$$

which is an assumption we can always make, because of the arbitrary way in which we can make points correspond on our Riemann surfaces.

Under these conditions, we can give y all possible values.

To each value there corresponds a position of the cycle Ω_i and, because of the invariance of this cycle, *corresponding to two infinitely close points on one of the cuts there correspond two infinitely close positions of the cycle Ω_i.*

The various positions of this cycle then generate a cycle of three dimensions.

§3. Two-dimensional cycles

To find all the two-dimensional cycles, it suffices to find all combinations of faces that are congruent to zero without being homologous to zero.

We can suppose first that the combination in question does not contain faces of category α since, by what we established in the preceding paragraph, each face of category α is homologous to faces of categories $\alpha\beta$ and β.

It is now a matter of finding whether these combinations can contain faces of category 1. First I observe the following: if C_1 and C_2 are two vertices of a polyhedron P we can always pass from one to the other by following certain edges of this polyhedron, so that we always have, on this polyhedron, the congruence

$$C_1 - C_2 \equiv \sum \zeta_q B_q,$$

where the right-hand side represents the set of edges along which one passes from C_1 to C_2. More generally, one can find integers ζ such that

$$\sum \varepsilon_k C_k \equiv \sum \zeta_q B_q, \tag{1}$$

where the ε are integers such that

$$\sum \varepsilon_k = 0,$$

since $\sum \varepsilon_k C_k$ can then be regarded as a sum of differences such as $C_1 - C_2$.

So, on the manifold V consider the combination of faces $\sum \varepsilon_k C_k$ and suppose that we have $\sum \varepsilon_k = 0$. We can then find integers ζ satisfying the congruence (1), so on the manifold V we have

$$\sum \zeta_q B_q \equiv \sum \varepsilon_k C_k + \sum \zeta_q \alpha_i \beta_i B_q - \sum \zeta_q \alpha_i \beta_{i+1} B_q$$

and, consequently,

$$\sum \varepsilon_k C_k \sim \sum \zeta_q \alpha_i \beta_{i+1} B_q - \sum \zeta_q \alpha_i \beta_i B_q,$$

which shows that the combination $\sum \varepsilon_k C_k$ is homologous to a combination of faces of category $\alpha\beta$.

Now suppose that we have a congruence of the form

$$\sum \varepsilon_k C_k + H = 0,$$

where H represents a combination of faces of categories $\alpha\beta$ and β. If we have $\sum \varepsilon_k = 0$, then we can replace $\sum \varepsilon_k C_k$ on the left-hand side by the combination of faces of categories β and $\alpha\beta$ to which it is homologous. The left-hand side then no longer contains faces of category 1.

If we have two congruences of the same form

$$\sum \varepsilon_k C_k + H \equiv 0, \quad \sum \varepsilon'_k C_k + H' \equiv 0,$$

then we can find two integers n and n' such that

$$n \sum \varepsilon_k + n' \sum \varepsilon'_k = 0,$$

and hence the congruence

$$\sum (n\varepsilon_k + n'\varepsilon'_k) C_k + nH + n'H \equiv 0,$$

which is a combination of the preceding two, can be reduced to one no longer containing faces of category 1.

To sum up, if there are congruences containing faces of category 1, *there cannot be two that are distinct.*

We therefore look specially at congruences that contain only faces of categories β and $\alpha\beta$. Let

$$\sum \theta'_k \alpha_i \beta_i B'_k + \sum \theta''_k \beta_i F''_k \equiv 0 \tag{2}$$

be one of these congruences.

We consider the points common to the manifolds that figure in this congruence and the surface $S(M)$. Reducing each of these manifolds to these common points, we get

$$\sum \theta'_k M B'_k \equiv 0. \tag{3}$$

Indeed, if the point M is any point on the cut OA_i, different from O, the surface $S(M)$ has no point in common with faces of category β, for which y can take only the value O. Likewise, the surface $S(M)$ has no point in common with faces of the form $\alpha_j \beta_j B'_k$, where the index j is different from i, because for these faces y must be on the cut OA_j, whereas M is on the cut OA_i.

Congruence (3) means that, on the surface $S(M)$, the set of edges B'_k of the polyhedron P'_i (weighted by the coefficients θ') must form a closed cycle. Let K_i be this cycle.

We observe that

$$\sum \theta'_k \alpha_i \beta_i B'_k \equiv \sum \theta'_k \alpha_i B'_k + H,$$

where H is the set of edges of categories β and $\alpha\beta$. Likewise,

$$\sum \theta''_k \beta_i F''_k \equiv H',$$

where H' is a combination of edges of category β.

We then have

$$\sum \theta'_k \alpha_i \beta_i B'_k + \sum \theta''_k \beta_i F''_k \equiv \sum \theta'_k \alpha_i B'_k + H + H',$$

so that the congruence (2) cannot hold unless we have identically

$$\sum \theta'_k \alpha_i B'_k = 0. \tag{4}$$

Since we do not have any relation between the edges $\theta'_k \alpha_i B'_k$ for different indices i, the identity (4) must hold when one gives the index i a definite value and we extend the summation only over the different values of the index k.

The identity (4) therefore means that, as the point A_i tends to A_i, the cycle K tends to zero.

Indeed, when M becomes A_i, the surface $S(M)$ degenerates and its genus decreases. Some of its cycles therefore vanish. We see how this happens by studying the Picard group. Let S_i be the substitution in this group that corresponds to the singular point A_i; it changes the cycle ω_h, for example, into

$$\omega'_h = m_1\omega_1 + m_2\omega_2 + \cdots + m_{2p}\omega_{2p}.$$

Then, for $M = A_i$, we have

$$\omega_h = \omega'_h,$$

that is, for $M = A_i$ the cycle $\omega'_h - \omega_h$ vanishes.

Do we obtain, in this way, all the cycles that vanish for $M = A_i$ (which I call the *vanishing cycles*)?

Picard showed (vol. 1, p. 82) that every algebraic surface can be reduced, by a birational transformation, to one with only ordinary singularities, that is, double curves with triple points.

He then showed (p. 95) how to determine, for such a surface, the singular points A_i and the corresponding substitutions in the Picard group.

One thereby sees, in this case, that there are no vanishing cycles except those generated as we have just seen.

If one wants to consider surfaces with more complicated singularities (for example, cone points) it is necessary to proceed differently.

Suppose, for example, that we have an ordinary surface with two singular points corresponding to the same substitution in the Picard group. If we allow this surface to vary in such a way that the two singular points come into coincidence, then the limiting surface will admit a vanishing cycle not generated in the above manner.

Then, if we consider any vanishing cycle $\sum \theta'_k M B'_k$, we have the congruence

$$\sum \theta'_k \alpha_i \beta_i B'_k \equiv -\sum \theta'_k \beta_i B'_k. \tag{5}$$

To each singular point A_i we associate in this way a vanishing cycle K_i and, consequently, a congruence of the form (5). By adding these congruences, one obtains

$$\sum\sum \theta'_k \alpha_i \beta_i B'_k \equiv -\sum\sum \theta'_k \beta_i B'_k.$$

Suppose that the set of cycles K_i is homologous to zero, such that we have, on the surface S,

$$\sum K_i \sim 0,$$

and on S_0

$$\sum\sum \theta'_k \beta_i B'_k \sim 0.$$

That is, we can find coefficients θ'' such that

$$\sum\sum \theta'_k \beta_i B'_k \equiv \sum \theta''_k \beta_i F''_k.$$

We then have

$$\sum\sum \theta'_k \alpha_i \beta_i B'_k + \sum \theta''_k \beta_i F''_k \equiv 0. \tag{2}$$

Thus, for each combination of vanishing cycles K_i such that $\sum K_i \sim 0$, there corresponds a congruence of the form (2).

To the congruence thus obtained it is convenient to adjoin the following:

$$\sum \beta F''_k \equiv 0,$$

which represents a two-dimensional cycle making up the entire surface S_0. All the congruences of the form (2) are manifestly combinations of those we have obtained.

Now, under what conditions are two congruences of the form (2) distinct? In other words, which are the congruences of this form whose left-hand side is homologous to zero?

To obtain all homologies of the form

$$\sum \theta'_k \alpha_i \beta_i B'_k + \sum \theta''_k \beta_i F''_k \sim 0$$

it is necessary to find all the congruences between cells and faces of the form

$$\sum \theta'_k \alpha_i \beta_i B'_k + \sum \theta''_k \beta_i F''_k \equiv \sum \varepsilon_k B_k + \sum \zeta'_k \alpha_i \beta_i F'_k. \tag{6}$$

If we reduce all the manifolds appearing in the congruence (6) to their common points with a surface S corresponding to a value of y not situated on one of the cuts, we obtain

$$\sum \varepsilon_k B_k \equiv 0,$$

which means that the set of edges $\sum \varepsilon_k B_k$ is a closed cycle on the polyhedron P. Let $K(y)$ be that cycle.

Let $K(M_i)$ (or $K'(M_i)$) be the limit to which this cycle tends as y approaches a point M on the cut OA_i from the first side of the cut (or from the second side), so that $K(M_i)$ (or $K'(M_i)$) represents the set of points common to the surface $S(M)$ and $\sum \varepsilon_k \alpha_i \beta_i B_k$ (or $\sum \varepsilon_k \alpha_i \beta_{i+1} B_k$).

We then have (since $\varepsilon_k B_k$ represents a closed cycle on the polyhedron P)

$$\sum \varepsilon_k B_k \equiv \sum \varepsilon_k \alpha_i \beta_i B_k - \sum \varepsilon_k \alpha_i \beta_{i+1} B_k, \tag{7}$$

whence

$$\sum \zeta'_k \alpha_i \beta_i F'_k \equiv \sum \theta'_k \alpha_i \beta_i B'_k + \sum \varepsilon_k \alpha_i \beta_{i+1} B_k - \sum \varepsilon_k \alpha_i \beta_i B_k + \sum \theta''_k \beta_i F''_k \tag{8}$$

or, reducing all manifolds to their common points with the surface $S(M_i)$,

$$\sum \zeta'_k M_i F'_k \equiv \sum \theta'_k M_i B'_k + K'(M_i) - K(M_i). \tag{9}$$

That is, on the polyhedron P'_i we must have

$$K_i \sim K(M_i) - K'(M_i), \tag{10}$$

since $\sum \theta'_k M_i B'_k$ is none other than the cycle we called K_i above.

Moreover, if condition (10) is satisfied, we can find numbers ζ' so as to satisfy the congruence (9). We then have, by letting M_i tend to A_i,

$$\sum \zeta'_k \alpha_i F'_k \equiv \sum \theta'_k \alpha_i B'_k + K'(A_i) - K(A_i).$$

But, by hypothesis, K_i is a vanishing cycle for the singular point A_i, so that

$$\sum \theta'_k \alpha_i B'_k = 0.$$

Moreover,

$$K'(A_i) = K(A_i) = \sum \varepsilon_k \alpha_i B_k.$$

Hence

$$\sum \zeta'_k \alpha_i F'_k \equiv 0,$$

which means that the set of faces $\sum \zeta'_k \alpha_i P'_k$ forms a closed surface. This cannot happen unless

$$\sum \zeta'_k \alpha_i F'_k = 0, \tag{11}$$

or if $\sum \zeta'_k \alpha_i F'_k$ represents the entire surface $S(K_i)$ or one of its components, in the case where the surface is decomposable (see above, p. 159).

Leaving aside the latter case, which will not occur for the surfaces with ordinary singularities to which Picard reduces the others (and which, moreover, we can treat as on p. 159), we see that we can always assume that

$$\sum \zeta'_k \alpha_i F'_k = nS(A_i),$$

where n is an integer. Whence

$$\sum (\zeta'_k - n) \alpha_i F'_k = 0.$$

But the congruence (9) continues to hold when we replace ζ'_k by $\zeta'_k - n$, since

$$\sum M_i F'_k \equiv 0,$$

because the surface $S(M_i)$ is closed.

We can therefore suppose that the identity (11) holds. It follows that

$$\begin{aligned} &\sum \zeta'_k \alpha_i \beta_i F'_k \\ &\equiv \sum \zeta'_k \alpha_i F'_k + \sum \theta'_k \alpha_i \beta_i B'_k + \sum \varepsilon_k \alpha_i \beta_{i+1} B_k - \sum \varepsilon_k \alpha_i \beta_i B_k - \sum \zeta'_k \beta_i F'_k. \end{aligned}$$

But, if we take account of the identity (11), and if we decompose the faces F' into faces F'', we can put

$$-\sum \zeta'_k \beta_i F'_k = \sum \theta''_k \beta_i F''_k.$$

This recovers the congruence (8) and, adjoining the congruence (7), we finally get the congruence (6).

Thus, to each system of homologies (10) there corresponds a homology of faces, and only one.

We verify similarly that the combination $\sum \beta_i F''_k$, which represents the entire Riemann surface S_0 and which, consequently, is congruent to zero, is not homologous to zero.

Indeed, if one has a congruence of the form (6) with all the θ' zero, then one must have a homology of the form (10) with the cycles K_i zero. This means that the cycle $K(y)$ will be invariant under all substitutions in the Picard group.

The cycles $K'(M_i)$ and $K(M_i)$ are none other than what we called Ω'_i and Ω_i in the preceding paragraph. We therefore recover the homology (5) of the preceding paragraph, which differs only in notation from the homology (9) of the present paragraph. Indeed, to pass from one to the other it suffices to annul the θ', change

$K'(M_i)$ to Ω'_i and $K(M_i)$ to Ω_i, and to write θ'_k in place of ζ'_k. To adopt these notations in the preceding paragraph, it suffices finally to write $\zeta_q B_q$ in place of $\varepsilon_k B_k$.

In this case, our cells satisfy the congruence (2) of the preceding paragraph, which is written

$$\sum \zeta_q B_q + \sum \theta'_k \alpha_i \beta_i F'_k \equiv 0,$$

or, returning to the notations of the present paragraph,

$$\sum \varepsilon_k B_k + \sum \zeta'_k \alpha_i \beta_i F'_k \equiv 0.$$

This shows that the left-hand side of (6) must be identically zero; that is, not only the θ', but also the θ'', must be zero.

Thus, all homologies between the faces $\beta F''$ reduce to the identity and, in particular, we do not have

$$\sum \beta_i F''_k \sim 0.$$

Q.E.D.

Before I conclude, I want again to examine the congruences involving cells of category 1. We have seen that we cannot have more than one such congruence, or rather, if there are two such congruences we can pass from one to the other by adjoining a homology.

Let us see whether there exists such a congruence

$$\sum \varepsilon_k C_k + H \equiv 0, \tag{12}$$

where H is a combination of faces of categories $\alpha\beta$ and β. We can assume that $\sum \varepsilon_k$ is nonzero, otherwise we are reduced to one of the congruences studied above.

I add that, if there exists a congruence of the form (12) with $\sum \varepsilon_k$ nonzero, this congruence is certainly distinct from the preceding ones, since there cannot be a homology of the form

$$\sum \varepsilon_k C_k + H \sim 0 \tag{13}$$

without $\sum \varepsilon_k$ being zero.

So, does there exist a congruence of the form (12)? To answer this without a long but not difficult discussion, I suppose that there are m vertices of the polyhedron P (which I call $C_1, C_2, \dots, C_m$), where m is the degree in z of the equation $F(x, y, z) = 0$, and which correspond to a given value of x, for example $x = x_0$ (see §5 below).

Then the combination $C_1 + C_2 + \cdots + C_m$, which I write $\sum C_k$ for short, is nothing but the Riemann surface represented by the equation in y and z

$$f(x_0, y, z) = 0.$$

We then have

$$\sum C_k \equiv \sum C_k \alpha_i \beta_i - \sum C_k \alpha_i \beta_{i+1}.$$

Since the m vertices $C_1, C_2, \dots, C_m$ are only permuted when we cross from one side of OA_i to the other by turning around A_i, we have

$$\sum C_k \alpha_i \beta_i = \sum C_k \alpha_i \beta_{i+1},$$

and consequently

$$\sum C_k \equiv 0.$$

This congruence is indeed of the form (12) and

$$\sum \varepsilon_k = m \neq 0.$$

We initially have two singular two-dimensional cycles which are the two Riemann surfaces, one corresponding to $y = 0$ and the other to $x = x_0$.

To form other two-dimensional cycles it suffices to consider q cycles

$$K_1, \quad K_2, \quad \ldots, \quad K_q,$$

corresponding to the q singular points

$$A_1, \quad A_2, \quad \ldots, \quad A_q,$$

each of which vanishes relative to its corresponding singular point. In addition, these cycles must satisfy, on the polyhedron P, the condition

$$\sum K_i \sim 0.$$

Two systems of cycles

$$\begin{array}{cccc} K_1, & K_2, & \ldots, & K_q, \\ K_1', & K_2', & \ldots, & K_q', \end{array}$$

give us two distinct two-dimensional cycles, at least if we have (on P)

$$K_i' - K_i \sim K'(M_i) - K(M_i),$$

where $K(y)$ is any cycle on the polyhedron P.

If we let U_i and U_i' denote the limits of the cycles K_i and K_i' as the point M_i tends to O, and let Ω_i denote the limit of the cycle $K(y)$ as the point y tends to O in the sector $A_{i-1}OA_i$, then on S_0 we have

$$\sum U_i \sim 0, \quad \sum U_i' \sim 0.$$

So, if we want two distinct cycles, we must not have

$$U_i' - U_i \sim \Omega_{i+1} - \Omega_i.$$

In this way we see how to obtain distinct cycles, and thereby distinct congruences of the form in question, and in the process find the number of distinct homologies.

We form the table of vanishing cycles relative to the different singular points A_i. We distinguish two kinds:

1° Those of the form $\Omega' - \Omega$, where Ω' is the transform of the cycle Ω under the transformation in the Picard group corresponding to the point A_i.

2° The vanishing cycles of the second kind are those not of the above form, nor combinations of cycles of this form.

Evidently, for each of the singular points, we must reduce the number of those vanishing as much as possible, so that the cycles in our table are linearly independent. Then if the cycles are

$$m_{1,1}\omega_1 + m_{1,2}\omega + 2 + \cdots + m_{1,2p}\omega_{2p},$$

$$\cdots\cdots\cdots\cdots\cdots\cdots\cdots\cdots\cdots\cdots$$

$$m_{k,1}\omega_1 + m_{k,2}\omega + 2 + \cdots + m_{k,2p}\omega_{2p},$$

the table of integers m, with k rows and $2p$ columns ($k < 2p$), is such that the determinants formed by suppressing $2p - k$ columns are not all zero.

Now we unite the tables relative to the q singular points; the united table has $2p$ columns and $\sum k = \mu$ rows. If the determinants formed by suppressing $\mu - 2p+r$ rows and r columns are all zero, but the determinants formed by suppressing $\mu - 2p + r + 1$ rows and $r + 1$ columns are not all zero, then the number of distinct congruences that we seek will be $\mu-2p+r+1$ (or $\mu-2p$ if the determinants formed by suppressing $\mu - 2p$ rows are not all zero).

To obtain the number of distinct homologies, it is necessary to find the number of combinations of the cycles $U_1, U_2, \dots, U_q$ such that we have

$$U_i \sim \Omega_{i+1} - \Omega_i.$$

Each cycle Ω evidently gives birth to such a combination; but if two cycles Ω and Ω' differ only by a cycle invariant under the Picard group they give birth to the same combination. The number of homologies is therefore equal to the total of cycles, $2p$, minus the number of distinct invariant cycles, which I call n.

The number of distinct two-dimensional cycles (including the two singular cycles) is therefore

$$2 + (\mu + 2p + 1) - (n - 2p).$$

We see how we can generate the nonsingular two-dimensional cycles by confining ourselves to the most general case, which is where there is no vanishing cycle of the second kind. In this case we have

$$K_i = \Gamma_i' - \Gamma_i,$$

where Γ_i is a cycle and Γ_i' is its transform under the substitution corresponding to A_i.

We then describe, in the plane of y, loops $L_1, L_2, \dots, L_q$ that begin at O and turn around the singular points $A_1, A_2, \dots, A_q$, respectively.

Let Γ_i be a cycle on the surface S_0. When y begins at O and describes the loop, the surface S and the cycle deform in such a way that, when y returns to O, the cycle becomes the cycle Γ_i' on S_0. In its movement it generates a surface σ_i bounded by the two closed curves Γ_i and Γ_i'.

We therefore have

$$\Gamma_1 + \Gamma_2 + \cdots + \Gamma_q = \Gamma_1' + \Gamma_2' + \cdots + \Gamma_q',$$

and the set $\sigma_1 + \sigma_2 + \cdots + \sigma_q$ will form a closed surface. This will be our two-dimensional cycle.

§4. One-dimensional cycles

The problem of one-dimensional cycles has been completely solved by Picard. I therefore have only to translate Picard's reasoning into our notation.

We are looking for congruences between edges. As we have seen at the beginning of paragraph 2, we can always assume that such a congruence contains no edges of category α. Our congruence must therefore be of the form

$$\sum \theta_k' \alpha_i \beta_i C_k' + \sum \theta_k'' \beta_i B_k'' \equiv 0. \tag{1}$$

We observe that

$$\sum \theta_k' \alpha_i \beta_i C_k' \equiv \sum \theta_k' \alpha_i C_k' + H,$$
$$\sum \theta_k'' \beta_i B_k'' \equiv H',$$

where H and H' are sets of edges of category β. We then have

$$\sum \theta'_k \alpha_i C'_k + H + H' = 0,$$

and, since the vertices $\alpha_i C'_k$ cannot be reduced, either to vertices of category β or to those of category α where the index of α is different from i, we must have

$$(2) \qquad \sum \theta'_k \alpha_i C'_k = 0,$$

where the summation is over the vertices $\alpha_i C'_k$ with the same index i, but different indices k.

What does the identity (2) mean? When y reaches A_i the polyhedron P'_i degenerates, several of its vertices come together, but none can disappear (although edges and faces can disappear). The algebraic sum of all the coefficients θ'_k relative to the different vertices $\alpha_i C'_k$ that come to coincide must then be zero; hence the sum of all the θ'_k is zero:

$$(3) \qquad \sum \theta'_k = 0.$$

Because of the relation (3), we can find a combination of edges $\sum \zeta'_k B_k$ on the polyhedron P_i such that

$$(4) \qquad \sum \zeta'_k B'_k \equiv \sum \theta'_k C'_k.$$

We then have

$$(5) \qquad \sum \zeta'_k \alpha_i \beta_i B'_k \equiv \sum \zeta'_k \alpha_i B_k - \sum \zeta'_k \beta_i B'_k + \sum \theta'_k \alpha_i \beta_i C'_k.$$

Moreover, when y arrives at A_i, the edges and vertices B'_k and C'_k of the polyhedron P'_i become the edges and vertices $\alpha_i B'_k$ and $\alpha_i C'_k$. We then have the congruence

$$\sum \zeta'_k \alpha_i B'_k \equiv \sum \theta'_k \alpha_i C'_k$$

or, by (2),

$$\sum \zeta'_k \alpha_i B'_k \equiv 0.$$

The latter congruence says that the combination $\sum \zeta'_k \alpha_i B'_k$ forms a closed cycle on the Riemann surface $S(A_i)$. But, when y arrives at A_i, the Riemann surface degenerates, cycles can disappear, but no new cycles appear. Therefore, to the cycle $\sum \zeta'_k \alpha_i B'_k$ there corresponds on the surface $S(M_i)$ at least one closed cycle $\sum \varepsilon'_i M_i B'_k$, which reduces to $\sum \zeta'_k \alpha_i B'_k$ for $M_i = A_i$, so that we have

$$(6) \qquad \sum \varepsilon'_k M_i B'_k \equiv 0,$$

$$\sum \varepsilon'_k \alpha_i B'_k = \sum \zeta'_k \alpha_i B'_k.$$

Then, if we replace ζ'_k by $\zeta'_k - \varepsilon'_k$, the congruence (4) continues to hold because of (6); the congruence (5) remains equally true, and we have

$$\sum (\zeta'_k - \varepsilon'_k) \alpha_i B'_k = 0.$$

Thus we can always assume that the ζ'_k have been chosen so that

$$(7) \qquad \sum \zeta'_k \alpha_i B'_k = 0.$$

If we combine the congruence (5) with the identity (7) we obtain the homology

$$(8) \qquad \sum \theta'_k \alpha_i \beta_i C'_k \sim \sum \zeta'_k \beta_i B'_k,$$

and, adjoining this homology to (1), we get

$$\sum \zeta'_k \beta_i B'_k + \sum \theta''_k \beta_i B''_k \equiv 0,$$

which involves only edges of category β.

We can therefore suppose that our congruence (1) contains only edges of category β. This is the theorem of Picard, according to which a one-dimensional cycle can always be brought to a position for which y is constant along the cycle.

To each of the closed cycles on the surface S_0 there corresponds a congruence of the form (1), but not all these congruences are distinct. This was shown by Picard. Let $\omega_1, \omega_2, \dots, \omega_{2p}$ be the $2p$ cycles of S_0, and suppose that a substitution S_h in the Picard group changes ω_i into

$$m_1\omega_1 + m_2\omega_2 + \cdots + m_{2p}\omega_{2p}.$$

Then we have the homology

$$\omega_i \sim m_1\omega_1 + m_2\omega_2 + \cdots + m_{2p}\omega_{2p}. \tag{9}$$

These homologies reduce the number of one-dimensional homologies, and Picard showed that, for the most general algebraic surface, the number is zero.

We use the term *subsisting cycles* for those that are not linear combinations of the different vanishing cycles with respect to the various singular points A_i. These are the cycles that remain distinct when the homologies (9) are taken into account. Indeed, the homology (9) says that the cycle

$$\omega - \sum M_k \omega_k$$

is nonvanishing with respect to the singular point corresponding to the substitution S_h in the Picard group.

Thus there are as many one-dimensional cycles as there are subsisting cycles. We have seen, on the other hand, that there are as many three-dimensional cycles as there are invariant cycles.

But, by the fundamental theorem on Betti numbers, the number of one-dimensional cycles must be the same as the number of three-dimensional cycles.

Thus the number of subsisting cycles must be the same as the number of invariant cycles.

We now verify this.

The verification is easy if there are no vanishing cycles of the second kind.

We recall that the Picard group has a particular form.

Let $\omega_1, \omega_2, \dots, \omega_{2p}$ be the fundamental cycles and consider the bilinear form

$$\Phi = \omega'_2\omega_1 - \omega'_1\omega_2 + \omega'_4\omega_3 - \omega'_3\omega_4 + \cdots + \omega'_{2p}\omega_{2p-1} - \omega'_{2p-1}\omega_{2p}.$$

Provided that the fundamental cycles have been conveniently chosen, when we subject the ω to one of the substitutions in the Picard group and at the same time subject the ω' to the same substitution, the form Φ will be unaltered. On the other hand, the number of subsisting cycles will be the same as the number of solutions of the system (A) of linear equations obtained by equating each fundamental cycle to its transform under each of the Picard substitutions.

Suppose then that $\sum m_i\omega_i$ is an invariant cycle. Since the expression $\sum m_i\omega_i$ can be assimilated in the linear form Φ, by setting

$$\omega'_2 = m_1, \quad \omega'_1 = -m_2, \quad \omega'_3 = m_3, \quad \omega'_4 = -m_4, \quad \dots,$$

and since, on the other hand, $\sum m_i\omega_i$ becomes $\sum m_i\omega_i$ when the ω are subjected to one of the substitutions in the Picard group, we conclude that the system of values

$$-m_2, \quad m_2, \quad -m_3, \quad m_4, \quad \dots$$

is its own transform under this linear substitution. It is therefore a solution of the system (A) just mentioned.

One sees, moreover, that two or more linearly independent cycles correspond in this way to two or more linearly independent solutions of the system (A), and conversely.

Thus there are as many subsisting cycles as there are invariant cycles. Q.E.D.

Now, what happens if there are vanishing cycles of the second kind?

As I have said, this case cannot occur for the surfaces with ordinary singularities to which Picard has reduced all the others (vol. 1, p. 85). It is true that it can occur for other surfaces if we want to study them without first applying Picard's transformation. But these surfaces have a special kind of singular point, and the four-dimensional manifold V generated by the surface also has a singular point.

The general theorems we are concerned with and which we have proved in *Analysis situs* and its supplements do not apply to manifolds with singular points. They fail to be true, in general, unless we make some special conventions.

The question whether there is a vanishing cycle of the second kind likewise depends on the conventions we make. It is true that such a cycle will be the boundary of a two-dimensional submanifold of V, but this submanifold contains a singular point of V.

To clarify the difficulty that arises, we take a very simple example. Imagine a surface, in ordinary space, with a conical point, or even more simply a cone of revolution and its prolongation. Let S be the apex of the cone, and C a circumference of the cone. In one sense, C is the boundary of a region on the cone, namely, the region between the circumference C and the apex S. But, on the other hand, a line drawn on the cone can leave this region without crossing C, by passing through the apex S to the other sheet of the cone.

Under these conditions, it seems preferable to leave aside this singular case and to confine ourselves to those surfaces with ordinary singularities to which all the others may be reduced.

§5. Various remarks

In the course of the proofs we have made various assumptions about our polyhedra. We recall them here:

1° We have assumed (see p. 153) that the polyhedron we call P' remains homeomorphic to itself when y follows the cut OA_i from A_i to O.

2° We have compared (see p. 159) the Riemann surface S_0 corresponding to the point O to the Riemann surface $S(M)$ corresponding to the point M, and we have said that we can make the two surfaces correspond point to point. We have used this correspondence to define the cycles U_i and U_i' that correspond on S_0 to Ω_i and Ω_i'.

3° We have next assumed (see p. 159) that, when the point M varies from A_i to M_0, the two (initially coincident) cycles U_i and U_i' sweep out a certain region R on S_0 which corresponds, on $S(M_0)$, to the region $\sum \theta_k' M_0 F_k'$.

4° We have claimed that, when the point M crosses successively the two sides of the cut OA_1, then the two sides of the cut OA_2, ..., then finally the two sides of the cut OA_q, the moving cycle U_i (which returns to its initial situation Ω_i^0 after occupying a continuous series of successive situations) does not sweep out the entire surface S_0.

5° We have said (p. 164) that, if the surface $f(x, y, z) = 0$ is reduced, by Picard's procedure, to one possessing only ordinary singularities, then each singular point A_i corresponds to only one vanishing cycle.

6° We have assumed (p. 167) that the vertices of P include m which correspond to a constant value of x, for example $x = 0$.

Since the legitimacy of these assumptions is almost evident, I have not wanted to interrupt the reasoning of the preceding paragraphs to give explicit proofs. Moreover, I have not constructed a particular polyhedron P, not wanting to make particular assumptions about the way in which the Riemann surface is subdivided.

I believe that it is now useful to return to these points and to give a proof that adopts particular assumptions about P.

We can, for example, construct the polyhedron P as follows.

We begin by reducing the surface $f = 0$ to one with only ordinary singularities.

We give y any value, and consider the corresponding Riemann surface S. This surface consists of m sheets over the plane of x (if the equation $f = 0$ is of degree m in z).

On the plane of x we mark the origin O and the singular points corresponding to the equations

$$f = 0, \quad \frac{df}{dz} = 0.$$

Let n be the number of singular points, and let $B_1, B_2, \ldots, B_n$ be these singular points. We join the point O to the n singular points $B_1, B_2, \ldots, B_n$ by n cuts $OB_1, OB_2, \ldots, OB_n$ which do not cross and which succeed each other around O in the order of their subscripts.

We obtain the Riemann surface as usual by joining the first side of a cut in one of the sheets with the second side of the cut on some other sheet. The cuts thereby subdivide the Riemann surface into our polyhedron P.

We see that the polyhedron has M faces (which are the F_k) and that each of the faces is a polygon with $2n$ edges.

Now what happens when y varies? The singular points B are displaced, the cuts OB are deformed, and we suppose that they are deformed in such a way that they continue not to cross each other and to succeed each other in the same order around O. When y describes a small closed contour, the deformation is one in which the cuts return to their original positions, unless there is s singular point inside the contour.

The possible singular points are of two kinds:

1° First, there are those corresponding to the case where two of the singular points B are exchanged (an argument of Picard shows that, if the surface $f = 0$ has only ordinary singularities, this cannot happen unless the plane $y =$const. is tangent to the surface $f = 0$).

2° Then there are those corresponding to the case where one of the singular points B comes to O.

The singular points of the second kind are not essential, and I would let them disappear if there were no advantage in retaining them.

I denote the singular points by $A_1, A_2, \ldots, A_q$ and draw the cuts $OA_1, OA_2, \ldots, OA_q$ in the plane of y values.

As long as y does not cross the cuts OA, the cuts OB can be deformed so that they do not cross each other and, consequently, in such a way that P remains homeomorphic to itself, and at the same time in such a way that, when y describes a closed contour, the cuts OB return to their initial positions.

Now we compare the configurations of the cuts OB when y is at two infinitely close points on opposite sides of a cut OA_i.

First suppose that A_I is a singular point of the first kind. If we assume that the surface $f = 0$ has only ordinary singularities and that, consequently, the plane $y = A_i$ is tangent to $f = 0$ at an ordinary point, we see that, when y turns around A_i, two singular points are exchanged, for example B_1 and B_3. Moreover, if the point B_1 permutes two sheets of the Riemann surface, then the point B_3 with which it is exchanged will permute the same two sheets. I call these two sheets the *first and second sheets of the surface.*

When y, after turning around A_i, returns infinitely closely to its initial position, but on the opposite side of the cut, we can assume that the cuts OB return to their initial position, with the exception of the cuts OB_1 and OB_3. The latter cuts, which initially occupy the positions of the solid lines OB_1 and OB_3 on Figure 1, finally occupy the positions of the dotted lines OB_3 and OB_1.

One sees that the Riemann surface can be subdivided into a polyhedron P in two ways. The two modes of subdivision differ because the solid lines OB_1 and OB_3 are replaced by the dotted lines OB_3 and OB_1.

If we superimpose the two modes of subdivision, we obtain the polyhedron that I call P'. One sees that two of the m faces of P, those corresponding to the first two sheets, have been subdivided into three subfaces, denoted on the figure by (α), (β), (γ).

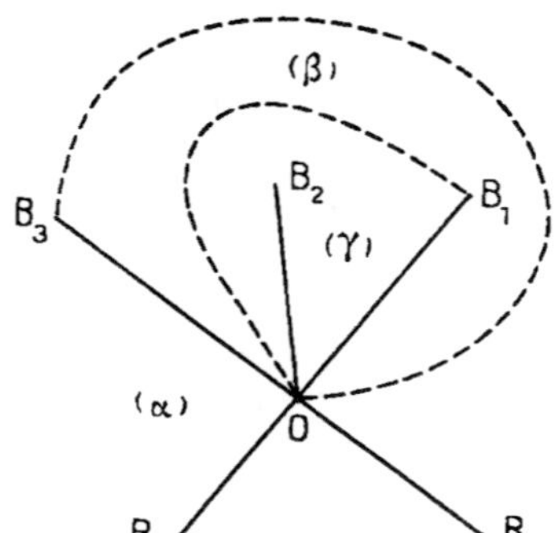

Fig. 1.

As y describes the cut OA_i, the points B are displaced in a continuous manner without coming into coincidence with each other or with the point O (except, of course, when y arrives at A_i). It follows that we can deform all our cuts OB (both solid and dotted) so as to avoid their meeting. In other words, we can say that the polyhedron P' constantly remains homeomorphic to itself. To say that P' remains homeomorphic to itself is to say that we can make a point-to-point correspondence between the surface $S(M)$ and another surface $S(M')$ corresponding to another position M' of y on OA_i, and with S_0 in particular. We remark that the correspondence can be made in such a way that a point at infinity on $S(M)$ corresponds to a point at infinity on $S(M')$.

When y arrives at A_i the polyhedron degenerates: the two points B_1 and B_3 come into coincidence. Likewise, the solid cut OB_1 comes to coincide with the dotted OB_3, and the dotted OB_1 comes to coincide with the solid OB_3. The subface (β) in Figure 1 disappears.

To go further, we remark that there are two cuts projected onto the solid line OB_1. When one follows the first of these (which we call B_1G) from O to B_1 one has the first sheet on the left and the second on the right; when one follows the second (which we call B_1D), one has the first sheet on the right and the second on the left. We define B_3G and B_3D similarly. If, instead of the solid line OB_1, we consider the dotted line OB_1, we similarly obtain two cuts that I call $B_1'G$ and $B_1'D$; and I again define $B_3'G$ and $B_3'D$ in the same way.

We see immediately that, when y turns around A_i, the cuts B_1G and $B_3'G$, B_1D and $B_3'D$, B_3G and $B_1'G$, B_3D and $B_1'D$ are exchanged; and that, for $y = A_i$, B_1G and $B_3'G$, ... come to coincide.

I denote by β_1 those subfaces β that belong to the first sheet if we adopt first subdivision, corresponding to the solid lines; the other subfaces β will be called β_2.

We then have the congruences

$$\beta_1 \equiv B_1'D - B_1D + B_3'D - B_3D,$$
$$\beta_2 \equiv B_1'G - B_1G + B_3'G - B_3G,$$

which come from the cuts forming the boundaries of β_1 and β_2.

Consider the combination

$$\omega = B_2D - B_1'D - B_3G + B_1'G.$$

This is a cycle on our Riemann surface. When y turns around A_i it changes to

$$B_1'D - B_3D - B_1'G + B_3G,$$

that is, to $-\omega$. This is therefore a vanishing cycle. It is easy to see that it is the only one.

We now return to the cycles that we called Ω_i and Ω_i' in paragraph 2. Let

$$\Omega_i = \zeta_1 B_1D + \zeta_2 B_1G + \zeta_3 B_3D + \zeta_4 B_3G + H,$$

where the ζ are integer coefficients and H is a combination of other edges of P_1. Since the cycle Ω_1 must be closed, we have

$$\zeta_1 + \zeta_2 = \zeta_3 + \zeta_4 = 0,$$

because the vertex B_1, for example, belongs only to the two edges B_1D and B_1G. It follows that we have

$$\Omega_i' = \zeta_1 B_3'D + \zeta_2 B_3'G + \zeta_3 B_1'D + \zeta_4 B"_1G + H.$$

This is because the edges of P, other than B_1D, B_1G, B_3D, B_3G, are not changed when y turns around A_i, so H does not change.

We then have

$$\begin{aligned}\Omega_i - \Omega_i' = &\quad \zeta_1(B_1D - B_1G - B_3'D + B_3'G)\\ &+ \zeta_3(B_3D - B_3G - B_1'D + B_1'G).\end{aligned}$$

But

$$\begin{aligned}(B_1D - B_1G - B_3'D + B_3'G) = &\quad (B_1'G - B_1G + B_3'G - B_3G)\\ &+ (B_1'D - B_1D + B_3'D - B_3D)\\ &+ (B_3D - B_1'D + B_3G - B_1'G)\end{aligned}$$

or

$$(B_1D - B_1G - B_3'D + B_3'G) + \omega \equiv \beta_2 - \beta_1;$$

whence

$$\Omega_i - \Omega_i' + (\zeta_1 - \zeta_3)\omega \equiv \zeta_1(\beta_2 - \beta_1),$$
$$\Omega_i - \Omega_i' \sim (\zeta_3 - \zeta_1)\omega.$$

But we have assumed that the cycle Ω_i is invariant, that is, that

$$\Omega_i \sim \Omega_i',$$

which requires $\zeta_1 = \zeta_3$. Under these conditions, we recover congruence (5) of paragraph 2, which can be written

$$\Omega_i = \Omega_i' \equiv \sum \theta_k' MF_k'.$$

Comparing this with the preceding congruence, we find

$$\sum \theta_k' MF_k' = \zeta_1(\beta_2 - \beta_1).$$

When y (or M) comes to A_i the subfaces β_1 and β_2 disappear, so that the right-hand side of this equation vanishes. The left-hand side reduces to $\sum \theta_k' \alpha_i F_k'$, whence one concludes

$$\sum \theta_k' \alpha_i F_k' = 0.$$

This justifies what we said on page 158 and those following.

Now suppose that A_i is a singular point of the second kind, and that when $y = A_i$ the singular point B_1 comes to O.

We can then draw a Figure 2 analogous to Figure 1. For the sake of simplicity I represent only three singular points, B_1, B_2, and B_3.

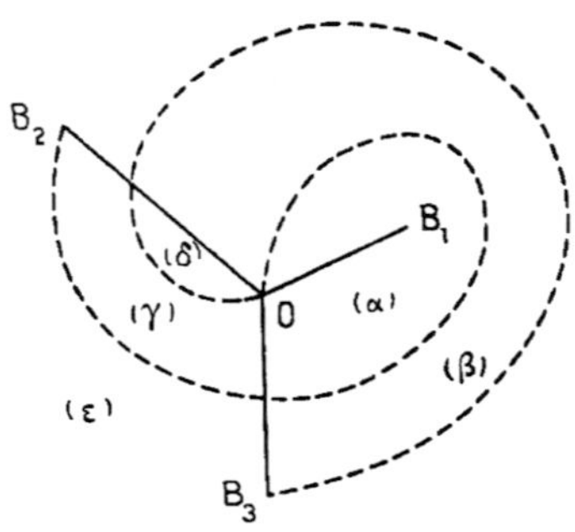

Fig. 2.

When y turns around A_i, the cut OB_1 returns to its original position, but the cuts OB_2 and OB_3 (drawn solid) are transformed to the cuts OB_2 and OB_3 (drawn dotted).

This gives two modes of subdivision and, by superimposition, we obtain the polyhedron P'. Each face of P' (that is, each sheet of S) is thereby subdivided into a certain number of subfaces. In the case of Figure 2, we have five, denoted on the figure by $(\alpha), (\beta), (\gamma), (\delta), (\varepsilon)$.

It is clear that, when y describes the cut OA_i, one can arrange for the preceding figure (and, consequently, the polyhedron P') to remain constantly homeomorphic to itself.

I let OB_2 and OB_3 denote the solid cuts, and OB_2' and OB_3' the corresponding dotted cuts. I observe that these cuts cross each other, with the pieces forming the edges of P'. On each of these cuts I distinguish the *terminal segment*, which is the one adjoining the point B_2 or B_3.

When y comes to A_i the polyhedron degenerates; several of its edges are reduced to zero, in particular the edge OB_1 and all segments of OB_2, OB_3, OB_2', OB_3', *except the terminal segments*. Moreover, the terminal segment of OB_2 comes to coincide with that of OB_2', and that of OB_3 with that of OB_3'. It follows that the four subfaces $(\alpha), (\beta), (\gamma), (\delta)$ disappear.

If there is no vanishing cycle here, the singular point A_i is not essential.

On the figure we have

$$\begin{aligned} OB_2 - OB_2' &\equiv (\alpha) + (\gamma) + (\delta), \\ OB_3 - OB_3' &\equiv (\alpha) + (\beta) + (\delta). \end{aligned}$$

Returning to our cycle Ω_i, we have

$$\Omega_i = \sum \zeta_1 OB_1 + \sum \zeta_2 OB_2 + \sum \zeta_3 OB_3.$$

I use the sign $\sum$ because there are several cuts (belonging to different sheets of the Riemann surface) that project onto OB_1. It follows that we have

$$\Omega_i' = \sum \zeta_1 OB_1 + \sum \zeta_2 OB_2' + \sum \zeta_3 OB_3';$$

whence

$$\begin{aligned} \Omega_i - \Omega_i' &= \sum \zeta_2 (OB_2 - OB_2') + \sum \zeta_3 (OB_3 - OB_3'), \\ \Omega_i - \Omega_i' &\equiv \sum \zeta_2 [(\alpha) + (\gamma) + (\delta)] + \sum \zeta_3 [(\alpha) + (\beta) + (\delta)], \end{aligned}$$

and, comparing with congruence (5) of paragraph 2,

$$\sum \theta_k' M F_k' = \sum \zeta_2 [(\alpha) + (\gamma) + (\delta)] + \sum \zeta_3 [(\alpha) + (\beta) + (\delta)].$$

When y comes to A_i, $(\alpha), (\gamma), (\delta)$ disappear, so what remains is

$$\sum \theta_k' \alpha_i F_k' = 0,$$

which again justifies what we said on page 158.

Supposing that the point y describes the different cuts $OA_1, OA_2, \ldots, OA_q$, the singular points B describe certain lines. We can assume that these lines are not indefinitely distant. Indeed, if this were so, we could always find a small circle in the plane of x values that did not cross any of these lines, and then, by a simple linear change of variables, we could send the center of this circle to infinity.

Thus, when the point y moves successively across the two sides of OA_i, the two sides of OA_2, etc., and finally those of OA_q, the singular points B remain at a finite distance. We can then control the deformation of the cuts OB so that the cuts remain at a finite distance. The cycle Ω_i, which is a combination of these cuts, therefore always remains at a finite distance. It will then be the same for the corresponding cycle U_i on S_0 (since we can choose the correspondence between $S(M)$ and S_0 in such a way that a point at infinity on $S(M)$ corresponds to a point at infinity on S_0). This cycle then cannot sweep out the entire surface S_0, which justifies what we said on p. 161.

We remark finally that, among the vertices of P, there are m that correspond to $x = 0$.

All our hypotheses are therefore justified.

Fifth Supplement to Analysis Situs

Rendiconti del Circulo matematico di Palermo 18 (1904), pp. 45–110.

§1

I have previously been involved with *analysis situs* on several occasions; I first published a memoir on that subject in the volume *Centenaire du Journal de l'École Polytechnique*; this memoir was followed by four supplements which have appeared in vol. XIII of the *Rendiconti del Circolo matematico di Palermo*, vol. XXXVII of the *Proceedings of the London Mathematical Society*, in the *Bulletin de la Société Mathématique de France* in 1901, and finally in the *Journal de Liouville* in 1901.

I now return to this same question, persuaded that it has not yet been exhausted and its importance is sufficient to justify further effort.

This time I confine myself to the study of certain three-dimensional manifolds, but the methods used without doubt are of more general applicability. I shall devote considerable space to certain properties of closed curves which can be traced on closed surfaces in ordinary space.

The final result which I have in view is the following. In the second supplement I showed that to characterize a manifold it does not suffice to know the Betti number; one also needs to know certain coefficients, which I have called torsion coefficients (2nd supplement, §5, p. 129) and which play an important role.

One could then ask whether the consideration of these coefficients is sufficient, e.g., whether a manifold, all Betti numbers and torsion coefficients of which equal 1, is simply connected in the true sense of the word, i.e., homeomorphic to a hypersphere; or whether, on the contrary, it is necessary, in order to affirm that a manifold is simply connected, to study its *fundamental group*, which I have defined in the *Journal de l'École Polytechnique*, §12, p. 38.

We can now answer this question; in fact, I construct an example of a manifold, all Betti numbers and torsion coefficients of which equal 1, but which is not simply connected.

§2

I consider a manifold V of m dimensions situated in the space of k dimensions. Then let

$$\varphi(x_1, x_2, \ldots, x_k) = t$$

be the equation of a hypersurface of $k-1$ dimensions situated in the same space, which I shall call the surface $\varphi(t)$; in this equation $x_1, x_2, \ldots, x_k$ are the coordinates of a point in the space of k dimensions and t an arbitrary parameter such that the surface $\varphi(t)$ deforms continuously as t varies continuously. I suppose that the function φ is uniform and such that only one surface $\varphi(t)$ passes through each point.

The surface $\varphi(t)$ intersects V in a certain number of $(m-1)$-dimensional manifolds

$$w_1(t), \quad w_2(t), \quad \ldots, \quad w_p(t),$$

the set of which constitutes a system $W(t)$.

When t varies continuously from $-\infty$ to $+\infty$, the system $W(t)$ varies continuously and *generates* the manifold V. If the manifold V is closed, the manifolds $w(t)$ are likewise.

That being given, I now define what I call the *skeleton* of the manifold V. To each of the partial manifolds $w_1(t), w_2(t), \ldots, w_p(t)$ I assign a point in ordinary space. One of the coordinates of this point, x for example, will equal the parameter t, the other two are chosen arbitrarily, subject only to the following conditions:

1° If two manifolds $w_i(t), w_i(t+\varepsilon)$ differ very little from each other, the two corresponding points will be very close together.

2° It may happen that for certain values, $t = t_0$ for example, that a manifold $w(t)$ decomposes into two; in this case, for example, the manifold $w_1(t_0-\varepsilon)$ will differ very little from the set $w_1(t_0+\varepsilon) + w(t_0+\varepsilon)$ of two manifolds. In that case, we must arrange the manner of correspondence so that the points representing the two manifolds $w_1(t_0+\varepsilon)$ and $w_2(t_0+\varepsilon)$ which result from the bifurcation of $w_1(t_0-\varepsilon)$ differ very little from each other and from the point representing $w_1(t_0-\varepsilon)$.

Under these circumstances, when t varies in a continuous manner, the points representing the p manifolds

$$w_1(t), \quad w_2(t), \quad \ldots, \quad w_p(t)$$

generate p continuous lines

$$L_1, \quad L_2, \quad \ldots, \quad L_p$$

at least as long as the number p does not change. But this number can change at $t = t_0$, if one of the manifolds decomposes into two, or if, on the contrary, two manifolds merge into one. In the first case one of the lines L bifurcates, in the second case two of the lines L combine into one.

In this way we obtain a sort of network of lines, and it is this network which I call the *skeleton* of V. I have drawn this network in the space of three dimensions and not the plane, because then we can always avoid lines meeting except at points of bifurcation.

If we follow one of these lines, L_1 for example, described by the point representing $w_1(t)$, we see that this manifold remains homeomorphic to itself (and in such a way that two manifolds $w_1(t)$ and $w_1(t+\varepsilon)$ corresponding to neighboring points differ very little from each other) *as long as we do not pass through a value t such that $w_1(t)$ has a singular point.*

We must then mark the lines of our network at points where the corresponding manifolds $w(t)$ have singular points. These will be the points of division which cut our lines into sections, but as long as we remain on one of these sections the corresponding manifold $w(t)$ will remain homeomorphic to itself.

We remark that if we consider one of the values t_0 which corresponds to a point of bifurcation where one of the manifolds w_i splits, then the manifold $w_i(t_0)$ itself will have a singular point. We are then obliged to study these singular points.

However, before proceeding on this study, I must make a few more remarks. If I have a closed manifold V, this means first of all that the manifolds $w(t)$ are all closed themselves. The second condition is that one of the lines L ends in a *cul-de-sac*, at $t = t_1$ for example, the corresponding manifold reduces to a point as t tends to t_1, i.e., on approaching the end of the *cul-de-sac.*

In the second place, I have said that I have chosen the function φ to be uniform, so that exactly one surface $\varphi = t$ passes through each point of the space. Apart from this condition, the system $W(t)$ can be arbitrary. The restriction does not prevent us from showing that any manifold V is susceptible to this mode of generation, however we can relax it and consider any system $W(t)$ of closed manifolds

$$w_1(t), \quad w_2(t), \quad \ldots, \quad w_p(t)$$

which vary continuously with t, allowing of course that for certain values of t one of the manifolds may reduce to a point or split into two.

Under these conditions, the system $W(t)$ again generates a manifold V and we can define the skeleton along the same lines as above, without changing anything.

Nevertheless, there is a case where it would be of advantage to make a slight change. I suppose that for two values of t, for example $t = 0$ and $t = 2\pi$, the system $W(t)$ is the same, or rather, I imagine that the two systems $W(t)$ and $W(t + 2\pi)$ are identical. It will then suffice to let t vary from 0 to 2π, and it will consequently be convenient to choose the point representing the manifold $w(t)$, not so that we have $x = t$ but so that $\arctan \frac{y}{x} = t$, that is, so that the representative points of $w(t)$ and $w(t + 2\pi)$ are identical.

If we suppose, for example, that $W(t)$ reduces to a single manifold $w(t)$, the skeleton of V will reduce to a closed curve under these conditions.

To take a thoroughly simple example, we consider a torus as our manifold V and regard it as being generated by a meridian circle which will be our manifold $w(t)$, identical to $w(t + 2\pi)$. With our new conventions, the skeleton of this torus will be a closed curve.

We now commence the study of singular points of the manifolds $w(t)$. The part of one of these manifolds close to the singular point could be represented by the following equations:

$$\begin{aligned} x_i &= \psi_t(y_1, y_2, \ldots, y_q) \quad (i = 1, 2, \ldots, k), \\ \varphi_h(y_1, y_2, \ldots, y_q) &= 0 \quad\;\; (h = 1, 2, \ldots, q - m), \end{aligned}$$

to which it is necessary to adjoin certain inequalities which do not concern us.

In the region envisaged, the functions φ and ψ are holomorphic. We can always assume that the singular point corresponds to

$$y_1 = y_2 = \cdots = y_q = 0$$

and at this value the first order partial derivatives of the functions φ are not all zero, *except for one among them,* φ_1.

Then the equations

$$\varphi_2 = \varphi_3 = \cdots = \varphi_{q-m} = 0$$

enable us to express each y as a function of $m+1$ of the others, depending on which functions remain holomorphic in the vicinity of the singular point. I then replace these y by the expressions derived for them, so that all expressions are in terms of $m+1$ of the quantities y, for example $y_1, y_2, \ldots, y_{m+1}$, and our equations take the form

$$x_i = \psi_i(y_1, y_2, \ldots, y_{m+1}),$$
$$\varphi_1(y_1, y_2, \ldots, y_{m+1}) = 0.$$

The function φ_1 can be expanded in powers of y and this expansion begins with terms of second degree, constituting a quadratic form $f(y_1, y_2, \ldots, y_{m+1})$.

It is not useful to envisage the singular points of other types—if they existed, they could be made to disappear by slightly changing the function $\varphi(x_1, \ldots, x_k)$ which defines the surfaces $\varphi(t)$, at least if we assume that the manifold v is itself free of singular points. We are therefore led to study the cone of second degree

$$f(y_1, y_2, \ldots, y_{m+1}) = 0$$

in the space of $m+1$ dimensions of the y, and its intersection with the hypersphere

$$y_1^2 + y_2^2 + \cdots + y_{m+1}^2 = 1.$$

Let C be that intersection. Everything will depend on the number of dimensions and the number of positive and negative squares in the decomposition of the form f as a sum of squares.

By change of coordinates we can always reduce f to the form

$$f = \sum A_i y_i^2 - \sum B_k y_k^2,$$

where the coefficients A and B are positive, so that we have q positive squares and $m+1-q$ negative squares, the index i running from 1 to q and the index k from $q+1$ to $m+1$.

I can then write

$$\sum A_i y_i^2 = \sum B_k y_k^2 = \lambda^2,$$

where

$$y_i = \eta_i \lambda, \quad y_k = \eta_k \lambda,$$

the η being any solutions of the equations

(1) $$\sum a_i \eta_i^2 = 1, \quad \sum B_k \eta_k^2 = 1.$$

We deduce

$$\lambda^2 \left(\sum \eta_i^2 + \sum \eta_k^2\right) = 1.$$

In studying the equations (1) we see that the $\sum \eta_i^2$ are bounded above and below by the inverses of the smallest and largest of the coefficients A. We similarly find upper and lower bounds for $\sum \eta_k^2$. We then conclude that λ is a continuous function of η (constrained by the equations (1)) and that this function cannot become zero or infinite. We may then assume that λ is always positive; λ will then be a continuous function completely determined by the η, and hence the same as the y.

Several cases can occur.

1° If q is zero or equal to $m+1$, so that all squares have the same sign, it is clear that the cone reduces to a point and C does not exist.

2° If q is not equal to 1 we can pass continuously from any solution of $\sum A_i \eta_i^2 = 1$ to any other solution; if, however, $q = 1$ this equation admits only two solutions: $\eta_1 = \pm \frac{1}{\sqrt{A_1}}$, and we cannot pass from one to the other in a continuous manner.

Likewise, if q is not equal to m we can pass continuously from any solution of the equation $\sum B_k \eta_k^2 = 1$ to any other; if, on the contrary, $q = m$, then the equation has only two solutions and such a passage is impossible.

The cases are therefore of three types:

If $1 < q < m$, C is a single piece.

If $1 = q < m$ or if $1 < q = m$, C consists of two pieces.

If $1 = q = m$, C consists of four pieces, or rather, it reduces to four discrete points.

In the latter case, we have $m = 1$ and the manifold V is of two dimensions; this alerts us to the fact that we shall encounter a difference between manifolds of two dimensions and those of more than two.

Suppose first of all that V has two dimensions ($m = 1$); we can then have:

- 1° $q = 0$ or $q = 2$. In this case C does not exist; when t passes through a value which corresponds to a similar singular point, we see that a new manifold $w(t)$ appears (or disappears); it reduces firstly to a point, then to a small closed curve. This singular point then corresponds to a *cul-de-sac* of the skeleton.
- 2° $q = 1$. In this case C reduces to four points which I can number 1, 2, 3, 4. The manifold $w(t)$ reduces to a curve which admits, as a singular point, an ordinary double point where two branches of the curve, $1 \cdot 3$ and $2 \cdot 4$, cross. I suppose that the singular point occurs for $t = 0$; I shall let $1', 2', 3', 4'$ denote the points of the manifold $w(t)$ which, for very small values of t, are respectively very close to the points $1, 2, 3, 4$ of the manifold $w(0)$.

I begin by observing that the branch of the curve which runs from the double point to the point 1 must return to the double point, because all our curves are closed; it can return via one of the other three points $2, 3, 4$. It follows that our points $1, 2, 3, 4$ are grouped in pairs, e.g., 1 with 2, 3 with 4, so that we can go from 1 to 2 and from 3 to 4 on the curve $w(0)$ without passing through the neighborhood of the double point. Similarly, we can go from $1'$ to $2'$, and from $3'$ to $4'$ on the curve $w(t)$ without passing through the neighborhood of the double point, and this is true for all sufficiently small values of t, positive or negative.

Now, if we consider the neighborhood of the double point, we see that for $t < 0$, for example, we can pass from $1'$ to $2'$ and from $3'$ to $4'$ on $w(t)$ also by passing near to the double point, whereas we cannot pass from $1'$ to $4'$ and from $2'$ to $3'$ in this way. On the other hand, for $t > 0$ we can pass not only from $1'$ to $4'$ and from $2'$ to $3'$, but also from $1'$ to $2'$ and $3'$ to $4'$.

It follows that for $t < 0$ the branches of $w(t)$ form two distinct closed curves, whereas for $t > 0$ they form only one.

Our singular point therefore corresponds to a bifurcation of the skeleton.

It is the same if 1 is associated with 4, and 2 with 3.

But now, suppose that 1 is associated with 3 and 2 with 4. We then see that for $t < 0$, as well as for $t > 0$, our manifold $w(t)$ reduces to a single closed curve; however, for $t < 0$ this curve passes through the points $1'3'4'2'1'$ in succession, whereas for $t > 0$ the order of points is $1'3'2'4'1'$. Our singular point does not therefore correspond to a bifurcation.

In this case I claim that the manifold V is nonorientable.

To prove this, we take any closed two-dimensional surface (orientable or nonorientable) and dissect it (leaving it as a single piece) so that it can be mapped on to the plane; we thus obtain a polygon, analogous to a Fuchsian polygon, whose edges are conjugate in pairs, two conjugate edges corresponding to two sides of the same cut.

Let AB and $A'B'$ be two conjugate edges, so that the vertex A is conjugate to A', and the vertex B to B'. If in moving from A to B the interior of the polygon is (say) on the left, while in moving from A' to B' it is on the right, we say that the conjugation is *direct.*

If, on the contrary, the interior is on the left when going from A to B and also in going from A' to B', we say that the conjugation is *inverse.* (This convention may seem surprising at first, but it will be justified on reflection.) That being given, if all pairs of edges are *directly* conjugate (as is the case for Fuchsian polygons) the corresponding surface is orientable. If the conjugation is inverse for one pair of edges, then the surface is nonorientable.

Suppose for example that our polygon is a rectangle whose vertices in cyclic order are $ABCD$ and whose opposite edges are conjugate. If AB is conjugate to DC, and AD to BC, the conjugation is direct, and the polygon can be regarded as a map of the torus, an orientable surface. If AB is conjugate to CD, and AD to BC, the conjugation is direct for one pair and inverse for the other, and the polygon is the map of a nonorientable surface analogous to that of Möbius. If AB is conjugate to CD, and AD to CD, the conjugation is inverse for both pairs and the polygon is the map of the *projective plane*, which is a nonorientable surface.

With this in mind, we apply these rules to our manifold V, dissected and mapped on to the plane so as to obtain our polygon. The figure presents only that part of the polygon which is of interest to us.

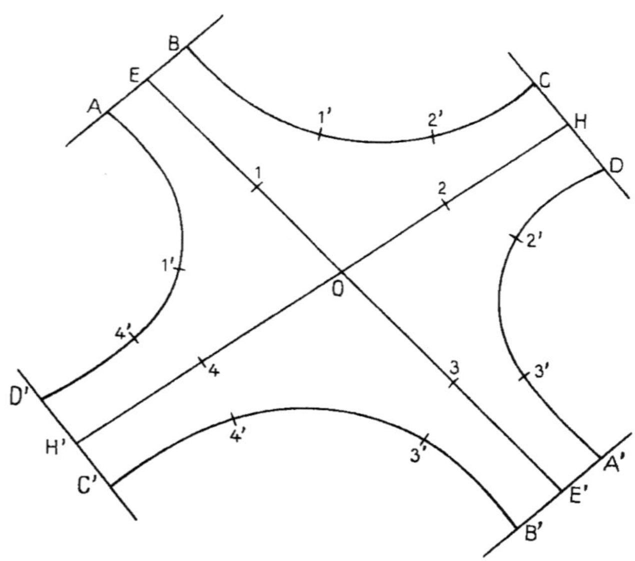

Figure 1

We have a singular point at O; the lines $CHD, AEB, C'H'D', A'R'B'$ represent part of the boundary of the polygon. The two lines $E103E'$ and $H204H'$ which cross at O are the maps of the curves $w(0)$. The lines $B1'2'C$, $B'3'4'C'$ are the map of the curve $w(t)$ for $t < 0$. The lines $D2'3'A'$ and $D'4'1'A$ are the map of the curve $w(t)$ for $t > 0$. By hypothesis, this curve $w(t)$ is closed onto itself in such a way that the branch $1'$ joins the branch $3'$ and the branch $2'$ joins the branch $4'$. Then A must be pasted to A', i.e., A is conjugate to A', and similarly B to B', C to C', D to D'. Thus in our polygon AB is conjugate to $A'B'$ and CD to $C''D'$. The conjugation is inverse, so our surface is nonorientable. Q.E.D.

Thus, if a two-dimensional V is orientable, its skeleton will not have singular points other than *culs-de-sac* and bifurcations. This is the secret of the relative simplicity of *analysis situs* on ordinary surfaces.

I now come to manifolds V of three dimensions ($m = 3$), to which I confine myself at present. Again we have to distinguish two cases:

1º $q = 0$ or $q = 3$; then C does not exist, and the singular point corresponds to a *cul-de-sac* of the skeleton.

2º $q = 1$ or $q = 2$; then C is composed of two pieces; the manifold $w(0)$ represents an ordinary conical point (if the singular point corresponds to $t = 0$); the parts of that manifold close to the singular point are assimilable in an ordinary cone of second degree. We now give t a very small value. The parts of $w(t)$ close to the singular point will be assimilable, e.g., in a hyperboloid of one sheet for $t < 0$ and a hyperboloid of two sheets for $t > 0$.

We consider the ellipse at the neck E of the hyperboloid of one sheet; for $t = -\varepsilon$ (where ε is positive and very small) this ellipse is a very small closed cycle traced on $w(t)$; for $t = 0$ it reduces to a point, and for $t = +\varepsilon$ it disappears.

What we have called C (intersection of the cone and the hypersphere) is composed here of two closed curves (as it is in the case of $q = 1$ or $q = m$), and we have to distinguish two cases: when we can pass from one of these two curves to the other on $w(t)$ without passing close to the singular point, and when this passage is impossible.

In the first case the ellipse E divides the manifold $w(-\varepsilon)$ into two parts, since we cannot pass from a neighborhood of one of the curves C to a neighborhood of the other curve C without passing through the neighborhood of the singular point, i.e., without crossing the ellipse at the neck E. We then have

$$E \sim 0$$

on $w(-\varepsilon)$.

In the second case, however, E does not separate $w(-\varepsilon)$.

In the first case, the sheets of $w(t)$ which pass in the neighborhood of the singular point form a single closed surface for $t = -\varepsilon$, because we can always pass from one point to another on these sheets and, in particular, from the neighborhood of one of the curves C to that of the other by crossing the ellipse E. On the other hand, for $t = +\varepsilon$ they form two closed surfaces, because we cannot pass from one of the curves C to the other.

In the first case, our singular point therefore corresponds to a bifurcation of the skeleton.

In the second case, on the other hand, these sheets of $w(t)$ always form a single closed surface, both for $t = -\varepsilon$ and $t = +\varepsilon$, because we can always pass from one curve C to the other without going near the singular point.

In the second case, our singular point does not correspond to a bifurcation.

This second case itself divides into subcases. We consider a line giving passage from one curve C to the other without going near the singular point. When we follow this line, if we are to conserve the form of the equations it will be necessary from time to time to change the variables and replace the variables y by other variables y', then replace these y' by new variables y'', etc. We assume always that these changes of variable are made in such a way that the Jacobian relating the new variables to the old is positive. Let $z_1, z_2, \ldots, z_{m+1}$ be the final variables when we return to the neighborhood of the singular point. We then have the x as functions of the z, but since, in the neighborhood of the singular point, our equations which give the x as functions of the y become valid again, the series becomes convergent again, we have the z as functions of the y; two cases can then occur according as the Jacobian of the z with respect to the y is positive or negative. In the first case the path is two-sided, in the second case, one-sided. This is as I have explained in *Analysis situs*, when defining nonorientable manifolds.

Thus, if there exist lines which permit us to pass from one curve C to the other without going near the singular point, it can be that all of these lines are two-sided, or some are two-sided and others one-sided, or all are one-sided.

For the moment we confine ourselves to the case where all the surfaces $w(t)$ are orientable. All the lines, if they exist, are therefore two-sided.

We know that for an orientable surface the Betti number is always odd. Then if an orientable surface is $(2p+1)$-tuply connected, it will admit $2p$ distinct cycles, any linear combination of which is homologous to zero. We then envisage the $2p$ cycles on the surface $w(-\varepsilon)$, and first distinguish between those which meet the ellipse at the neck E and those which do not. If a cycle K meets E, we must divide the points of intersection into two categories according to the sign of a certain determinant, as I explained in *Analysis situs*, p. 24. In this way we define the number $N(K, E)$ (cf. *Analysis situs*, p. 25) which will be the difference of the numbers of points of intersection in the two categories. If the number N relative to the cycle K is zero, K will be homologous to a cycle which does not meet E at all. If, on the contrary, the number N is not zero, all cycles homologous to K will meet E.

What happens then, when t varies continuously from the value $-\varepsilon$ to the value $+\varepsilon$? We could find a line K' on $w(+\varepsilon)$ differing infinitely slightly from the cycle K traced on $w(-\varepsilon)$ only if K does not meet E, the line K' is closed and constitutes a new cycle on $w(\varepsilon)$; if, on the contrary, K meets E, the line K' cannot be closed. Then, in order to have cycles on $w(\varepsilon)$ differing infinitely little from a cycle homologous to K, it is necessary and sufficient that the number $N(K, E)$ be zero.

In other words, all the cycles for which this number is nonzero disappear when t passes from $-\varepsilon$ to $+\varepsilon$, and all the others survive.

Let K be a cycle of $w(-\varepsilon)$ which survives and K' the cycle corresponding to it on $w(\varepsilon)$.

Under what circumstances do we have $K' \sim 0$?

If we have $K' \sim 0$ there exists an area A' on $w(\varepsilon)$ bounded by K'; on $w(-\varepsilon)$ we can find an area A differing infinitely little from A', and this area will be bounded by K only or else by K and the ellipse at the neck E, so that

$$K \sim 0 \text{ or } K \sim E.$$

I might add that if we trace any cycle K' on $w(\varepsilon)$ we can always find a cycle K on $w(-\varepsilon)$ which differs from it arbitrarily little, and we cannot have $K \sim 0$ without having $K' \sim 0$; for if there is an area A on $w(-\varepsilon)$ bounded by K there is an area A' on $w(\varepsilon)$ bounded by K' and which differs infinitely little from A.

Thus, when t passes from $-\varepsilon$ to $+\varepsilon$, certain cycles can disappear, but new cycles cannot appear; certain cycles can become homologous to zero, but no cycle can lose this property, so that the Betti number can decrease, but not increase.

It follows that only two cases can occur:

1° $E \sim 0$ on $w(-\varepsilon)$; in this case we have seen that when t passes from $-\varepsilon$ to $+\varepsilon$ the manifold $w(t)$ will decompose into two others.

The cycles on $w(-\varepsilon)$ cannot disappear; in fact, since we have $E \sim 0$, we have $N(K, E) = 0$ for all cycles K. Also, none of these cycles K can become homologous to zero. We have seen which new homologies can be introduced between the cycles K in the passage from $-\varepsilon$ to $+\varepsilon$; all of them can be deduced from the new homology $E \sim 0$; and, in fact, we have said that when we have $K' \sim 0$ without having $K \sim 0$ it is necessary that $K \sim E$. But in the case we are concerned with we have $E \sim 0$ on $w(-\varepsilon)$ as well as on $w(\varepsilon)$. There are therefore no new homologies.

The total number of distinct cycles then remains the same; if $w(-\varepsilon)$ is $(2p+1)$-tuply connected, $w(\varepsilon)$ consists of two surfaces, respectively $(2p'+1)$-tuply and $(2p''+1)$-tuply connected, where $p' + p'' = p$.

2° In the second case, we do not have $E \sim 0$ on $w(-\varepsilon)$; we have seen that in that case $w(t)$ does not decompose; we suppose that $w(-\varepsilon)$ is $(2p+1)$-tuply connected, with $2p$ distinct cycles. At the moment when t becomes positive, certain cycles K disappear; they are those for which $N(K, E)$ is not zero. But if we have two cycles K_1 and K_2 such that

$$N(K_1, E) = m_1, \quad N(K_2, E) = m_2,$$

then we have

$$N(m_2K_1 - m_1K_2, E) = 0$$

so that the cycle $m_2K_1 - m_1K_1$ does not disappear. It follows that all the cycles which disappear are linear combinations of *one* among them and the cycles which do not disappear. The number of distinct cycles then diminishes, *for this reason*, by one and one only.

On the other hand, one and only one new homology is introduced among the cycles which survive,

$$E \sim 0,$$

so that the number of cycles again diminishes by one.

In summary, the total number of distinct cycles diminishes altogether by two, so that $w(\varepsilon)$ is $(2p-1)$-tuply connected.

§3

Before going further, we must briefly review what we know of two-dimensional surfaces, or rather those properties of surfaces we shall require in what follows; I commence with orientable surfaces.

We know that a $(2p+1)$-tuply connected closed orientable surface admits $2p$ distinct cycles. Let $C_1, C_2, \ldots, C_{2p}$ be $2p$ fundamental cycles on the surface, chosen in such a way that each cycle on the surface is homologous to a linear combination of them.

Now let

$$X = \sum x_i C_i, \quad Y = \sum y_i C_i$$

be two of these linear combinations where the coefficients x and y are integers. We consider the number $N(X, Y)$ relating intersections of the two cycles X and Y; this number is equal to

$$N(X, Y) = F(x, y),$$

where $F(x, y)$ is a bilinear form in the variables x and y.

This form and all its coefficients are integers, and it changes sign when we permute x with y, so that

$$F(x, y) = -F(y, x).$$

Finally, its discriminant is equal to 1.

We can choose the fundamental cycles C (in infinitely many ways) so that the form $F(x, y)$ is *reduced*; i.e., it reduces to

$$x_1 y_2 - x_2 y_1 + x_3 y_4 - x_4 y_3 + x_5 y_6 - x_6 y_5 + \cdots .$$

We see that if the form F is reduced, $N(C_i, C_k)$ will be zero if the indices i and k have the same parity; that is, if i and k have the same parity the cycles C_i and C_k will not intersect, or else will be homologous to cycles which do not intersect.

In what follows it will often be useful to us to replace our surface by a Fuchsian polygon; this can be done in two ways. Assuming that the form F is reduced, we consider the p cycles of odd rank

$$C_1, \quad C_3, \quad \ldots, \quad C_{2p-1},$$

which we may assume do not intersect.

In addition, here is how we can give account of the generation of these cycles. We form the skeleton of our surface; this skeleton will be a network on which we can describe p distinct closed lines; by suitable choice of p points on the network the p closed lines are cut off, the network being reduced to a single piece. Each of these points represents a closed curve (which will be the manifold $w(t)$ of the preceding paragraph). We then have p closed curves which have no common point and which are our p cycles

$$C_1, \quad C_3, \quad \ldots, \quad C_{2p-1}.$$

We cut our surface along these p curves; it remains in a single piece, but it can now be mapped on to the plane, and, after the mapping it reduces to a plane region bounded by $2p$ closed curves. One of the $2p$ curves bounds it from the exterior and the others from the interior. These $2p$ curves are conjugate in pairs, two conjugate curves corresponding to two sides of the same cut. The region will therefore be capable of assimilation in the third family of Fuchsian polygons; and we have the advantage of being able to envisage the Fuchsian group it generates, and the decomposition of the plane into an infinity of polygons congruent to the generator polygon.

Suppose now that we have cut our surface along the $2p$ cycles. The $2p$ cycles have been traced so that they all pass through the same point and have no other point in common. If, after cutting, we map the surface on to a plane, we obtain a polygon of $4p$ sides conjugate in pairs, assimilable in a Fuchsian polygon (which does not have a single cycle of vertices) which has the sum of its angles equal to 2π. If the form F is reduced, the rule of conjugation of the sides will be the following: 1 with 3, 2 with 4, 5 with 7, 6 with 8, 9 with 11, 10 with 12, etc. This allows us to envisage the Fuchsian group and the decomposition of the fundamental circle into congruent polygons.

This now brings us to a question which will detain us for some time. The Fuchsian group in question is none other than what was called, in *Analysis situs* §12, the fundamental group of the surface. The definition of this group is based on the notion of *equivalence* of cycles and on the distinction between this notion and that of homology (cf. p. 15 and p. 40 of *Analysis situs*).

We consider two cycles K and K', beginning and ending at the same point; I shall write the *equivalence*

$$K \equiv K'$$

if we can pass from one to the other by a continuous deformation without leaving the manifold in question. It may be that the cycle K' passes several times through the initial point M and is therefore equivalent to a number of consecutive cycles K_1, K_2, K_3 which occur in the order indicated; we can then write

$$K \equiv K_1 + K_2 + K_3,$$

but now we no longer have the right to change the order of the terms and write, for example,

$$K \equiv K_1 + K_3 + K_2.$$

This is precisely what distinguishes equivalences from homologies; in the latter we have the right to change the order, so that the equivalence

$$K \equiv K_1 + K_2 + K_3$$

allows us to deduce not only the homology

$$K \sim K_1 + K_2 + K_3$$

but also the homology

$$K \sim K_1 + K_3 + K_2.$$

Likewise, suppose that K', instead of starting at point M, begins and ends at some other point M'. Let L be any line from M to M'. Then the cycle $L + K' - L$, like K, runs from M to M. Suppose that we have the equivalence

$$K \equiv L + K' - L.$$

Since we do not have the right to change the order of terms, we cannot deduce the equivalence $K \equiv K'$, but only the homology $K \sim K'$.

Thus with homologies, terms are composed according to the rules of ordinary addition; with equivalences the terms are composed according to the rules of substitutions in a group; this is why the set of equivalences can be said to be symbolized by a group, which is the fundamental group of the manifold.

In the case we are concerned with, we envisage the fundamental circle decomposed, as we have said, into congruent Fuchsian polygons. Each of these polygons has $4p$ sides. Any cycle will be represented by a closed line, or a line from one point of the plane to another "congruent" point (i.e., a transform of the former point under one of the transformations of the Fuchsian group).

In the first case the cycle is equivalent to zero; in the second case it is not.

The fundamental group is then the Fuchsian group itself, the group derived from the $2p$ substitutions corresponding to the $2p$ fundamental cycles C_i; in the case where the form F is reduced, the rule of conjugation of the sides of the polygon is what I have said above; between the $2p$ cycles there is a single equivalence which is written

$$0 \equiv C_1 + C_2 - C_1 - C_2 + C_3 + C_4 - C_3 - C_4 + C_5 + C_6 - C_5 - C_6 + \cdots .$$

This equivalence suffices to define the fundamental group.

We can find the form of a cycle which is homologous to zero without being equivalent to zero. We see first of all that we cannot have $p = 1$, since then the equivalence I have just written becomes

$$C_1 + C_2 \equiv C_2 + C_1,$$

which signifies that any two cycles commute (from the point of view of equivalence). Moreover, we know that in this case the Fuchsian functions reduce to elliptic functions, and the Fuchsian group of these substitutions is commutative.

If $p > 1$ we suppose that the rule of conjugation of the sides is as expressed above, i.e., 1 with 3, 2 with 4, etc. Let 0 and 1 be the two vertices of edge 1; 1 and 2 those of edge 2 etc.; finally let $4p - 1$ and $4p = 0$ be those of edge $4p$. We join the vertices 0 and 4 by a line through the interior of the polygon. This line represents a cycle which will be equivalent to

$$C_1 + C_2 - C_1 - C_2.$$

It will therefore be homologous to zero, but not equivalent to zero.

We now have recourse to another mode of representation; we represent our surface by a Fuchsian polygon of the third family bounded by $2p$ closed curves conjugate in pairs, one of them bounding the polygon from the exterior, the others from the interior.

Let k be one of these closed curves, corresponding to the cycle C_2, let k' be its conjugate. Let M be a point of k, M' the corresponding point of k'; we join M and M' by a line MPM' which corresponds to the cycle C_1. We now trace a closed curve K in the interior of the polygon enveloping k and k', but not enveloping any of the other curves which form the boundary of the polygon.

Let Q be a point on k where this cycle begins and ends. Let L be a line from Q to M. It is clear that we have the equivalence

$$K \equiv L + k + MPM' + k' - MPM' - L,$$

that is,

$$K \equiv L + C_2 + C_1 - C_2 - C_1 - L;$$

from which we can conclude $K \sim 0$, i.e., that K is homologous to zero without being equivalent to zero.

Finally, we pass to the surface itself, and confine ourselves to *simple* cycles, i.e., *to cycles which do not intersect themselves.*

Let C be such a cycle which is homologous to zero; it decomposes the $(2p+1)$-tuply connected surface into two parts, so that if we strangle the surface in such a way as to reduce the cycle C to a point, the surface decomposes into two. If one of the two surfaces obtained in this way is simply connected, then the cycle C will be equivalent to zero. If the two surfaces are $(2p' + 1)$-tuply connected and $(2p'' + 1)$-tuply connected, where $p' > 0, p'' > 0, p' + p'' = p$, the cycle C, on the contrary, is homologous to zero without being equivalent to zero.

It is easy to see, and of course we already know, that two orientable surfaces with same Betti number are always homeomorphic. It suffices to remark that each of them can be replaced by a Fuchsian polygon of the third family, bounded by $2p$ closed curves, and that two similar polygons, i.e., two regions bounded from the exterior by one closed curve and from the interior by $2p-1$ closed curves are evidently homeomorphic to each other.

But we can go further. Let S be a $(2p+1)$-tuply connected surface, and trace on it two systems of $2p$ cycles:

$$C_1, \quad C_2, \quad \dots, \quad C_{2p}; \qquad C'_1, \quad C'_2, \quad \dots, \quad C'_{2p}.$$

The surface S can always be regarded as homeomorphic to itself; i.e., point M on the surface can be made to correspond to any other point M' in such a way that the correspondence extends to a one-to-one bicontinuous map of the whole surface. In that case the correspondence can be chosen so that, when the point M describes the cycles

$$C_1, \quad C_2, \quad \dots, \quad C_{2p}$$

the point M' describes either

1° the cycles $C'_1, C'_2, \dots, C'_{2p}$, or
2° cycles equivalent to $C'_1, C'_2, \dots, C'_{2p}$, or
3° cycles homologous to $C'_1, C'_2, \dots, C'_{2p}$.

I now deal with the third question, which is the easiest. I consider the form $F(x,y)$ which represents the number N relative to the intersection of the cycles $\sum xC$ and $\sum yC$. I also consider the form $F'(x',y')$ relative to the intersection of the cycles $\sum x'C'$ and $\sum y'C'$.

It is clear first of all that if the correspondence is possible in such a way that M' describes a cycle homologous to C'_i when M describes a cycle homologous to C_i then the two forms will be identical, i.e., they will differ only by a substitution of the variables x', y' for the variables x and y. If, for example, we suppose that F is reduced and that

$$F = x_1y_2 - x_2y_1 + x_3y_4 - x_4y_3 + \cdots,$$

then we must have

$$F' = x'_1y'_2 - x'_2y'_1 + x'_3y'_4 - x'_4y'_3 + \cdots.$$

The cycles C' are linear combinations of the cycles C, and if we suppose

$$\sum xC = \sum x'C', \quad \sum yC = \sum y'C',$$

the new variables x' and y' will also be linear combinations of the x and the y with integer coefficients. It is the case then, that the form F is not altered by the linear substitution that converts the x to x' and the y to y'.

I claim that this necessary condition is also sufficient.

To prove this, we look at those linear substitutions which do not alter the form F, which we assume to be reduced.

1° If we suppose that we have

$$\begin{aligned} x'_1 &= x_1 + x_2, & x'_i &= x_i \quad (i > 1), \\ y'_1 &= y_1 + y_2, & y'_i &= y_i \quad (i > 1), \end{aligned}$$

it is clear that we have

$$x'_1y'_2 - x'_2y'_1 + \cdots = x_1y_2 - x_2y_1 + \cdots.$$

It is clear that this will also be the case if we put

$$x'_2 = x_1 + x_2 \quad \text{or}$$
$$x'_3 = x_3 + x_4 \quad \text{or}$$
$$x'_4 = x_4 + x_3,$$

or more generally

$$x'_{2k-1} = x_{2k-1} + x_{2k} \quad \text{or}$$
$$x'_{2k} = x_{2k-1} + x_{2k},$$

where all the remaining x'_i equal the corresponding x_i.

It will still remain the same if the x are subject to the inverses of the preceding substitutions, i.e., if we put

$$x'_{2k-1} = x_{2k-1} - x_{2k} \quad \text{or}$$
$$x'_{2k} = x_{2k-1} - x_{2k},$$

while all the other x'_i remain equal to the corresponding x_i.

It goes without saying that the linear substitution by which we pass from the y to the y' is identical to that by which we pass from the x to x'.

We see then, a first type of linear substitution which does not alter the reduced form F.

2° Now here is a second type.

Suppose that we put

$$x'_1 = x_1 + x_3, \quad x'_4 = x_4 - x_2,$$
$$x'_i = x_i \quad (\text{except for } i = 1 \text{ and } i = 4),$$

or more generally

$$x'_{2k-1} = x_{2k-1} + x_{2j-1}, \quad x'_{2j} = x_{2j} - x_{2k},$$
$$x'_i = x_i \quad (\text{except for } i = 2k-1 \text{ and for } i = 2j).$$

It is clear that the reduced form F is not altered.

It is also not altered by the inverse substitution

$$x'_{2k-1} = x_{2k-1} + x_{2j-1}, \quad x'_{2j} = x_{2j} + x_{2k},$$
$$x'_i = x_i \quad (\text{except for } i = 2k-1 \text{ and for } i = 2j).$$

This is our second type.

But it is easy to see that every substitution which does not alter the reduced form F can be considered as a combination of substitutions of these two types.

It therefore suffices to demonstrate the theorem for substitutions of these two types.

To deal with the second type first, we can represent our surface by a Fuchsian polygon of the third family, bounded from the exterior as well as the interior by $2p$ closed curves

$$A_1,\ A'_1; \quad A_3,\ A'_3, \quad \dots; \quad A_{2p-1},\ A'_{2p-1}$$

conjugate in pairs and corresponding to p cycles of odd index

$$C_1, \quad C_3, \quad \dots, \quad C_{2p-1}.$$

To construct the p cycles of even index, it suffices to operate in the following fashion.

Let

$$P_1, \quad P_3, \quad \dots, \quad P_{2p-1}$$

be p points taken arbitrarily on the curves $A_1, A_3, \ldots, A_{2p-1}$; let $P'_1, P'_3, \ldots, P'_{2p-1}$ be the corresponding points on the conjugate curves $A'_1, A'_3, \ldots, A'_{2p-1}$. We join P_1 to P'_1, P_3 to $P'_3, \ldots, P_{2p-1}$ to P'_{2p-1} by p lines

$$L_1, L_3, \ldots, L_{2p-1},$$

traced so that they do not mutually intersect. These p lines will be homologous to p cycles of even index

$$C_2, \quad C_4, \quad \ldots, \quad C_{2p}.$$

To simplify, we suppose that it is the curve A_5 (which does not play any role in what follows) which bounds our Fuchsian polygon from the outside. We construct a curve B which envelopes the two curves A_1 and A_3 and does not envelope the others; this curve B will represent a cycle homologous to

$$C_1 + C_3.$$

Let R be the region bounded from the outside by this curve B and from the inside by A_1 and A_3. We envisage the substitution of the Fuchsian group which changes A_1 into A'_1 (and which corresponds to the cycle C_2). Let R' be the image of the region R under this substitution; it will be bounded from the exterior by A'_1 and from the interior by two closed curves B' and A''_3 which are images of B and A_3.

We modify the Fuchsian polygon by cutting away the region R and adjoining the region R'. Our new Fuchsian polygon will be bounded from the outside by A_5 and from the inside by

$$A'_5; \quad B; \quad B'; \quad A'_3,\ A''_3; \quad A_7,\ A'_7; \quad \ldots; \quad A_{2p-1},\ A'_{2p-1}.$$

Two points of the plane which transform into each other by a substitution of the Fuchsian group evidently correspond to the same point on the surface S. Our new Fuchsian polygon will therefore correspond, like the old, to the whole surface S, because the region R removed has been replaced by its image R'. Moreover, these two polygons, which are both plane regions bounded by $2p$ closed curves, are homeomorphic to each other in such a way that

$$A_1,\ A'_1; \quad A_3,\ A'_3; \quad A_5,\ A'_5; \quad \ldots; \quad A_{2p-1},\ A'_{2p-1}$$

correspond to

$$B,\ B'; \quad A_3,\ A'_3; \quad A_5,\ A'_5; \quad \ldots; \quad A_{2p-1},\ A'_{2p-1}.$$

The result is that for this homeomorphism the odd cycles

$$C_1, \quad C_3, \quad C_5, \quad \ldots, \quad C_{2p-1}$$

correspond to cycles homologous to

$$C_1 + C_3, \quad C_3, \quad C_5, \quad \ldots, \quad C_{2p-1}.$$

To what do the cycles of even order correspond? Each of these cycles corresponds to a substitution of the Fuchsian group, for example C_2 corresponds to the substitution T, which changes A_1 into A'_1, C_4 to the substitution T_3 which changes A_3 into A'_3 etc. It is also evident that, in the homeomorphism in question, the substitution T_1 is replaced by that which changes B into B' (cycles which correspond to A_1 and A'_1), which is again T_1, the substitution T_3 is replaced by that which changes A_3'' into A_3', which is $T_1^{-1}T_3$, and the other substitutions remain the same. We could have already foreseen that the cycles

$$C_2, \quad C_4, \quad C_6, \quad \dots$$

would correspond to

$$C_2, \quad C_4 - C_2, \quad C_6, \quad \dots .$$

But a doubt may persist, since the substitution T_1 corresponds not only to the cycle C_2, but to all the cycles $C_2 + K$, where K is any linear combination of cycles of odd order.

We should therefore return to the lines L we have defined; we can always suppose that none of these lines cut B, with the exception of the lines L_1 and L_3 which cut B at N_1 and N_3. I let M_1 and M_1', M_3 and M_3' designate the points of intersection, of A_1 and A_1' with L_1, A_3 and A_3' with L_3; and let N_1' and N_3' be the images of N_1 and N_3 under T which are on B', then I envisage the segments $N_1'M_1'$, $N_3'M_3''$ which are the images under T, of the segments N_1M_1 and N_3M_3 of the lines L_1 and L_3. The point M_3'' is on A_3''. The lines L do not meet the new segments $N_1'M_1', N_3'M_3''$ at any point.

We consider again the segments N_3N_1 on B and N'_3N_1' on B', or rather, the segments $N_3N_1^0$ and $N_3'N_1'^0$ adjoining the points N_1^0 and $N_1'^0$ situated on B and B' infinitely close to N_1 and N_1' and such that N_1 and N_1' are not on the arcs $N_3N_1^0$ and $N_3'N_1'^0$; in addition, I envisage a line $N_1'^0M_1'^0N_1^0$ infinitely close to $N_1'M_1'N_1$ and not intersecting it; with the aid of these various segments I can construct the lines

$$L_1', \quad L_3', \quad \dots,$$

which correspond under our homeomorphism to the lines

$$L_1, \quad L_3, \quad \dots .$$

The first will be the line $N_1M_1'N_1'$; the second, which must go from M_3'' to M_3', will be

$$M_3''N_3' + N_3'N_1'^0 + N_1'^0M_1'^0N_1^0 + N_1^0N_3 + N_3M_3'.$$

Otherwise, L_5' will be identical to L_5, L_7', etc. It will suffice to verify that these various lines do not intersect.

We see then that these lines are homologous to

$$L_1, \quad L_3 - L_1, \quad L_5, \quad \dots$$

which in terms of cycles $C_2, C_4, \dots$ correspond to

$$C_2, \quad C_4 - C_2, \quad C_6, \quad \dots .$$

In summary, we then have $C_i' \sim C_i$ except for $i = 1$ and 4, when

$$C_1' \sim C_1 + C_3, \quad C_4' \sim C_4 - C_2.$$

If we now suppose that $\sum xC = \sum x'C'$, we shall have $x_i = x_i'$ except when $i = 2$ and 3, and then

$$x_3' = x_3 - x_1, \quad x_2' = x_2 + x_4.$$

If we had required

$$x_3' = x_3 + x_1, \quad x_2' = x_2 - x_4,$$

it would have been necessary to trace B around A_1 and A_3 and transform the region R, not by T_1, but by T_3.

This procedure is applicable to all substitutions of the second type.

We operate in an analogous fashion for substitutions of the first type; I suppose, for example

$$x_1' = x_1 + x_2, \quad x_i' = x_i \quad (i > 1),$$

that is,

$$C_i \sim C_i', \quad C_1' \sim C_2 - C_1.$$

I do not represent our surface by a Fuchsian polygon of the third family, nor by a Fuchsian polygon of the first family as I have above, instead I employ a mode of representation in some sense intermediate between the two.

We remark, in fact, that nothing obliges us to confine ourselves to Fuchsian polygons in the strict sense, i.e., polygons bounded by arcs of circles orthogonal to the fundamental circle. In the question we are concerned with, nothing prevents us from replacing, e.g., a Fuchsian polygon of the first family by another curvilinear polygon which is homeomorphic to it, but otherwise arbitrary.

We can profit from this flexibility by adopting the following mode of representation.

Our polygon will be bounded on the outside by a curvilinear quadrilateral and on the inside by $2p-2$ closed curves. The opposite edges of the quadrilateral will be conjugate, and the $2p-2$ closed curves will be conjugate in pairs in an arbitrary manner.

Let a, b, c, d be the four vertices of the quadrilateral, and let

$$A_3, \ A_3'; \quad A_5, \ A_5'; \quad \ldots; \quad A_{2p-1}, \ A_{2p-1}'$$

be the $2p-2$ closed curves in their conjugate pairs; these closed curves correspond to the $p-1$ cycles of odd order

$$C_3, \quad C_5, \quad \ldots, \quad C_{2p-1}.$$

The sides ab and dc of the quadrilateral correspond to the cycle C_1; the sides ad and bc to the cycle C_2.

We choose any points P_i on the closed curves A_i, and join each of these points P_i to the corresponding point, P_i' on the curve A_i'. The line L_i which connects P_i to P_i' corresponds to a cycle C_{i+1}, under the condition that these lines L have been traced in such a way as not to cut each other or the sides of the quadrilateral.

We join the opposite vertices a and c of the quadrilateral by a curvilinear diagonal ac which divides the quadrilateral into two triangles acb and acd and which is traced in such a way that all the closed curves A and A' lie in the interior of the first triangle acb.

The group which will play the role of our Fuchsian group in a moment will be derived from the following substitutions: T_1 changes ad into bc; T_2 changes ab into dc; T_i $(i = 3, 5, \ldots, 2p-1)$ changes A_i into A_i'.

Then T_1 transforms adc into bcf.

We replace the triangle adc by the triangle bcf and, as a consequence, we can replace our generator polygon by a new polygon bounded on the outside by the quadrilateral $abfc$ and on the inside by the $2p-2$ closed curves A and A'; these closed curves are again conjugate in pairs and the opposite edges of the quadrilateral are conjugates.

These two polygons (both of which correspond to the whole surface S) are homeomorphic to each other in such a way that

$$ab, \quad bc, \quad cd, \quad da, \quad A_i, \quad A_i', \quad L_i,$$

respectively, correspond to

$$ab, \quad bf, \quad fc, \quad ca, \quad A_i, \quad A_i', \quad L_i.$$

The cycles

$$C_1, \quad C_2, \quad C_i, \quad C_{i+1}$$

correspond, under this homeomorphism, to

$$C_1, \quad C_2 + C_1, \quad C_i, \quad C_{i+1}.$$

Then the surface S is homeomorphic to itself in such a way that these cycles correspond

$$\begin{aligned} C_k' &= C_k \qquad\qquad (k = 1, 3, 4, 5, \ldots, 2p), \\ C_2' &= C_2 + C_1. \end{aligned}$$

In that case we have

$$x_1' = x_1 - x_2, \quad x_2' = x_2, \quad x_3' = x_3, \quad \ldots, \quad x_{2p}' = x_{2p}.$$

This is now a substitution of the first type for which the theorem has been proved, and it is clear that we can similarly prove it for all other substitutions of the first type.

In summary, we can now say:

The necessary and sufficient condition for the surface S to be homeomorphic to itself in such a way that cycles C_i correspond to homologous cycles C_i' is that the form $F(x, y)$ relative to the cycles C be identical to the form $F(x', y')$ relative to the cycles C'.

It is easy to deduce that a cycle $\sum a_i C_i$ is always homologous to a *simple* cycle, i.e., one which does not intersect itself, if the integers a_i are relatively prime. If, on the contrary, the integers a_i are not relatively prime it cannot be homologous to a simple cycle.

We begin by establishing the first point.

It will suffice to show that we can find $2p$ cycles

$$C_1', \quad C_2', \quad \ldots, \quad C_{2p}'$$

such that

$$C_1' = \sum a_i C_i$$

and such that the form $F(x', y')$ relative to the cycles C' is identical to the form $F(x, y)$ relative to the cycles C; in such a way that we have

$$F(x', y') = x_1' y_2' - x_2' y_1' + x_3' y_4' - x_4' y_3' + \cdots$$

if we assume, as usual, that the cycles C have been chosen so that the form $F(x, y)$ is reduced.

In fact, if the cycles C' satisfy this condition, the surface S will be homeomorphic to itself in such a way that the cycle C_i corresponds to a homologous cycle C_i'. Then we can find a cycle homologous to C_1' which corresponds to C_1 under this homeomorphism and which consequently will not intersect itself, since C_1 does not intersect itself.

We remark initially that for relatively prime integers a_i we can find $2p$ cycles

$$C_1'', \quad C_2'', \quad \ldots, \quad C_{2p}''$$

such that

$$C_1'' = C_1' = \sum a_i C_i, \quad C_k'' = \sum b_{ik} C_i,$$

where the a_i and b_{ik} are integers whose determinant equals 1.

Let $F(x'', y'')$ be the form relative to the cycles C''.

What kind of relation do we have between the variables x' and x''? Since the cycle C'_1 must be identical to C''_2, it is clear that

$$x'_2, \quad x'_3, \quad \ldots, \quad x'_{2p}$$

must be linear combinations of

$$x''_2, \quad x''_3, \quad \ldots, \quad x''_{2p},$$

and the same must be true of the difference $x'_1 - x''_1$.

It remains to show that there is a linear transformation of variables which satisfies this condition and which, at the same time, is such that $F(x', y')$ is reduced.

But we can write

$$F(x'', y'') = x''_1 Y_2 - y''_2 X_2 + \Phi(x'', y''),$$

where X_2 is a linear combination of $x''_2, x''_3, \ldots, x''_{2p}$ with relatively prime integer coefficients, where Y_2 is the same combination of $y''_2, y''_3, \ldots, y''_{2p}$, and where finally Φ is a bilinear form in

$$x''_2, \quad x''_3, \quad \ldots, \quad x''_{2p}; \qquad y''_2, \quad y''_3, \quad \ldots, \quad y''_{2p}.$$

I claim that the coefficients of X_2 are relatively prime and that, in fact, if the greatest common divisor a exceeds 1 the determinant of the form $F(x'', y'')$ will be divisible by a^2, which is impossible because the determinant is equal to 1.

Since the coefficients are relatively prime, we can find $2p-1$ linear combinations

$$X_2, \quad X_3, \quad \ldots, \quad X_{2p}$$

of $x''_2, x''_3, \ldots, x''_{2p}$, the first of which is precisely X_2 and whose coefficients are integers whose determinant equals 1.

Under these conditions Φ will be a bilinear form in

$$X_2, \quad X_3, \quad \ldots, \quad X_{2p}$$

and the corresponding combinations

$$Y_2, \quad Y_3, \quad \ldots, \quad Y_{2p}$$

formed with the y''. We can then write

$$\Phi = X_2 Y' - Y_2 X' + \psi(X, Y),$$

where X' is a linear combination of $X_3, X_4, \ldots, X_{2p}$, where Y' is the same combination of the Y, and where ψ is a bilinear form of $qp - 4$ variables

$$X_3, \quad X_4, \quad \ldots, \quad X_{2p}; \qquad Y_3, \quad Y_4, \quad \ldots, \quad Y_{2p}.$$

The determinant of this form ψ must divide that of F; it therefore equals 1. Since the form ψ has determinant 1 we can find $2p-2$ linear combinations

$$x'_3, \quad x'_4, \quad \ldots, \quad x'_{2p}$$

of the variables $X_3, X_4, \ldots, X_{2p}$ such that the form ψ is reduced when we take the x' as new variables, with the corresponding combinations y' of the Y.

If we now put

$$\begin{aligned} x'_1 &= x''_1 - X, \qquad & y'_1 &= y''_1 - Y', \\ x'_2 &= X_2, & y'_2 &= Y_2, \end{aligned}$$

we will get

$$F = x'_1 y'_2 - x'_2 y'_1 + \psi.$$

Since the form ψ is reduced, the same is true of F, so that the new variables x' meet our needs.

The first point is therefore established.

Now suppose that the a_i are not relatively prime; I claim that every cycle homologous to $\sum a_i C_i$ intersects itself. In fact, let

$$a_i = b_i d,$$

where d is the greatest common divisor of the a_i and the b_i are relatively prime. Then let

$$\sum b_i C_i = C'_1, \quad \sum a_i C_i = dC'_1.$$

According to the results above, the surface S is homeomorphic to itself under a mapping which makes C_1 correspond to C'_1. It will then suffice to show that every cycle homologous to dC_1 intersects itself, because under our homeomorphism every cycle homologous to $\sum a_i C_i$ corresponds to a cycle homologous to dC_1.

For this, we recall the representation of our surface S by a Fuchsian polygon R_0 of the first family with $4p$ edges, a polygon which, together with all its transforms, fills the fundamental circle.

Let K be our cycle which we assume to be homologous to dC_1. This cycle will be represented by a certain line amb, where a is a certain point in the interior of the fundamental circle and b is its transform under some substitution of the Fuchsian group. As well as this line, we have to consider all of its transforms by different substitutions of the Fuchsian group, since any one of these transforms represents the same line.

In particular, we visualize those arcs of the line amb whose transforms are interior to the polygon R_0. I shall call the set of these arcs the *image* of the cycle K.

This image is composed of a certain number of arcs

$$A_1B_1, \quad A_2B_2, \quad \ldots, \quad A_nB_n$$

between certain points $A_1, A_2, \ldots, A_n$ on the perimeter of R_0 and other points $B_1, B_2, \ldots, B_n$ also on the perimeter of R_0. For the cycle to be continuous and closed it is necessary that the points B_1 and A_2, B_2 and $A_3, \ldots, B_{n-1}$ and A_n, B_n and A_1 correspond to the same points of S and consequently that these be conjugate points on the perimeter of R_0, i.e., corresponding points on conjugate sides.

We assign a number to each point on the perimeter of R_0 in the following way:

1º The numbers corresponding to points A_i and B_i will be integers.

2º The numbers corresponding to other points on the perimeter will equal integers $+\frac{1}{2}$.

3º If the perimeter is described in the positive sense, our number will not vary except when we pass a point A_i or a point B_i.

4º It will increase by one when we pass through one of the points A_i and decrease by one when we pass through one of the points B_i.

5º The value of the number at one of the points A_i or one of the points B_i will be the arithmetic mean of the values on either side of it.

If, for example, we encounter the points

$$A_3, \quad A_2, \quad B_1, \quad B_2, \quad A_1, \quad B_3$$

in that order when traversing the perimeter in the positive sense, the number will be equal to 0 at A_3, $\frac{1}{2}$ on the arc A_3A_2, 1 at A_2, $1+\frac{1}{2}$ on the arc A_2B_1, 1 at B_1, $\frac{1}{2}$ on the arc B_1B_2, 0 at B_2, $-\frac{1}{2}$ on the arc B_2A_1, 0 at A_1, $\frac{1}{2}$ on the arc A_1B_3, 0 at B_3 and finally $-\frac{1}{2}$ on the arc B_3A_3.

Since we have equal numbers of A points and B points, we always return to the initial value after traversing the whole perimeter.

Having done that, I claim first of all that if the cycle does not intersect itself or, what comes to the same thing, if none of the arcs A_iB_i intersect each other, then the two points A_i and B_i correspond to the same number. In fact, let α be one of the two arcs on the perimeter of R_0 going from A_i to B_i. If the point A_k is on this arc α, then the point B_k must be also, otherwise the arcs between A_i, B_i and A_k, B_k would cross. It follows that the arc α has as many A points as B points, which means that the points at the extremities, A_i and B_i, correspond to the same number. Q.E.D.

We now compare the numbers corresponding to points B_i and A_{i+1} (for greater symmetry of notation I designate the point A_1 both by A_1 and A_{n+1}).

As we have said, the points A_i and B_{i+1} are *conjugate* on the perimeter of R_0.

How do we express the fact that our cycle K is homologous to dC_1? This must say that if we consider the intersections of the cycle K with the different fundamental cycles C_i, and if we agree to regard intersections as positive or negative according to the sense in which the two cycles intersect (cf. *Analysis situs, Journal de l'École Polytechnique*) the number of positive intersections will be the same as the number of negative intersections for all the cycles C_i except C_2, and for C_2 the former exceeds the latter by d. (I say C_2 because C_1 cuts C_2 at a point, it does not cut the other cycles C_2 if we choose the fundamental cycles in such a way that the form F is reduced.)

In other words, let

$$P_1, \quad P_2, \quad P_1', \quad P_2', \quad P_3, \quad P_4, \quad P_3', \quad P_4'$$

be the successive edges of R_0; I assume $p = 2$ to fix ideas; in this case the edges P_1, P_2, P_3, P_4 are, respectively, conjugate to P_1', P_2', P_3', P_4'. The edge P_i corresponds to the cycle C_i and the edge P_i' to the cycle C_i traversed in the opposite sense. This represents the law of conjugation of the edges when the form F is reduced.

Then let N_i be the number of points A found on P_i minus the number of points B found on the same edge P_j; let N_i' be the corresponding difference for the edge P_i'.

We then have

$$N_2 = d, \quad N_2' = -d, \quad N_1 = N_3 = N_4 = N_1' = N_3' = N_4' = 0.$$

These are the conditions that express the fact that the cycle K is homologous to dC_1.

We let Q_i, S_i represent the vertices of P_i, where the positive sense of P_i is from Q_i to S_i; similarly, let A_i', S_i' be the two vertices of P_i'. It is clear from this definition that S_1 will be identical with Q_2, S_2 with Q_1' etc.

The number corresponding to S_i will equal that corresponding to Q_i, plus N_i; and since all the N and N' are multiples of d, we are forced to conclude that the numbers corresponding to the various vertices of R_0 differ from each other by multiples of d.

We now consider two conjugate edges P_i and P_i' and imagine two points, the first traversing P_i from Q_i and S_i and the second traversing P_i' from S_i to Q_i', *in such a way as to remain conjugate throughout.* When the first passes a point A, the second will pass a conjugate which will be a B point; the number relative to the first will increase by one, and similarly relative to the second, for in the latter case we are passing a B point *in the reverse direction.* Similarly when the first point passes through B and the second through A, the two numbers will be diminished by one.

The difference between the two numbers therefore remains constant, and since it was originally a multiple of d, it will always be a multiple of d.

Thus, *the two numbers relative to two conjugate points A_i and B_{i+1} differ by a multiple of d*; and since the number relative to B_i is equal to the number relative to A_i we finally conclude that the numbers relative to the $2n$ points

$$A_1, \quad A_2, \quad \ldots, \quad A_n; \qquad B_1, \quad B_2, \quad \ldots, \quad B_n$$

differ from each other by multiples of d.

We now traverse the perimeter of R_0 in the positive sense and look at consecutive points A_iA_k or A_iB_k or B_iB_k; according to our definition, the numbers relative to these points will be equal or they will differ by one. Since the difference has to be a multiple of d, we have to conclude that all the numbers relative to the points A and B are equal, zero for example.

But then at other points of the perimeter our number will be $\pm\frac{1}{2}$, so if we consider two vertices of R_0 in particular, the difference between their two numbers will be 0 or ± 1. But for the two vertices P_2 and Q_2 this difference is $N_2 = d$.

We have therefore arrived at a contradiction, from which we conclude that our initial hypothesis was absurd, and that *the cycle K intersects itself.* Q.E.D.

§4

In the preceding paragraph we have seen that it is relatively easy to recognize when a given cycle is homologous to a simple cycle, or if two given cycles are respectively homologous to two cycles which do not intersect. In the present paragraph we have to examine an analogous question:

How can we recognize when a given cycle is *equivalent* to a simple cycle, or if two given cycles are equivalent to two cycles which do not intersect?

However, before starting on this question, we return to the definition of equivalence.

Until now we have always understood equivalence in the following manner:

When we write

$$C \equiv C',$$

we understand that the initial and final point of the closed cycle C is the same as that of C', and that there is a simply connected area between C and C' whose boundary consists of C and C'. In other words, we can pass from C to C' by making C vary in a continuous manner so that it always forms a single closed curve with fixed initial and final point. This is what we may call *proper equivalence.*

It will be useful to write

$$C \equiv C' \quad \text{(impr.)}$$

if it is possible to pass from C to C' by varying C continuously, so that it always forms a single closed curve *but allowing the initial and final point to vary.* In other words, we have improper equivalence

$$C \equiv C' \quad \text{(impr.)}$$

when we have a proper equivalence

$$C \equiv -\alpha + C' + \alpha,$$

where α is any arc whose initial point is the final point of C' and whose initial point is the final point of C.[28]

We therefore have four sorts of relations: proper equivalences, in which we cannot change the order of terms; improper equivalences, where we can change the order of terms on condition that we respect *cyclic order*; homologies without division, which are subject changes in order and which can be added, subtracted and multiplied; and finally homologies with division, which can be divided as well.

Unless the contrary is explicitly stated, when we speak of an equivalence we always refer to a proper equivalence.

In the study of the question before us, equivalence can be viewed from several different vantage points. Firstly, if we represent our surface by a Fuchsian polygon R_0 of the first family, we construct the different transforms of this polygon by the transformations of the corresponding Fuchsian group G; these transforms fill the fundamental circle. An arbitrary cycle C will then be represented by an arc of the curve MM' going from the point M to one of its transforms, M'. Two properly equivalent cycles will be represented by arcs MPM' and MQM' with the same extremities, and conversely, two arcs with the same extremities will represent two equivalent cycles. An arc M_1QM_1' will represent a cycle improperly equivalent to the cycle represented by the arc MM' if the substitution of the group G which changes M into M' also changes M_1 into M_1'.

Let arc MPM' represent a cycle C; we consider the various transforms of this arc under the substitutions of the group G; all these transforms also represent the cycle C. The condition for the cycle C not to intersect itself is that the arc MPM' does not meet any of its transforms.

Similarly, let MPM', M_1QM_1' be two arcs representing two cycles C and C'; the condition for the two cycles C and C' not to intersect is that the arc MPM' does not cut the arc M_1QM_1', or any of its transforms.

That being given, we look among the cycles improperly equivalent to C for one which does not intersect itself, going back to the arc MPM' and the substitution S of the group G which changes M into M'. This substitution is hyperbolic; in fact, in the case we are dealing with, which is that of a polygon R_0 of the first family whose vertices form a unique cycle and whose angles sum to 4π, all the substitutions of G are hyperbolic.

The substitution S therefore has two fixed points α and β on the fundamental circle.

We connect these two points by a non-Euclidean line, i.e.,, following the terminology adopted in the theory of Fuchsian functions, by a circle orthogonal to the fundamental circle. Let M_1 be any point on that non-Euclidean line $\alpha\beta$; its

---

[28]Thus improperly equivalent elements are what Poincaré has called "transforms" of each other on p. 50, and are what we now call "conjugates", in the group-theoretic sense of the word "conjugate" mentioned on p. xvi. (Translator's note.)

transform M_1' under the substitution S will likewise lie on the line $\alpha\beta$. Let M_1QM_1' be the arc of the non-Euclidean line between M_1 and M_1'. It will represent a cycle improperly equivalent to the cycle MPM'.

We now consider the transforms of M_1QM_1' under the various substitutions of G, which will also be arcs of non-Euclidean lines. The transforms under S and all its multiples give the whole line $\alpha\beta$; the other transforms give other non-Euclidean lines, namely the lines $\alpha'\beta'$ joining the fixed points α' and β' of the various substitutions of G conjugate to S, i.e., the various hyperbolic substitutions $T^{-1}ST$ where T is any substitution of G.

Then the cycle M_1QM_1' will not intersect itself if its various non-Euclidean lines do not intersect; and for this to happen it is necessary and sufficient that for any of the substitutions $T^{-1}ST$ the two fixed points α' and β' do not separate the fixed points α and β, i.e., the points do not recur in the order $\alpha\alpha'\beta\beta'$ or its inverse.

Conversely, I suppose that two of the non-Euclidean lines intersect; then I claim that all cycles improperly equivalent to MPM' are self-intersecting. If they intersect, in fact, it is because the double points α, β and α', β' of S and $T^{-1}ST$ *separate* each other. Now consider any arc M_2M_2' improperly equivalent to MPM'; then M_2' is the transform of M_2 under S. We consider first of all the transforms of the arc M_2M_2' under multiples of the substitution S; then connect the successive transforms of M_2 by multiples of S, and therefore form a continuous path from α to β.

For the same reason, the transforms of the arc M_2M_2' under the substitutions S^mT (where m is a positive or negative integer) form a continuous path from α' to β'; since α, β and α', β' separate each other, the two paths necessarily cross; i.e., the two transforms of the arc M_2M_2' intersect, which means that the cycle M_2M_2' is self-intersecting. Q.E.D.

It is the same if MPM' and NQN' are two arcs representing closed cycles.

Among the cycles improperly equivalent to MPM' and NQN', are there any which do not intersect? Let S and S_1 be the substitutions which change M into M' and N into N'. Let α and β be the fixed points of S; α_1 and β_1 the fixed points of S_1. We trace the two non-Euclidean lines $\alpha\beta$ and $\alpha_1\beta_1$ and take any two points M_1 and N_1 on these lines; let M_1' be the transform of M_1 under S and N_1' that of N_1 by S_1; the point M_1' will be on the line $\alpha\beta$ and the point N_1' on the line $\alpha_1\beta_1$.

We consider the arcs M_1M_1' and N_1N_1' of the non-Euclidean lines; they represent two cycles improperly equivalent to MPM' and NQN'.

By reasoning similar to that above, we see that if the fixed points α and β of S do not separate the fixed points α_1 and β_1 of S_1, nor the fixed points of the various transforms $T^{-1}S_1T$ of S_1, then the cycles M_1M_1' and N_1N_1' do not intersect. And if, on the contrary, α and β separate α_1 and β_1, or the fixed points of one of the transforms $T^{-1}S_1T$, then not only do the cycles M_1M_1' and N_1N_1' intersect, but it is the same for any two cycles improperly equivalent to MPM' and NQN'.

We can again present this fact under another form. Suppose that the cycle M_1M_1' does not intersect itself; then the non-Euclidean line $\alpha\beta$ and its various transforms do not intersect; these non-Euclidean lines then partition the surface of the fundamental circle into infinitely many regions. If the point N belongs to one of these regions and the transform N_1 of N under S belongs to another, the cycles M_1M_1' and NN' intersect, so that the cycles are improperly equivalent; if the two points N and N' belong to the same region, the cycles do not intersect.

We now adopt a different point of view. We envisage a cycle C represented by an arc MPM' and all the transforms of that arc. The cycle will be self-intersecting if two of these transforms intersect; but if there is an intersection between two of the transforms, there will be infinitely many such, resulting from each other by the substitutions of G and, in particular, there will be one in the interior of R_0.

It therefore suffices to consider the portions of the arc MPM' and its transforms which are in the interior of R_0. Our cycle will then be represented by a certain number of arcs A_iB_i running from one point on the perimeter of R_0 to another.

When a point describes the closed cycle C on the closed surface S the corresponding point on R_0 will describe the successive arcs

$$A_1B_1, \quad A_2B_2, \quad \dots, \quad A_nB_n.$$

The points A and B belong to the perimeter of R_0, we know that this perimeter consists of a certain number of sides conjugate in pairs; it is clear that the points B_1 and A_2, B_2 and $A_3, \dots, B_{n-1}$ and B_n, B_n and A_1 must be conjugate.

If none of the arcs A_iB_i intersect each other, the cycle is simple.

Similarly, if instead of one cycle we have two or more, and if the arcs representing the different cycles do not intersect each other, the different cycles are disjoint.

To fix ideas, suppose that $p = 2$. Then the polygon R_0 is an octagon whose consecutive sides represent the respective cycles

$$+C_1, \quad +C_2, \quad -C_1, \quad -C_2, \quad +C_3, \quad +C_4, \quad -C_3, \quad -C_4,$$

which shows first of all that the four fundamental cycles satisfy the equivalence

$$C_1 + C_2 - C_1 - C_2 + C_3 + C_4 - C_3 - C_4 \equiv 0 \tag{26}$$

because the polygon R_0 is a simply connected area.

Let M be a point in the interior of R_0, let N a point situated on one of the sides of R_0, and let N' the corresponding point on the conjugate side. We see immediately that the cycle $MN' + NM$ is improperly equivalent to

$$\left\{\begin{array}{ll} +C_2 & \text{if } N \text{ is on the side } +C_1, \\ -C_1 & \text{if } N \text{ is on the side } +C_2, \\ -C_2 & \text{if } N \text{ is on the side } -C_1, \\ +C_1 & \text{if } N \text{ is on the side } -C_2, \\ +C_4 & \text{if } N \text{ is on the side } +C_3, \\ -C_3 & \text{if } N \text{ is on the side } +C_4, \\ -C_4 & \text{if } N \text{ is on the side } -C_3, \\ +C_3 & \text{if } N \text{ is on the side } -C_4. \end{array}\right. \tag{27}$$

That being given, we see that the arc A_iB_i is equivalent to the arc A_iMB_i; consequently our cycle

$$C \equiv A_1B_1 + A_2B_2 + \cdots + A_nB_n$$

is equivalent to

$$A_1MB_1 + A_2MB_2 + \cdots + A_nMB_n$$

and consequently, improperly equivalent to

$$(MB_n + A_1M) + (MB_1 + A_2M) + \cdots + (MB_{n-1} + A_nM).$$

But each bracketed expression, for example $MB_1 + A_2M$, is analogous to a cycle $MN' + NM$ of the type just discussed. It is therefore equivalent to one of the fundamental cycles $\pm C_1, \pm C_2, \pm C_3, \pm C_4$, and to know which one it suffices to examine which side of R_0 contains the point A_i and consult the table (27).

We see in this way that our cycle C is improperly equivalent to a combination of fundamental cycles and we have a means of determining this combination. The combination found is not the only one to which C is equivalent, since we can transform any equivalence obtained by means of the equivalence (26), which is the only one which holds among the fundamental cycles.

Conversely, given any combination K of fundamental cycles, we have the means of forming an equivalent cycle represented by a series of arcs $A_1B_1, A_2B_2, \ldots, A_nB_n$.

For example, suppose our combination K is written

$$K = +C_1 + C_1 + C_2 - C_3 + C_4 + C_4 - C_1 - C_2 - C_2$$

or some analogous form; each of the terms of the combination will be one of the fundamental cycles C_i with a coefficient $+1$ or -1. The set of two consecutive terms will be called a *sequence*, and I shall also call the set consisting of the last and the first term a sequence, so that our combination will contain as many sequences as terms.

Each sequence corresponds to an arc A_iB_i; the point A_i will be found on the side

$$+C_1, \quad +C_2, \quad -C_1, \quad -C_2, \quad +C_3, \quad +C_4, \quad -C_3, \quad -C_4$$

if the first term of the sequence is, respectively,

$$+C_2, \quad -C_1, \quad -C_2, \quad +C_1, \quad +C_4, \quad -C_3, \quad -C_4, \quad +C_3,$$

and the point B_i will be found on the side

$$-C_1, \quad -C_2, \quad +C_1, \quad +C_2, \quad -C_3, \quad -C_4, \quad +C_3, \quad +C_4$$

if the second term of the sequence is, respectively,

$$+C_2, \quad -C_1, \quad -C_2, \quad +C_1, \quad +C_4, \quad -C_3, \quad -C_4, \quad +C_3.$$

Two arcs A_iB_i and A_kB_k intersect of necessity if the points A_i, B_i *separate* the points A_k, B_k on the perimeter of R_0. We then say that the two corresponding sequences are *incompatible*. If, on the contrary, these four points do not separate each other, we can trace the two arcs in such a way that they do not meet.

How can we recognize when two sequences are incompatible? This will not present any difficulty when the four points A_i, B_i, A_k, B_k are on four different sides; the circular order of the four points will be that of the four sides, which is known.

However, if, for example, A_i and A_k are on the same edge, it is necessary to look at two consecutive sequences, $A_{i-1}B_{i-1} + A_iB_i$ and $A_{k-1}B_{k-1} + A_kB_k$. On the one hand we want A_{i-1}, B_{i-1} not to separate A_{k-1}, B_{k-1} and on the other hand we do not want A_i, B_i to separate A_k, B_k either. To designate the side containing one of the points $A_i, \ldots$, we employ the same letter A_i.

By hypothesis, the points A_i, A_k occur on the same side A_iA_k, and it follows that the points B_{i-1}, B_{k-1} are likewise on the same side $B_{i-1}B_{k-1}$, conjugate to A_iA_k.

That being established, we traverse the perimeter of R_0 so as to successively encounter the sides $A_{i-1}, B_{i-1}B_{k-1}, A_{k-1}$; if the points A_{i-1}, B_{i-1} do not separate A_{k-1}, B_{k-1} we shall encounter B_{i-1} before B_{k-1}.

Now if we traverse the perimeter of R_0 *in the opposite direction*, in traversing the side conjugate to A_iA_k we must encounter A_i before A_k, because A_1 is conjugate to B_{i-1} and A_k to B_{k-1}; and since the points A_i, B_i are not to separate A_k, B_k we shall encounter the sides B_i, A_iA_k and B_k in the order indicated.

Thus, for the sequences to be compatible, it is necessary to encounter the sides $A_{i-1}, B_{i-1}B_{k-1}, A_{k-1}$ successively, or else the sides B_i, A_iA_k, B_k if R_0 is traversed in the opposite sense.

The other doubtful case reduces to the preceding by reversing one of the two arcs A_iB_i or A_kB_k.

That being given, a cycle, all sequences of which are compatible, will be equivalent to a simple cycle; a cycle which has incompatible sequences will not be equivalent to a simple cycle, unless we can make these sequences disappear by means of the equivalence (26). In the same way we recognize if two or more cycles are equivalent to cycles which do not intersect.

Application of these rules shows us, for example, that of all the combinations of odd cycles C_1 and C_3, the only ones equivalent to simple cycles are the following

$$C_1, \quad C_3, \quad C_1 + C_3, \quad C_3 + C_1.$$

However, for what follows, I want to take yet another point of view.

We represent our surface by a Fuchsian polygon R_0' of the third family which, for $p = 2$, will be bounded from the outside by a circle and from the inside by three other circles. We construct the different transforms of R_0' under the substitutions of the corresponding group G'; this time it fills the whole plane with the exception of an infinity of singular points corresponding to the boundary points of the fundamental circle.

A cycle is again represented by an arc MM' between a point M and one of its transforms M'.

However, two arcs with the same extremities do not always represent two equivalent cycles; it is necessary in addition that the area between them not contain any singular point. Apart from this, all we have said about Fuchsian polygons of the first family remains valid.

We construct the various transforms of the arc MM' under the substitutions of G'; and retain just the portions of these transforms which lie inside R_0'; our cycle is then represented by a series of arcs

$$A_1B_1, \quad A_2B_2, \quad \ldots, \quad A_nB_n$$

going from one point on the perimeter of R_0' to another, and such that the points B_{i-1} and A_i, B_n and A_i, are conjugate.

In order that a cycle be simple, or that cycles be disjoint, the condition is that the arcs A_iB_i which represent the single cycle or the set of them do not intersect; this can be recognized by means analogous to those above.

However, one question remains to be dealt with. Given a cycle represented by a certain number of arcs A_iB_i, to which combination of fundamental cycles is it equivalent?

To resolve this question, we return to the case of the polygon of the first family R_0.

Our polygon R_0' is bounded on the outside by a circle $+A$ and on the inside by a circle $-A$ conjugate to $+A$, and circles $+B$ and $-B$ conjugate to each other. I want

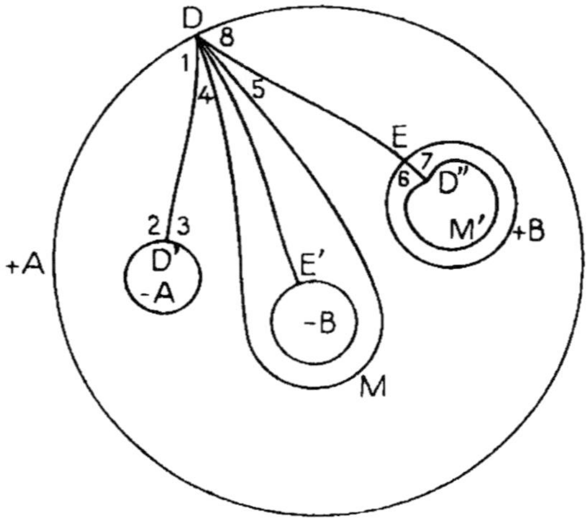

Figure 2

to modify R_0' in such a way as to transform it into a polygon R_0'' homeomorphic to R_0.

Let D and D' be two conjugate points on $+A$ and $-A$; let E and E' be two conjugate points on $+B$ and $-B$. We join DD', DE, DE' by arcs which are regarded as cuts. We envelope the cut DE and the circle $-B$ by a closed curve DMD very little distance from it. We consider the figure $DE'M$ bounded by this closed curve and $-B$, together with its transform $D''EM'$ under the substitution of G' which changes $-B$ into $+B$.

If we subtract the figure $DE'M$ from the polygon R_0' and annex the figure $D''EM'$ in return, we obtain a new polygon R_0'' which represents the closed surface as justifiably as R_0'; this polygon will be bounded on the outside by the circle $+A$ and on the inside by the circle $-A$ and the closed curves DMD and $D''M'D''$; but, thanks to the cuts $DD', DE, DE', D''E$, this polygon R_0'' will be simply connected; the numerals $1, 2, \ldots, 8$ indicate the sequence of vertices of this polygon when we traverse its perimeter. We see, from the order of conjugation of the sides, that this polygon is homeomorphic to R_0 in such a way that the sides of R_0''

$$81, \quad 12, \quad 23, \quad 34, \quad 45, \quad 56, \quad 67, \quad 78$$

correspond to the sides of R_0

$$+C_1, \quad +C_2, \quad -C_1, \quad -C_2, \quad +C_3, \quad +C_4, \quad -C_3, \quad -C_4.$$

Having thus reduced to the case of R_0, it is easy to enunciate the rule.

For each of the arcs A_iB_i crossing R_0' there may be several corresponding arcs crossing R_0'', because a primitive arc may be divided into several pieces by the cuts. For each of the partial arcs crossing R_0'' there corresponds, according to the rule demonstrated in the case of R_0, one and only one term in the combination of fundamental cycles found. Each of our primitive arcs A_iB_i will therefore correspond to one or more terms of this combination.

The first of these terms will depend on the position of the initial point A_i; if this point is on the circle

$$+A, \quad -A, \quad +B, \quad -B,$$

then the first term will be, respectively,

$$+C_2, \quad -C_2, \quad -C_4, \quad +C_4.$$

The terms following depend on the cuts DD', DE, DE' and the order in which they are encountered by the arc A_iB_i; if this arc, going from left to right, encounters

$$DD' \text{ or } DE \text{ or } DE',$$

then the corresponding terms will be, respectively,

$$+C_1, \quad +C_3, \quad -C_4 - C_3 + C_4$$

and if the cuts are encountered from right to left,

$$-C_1, \quad -C_3, \quad -C_4 + C_3 + C_4.$$

It is then easy, following this rule, to form the desired combination.

Throughout this chapter, I have made my examination from the point of view of improper equivalence; if one wanted to deduce the analogous theorems for proper equivalence it would suffice to observe that the whole closed surface is homeomorphic to itself in such a way that any point A on the surface can correspond to any other point A' on the same surface.

§5

We envisage, in particular, a three-dimensional manifold V defined as in paragraph 2; its *skeleton* will reduce to a simple line segment along which the variable we have called t will vary from 0 to 1. The system $W(t)$ will be composed of a unique manifold; this manifold will be an ordinary closed surface which we can assume to be orientable, reducing to a point for $t = 0$ and with connectivity increasing monotonically to $2p+1$ as t increases from 0 to 1.

It follows from this definition that *the manifold V is not closed.*

From what we have seen in paragraph 2, the line comprising the skeleton of V will divide into segments at certain critical values of t.

Let

$$t_1, \quad t_2, \quad \ldots, \quad t_p$$

be the critical values; they are those for which the surface $W(t)$ has a singular point, and hence by our hypothesis, those for which the connectivity of the surface increases by two.

Thus W will be 1-tuply connected for t between 0 and t_1, 3-tuply for t between t_1 and $t_2, \ldots, (2q+1)$-tuply for t between t_q and t_{q+1}, and finally $(2p+1)$-tuply for t between t_p and 1.

The surface W remains homeomorphic to itself as long as the variable t remains within the same segment.

Suppose that we allow t to decrease and that t passes through a critical value; then, as we have seen in paragraph 2, one of the cycles C of the surface S reduces to a point; all cycles equivalent to C will become equivalent to zero and the parts of all cycles cut off by C will cease to exist.

This is, however, the number of distinct cycles, and hence the connectivity is reduced by two.

We now have to define the cycle K_q. For $t = t_q + \varepsilon$ there is an infinitely small cycle on $W(t_q + \varepsilon)$ which reduces to a point when $t = t_q$.

This is the cycle that I call K_q. The surface $W(t)$ remains homeomorphic to itself as t varies from t_q to t_{q+1} (and we can assume that the homeomorphism is such that for two infinitely close values t and t' the two corresponding points on $W(t)$ and $W(t')$ differ from each other infinitely little). The surface $W(t)$ therefore remains homeomorphic to $W(t_q + \varepsilon)$ and the cycle K_q will correspond to a cycle on $W(t)$ which I shall again call K_q. To define K_q on the surface $W(t_{q+1} + \varepsilon)$ it suffices to say that this cycle differs infinitely little from the same cycle on the surface $W(t_{q+1} - \varepsilon)$; since $W(t)$ remains homeomorphic to itself for all values of t between t_{q+1} and t_{q+2}, we can define K_q as above for all values of t, and so on.

Having defined K_q in this way, I arrive at an essential property concerning two cycles K_α and K_β which do not intersect. Let $\beta > \alpha$ and let $t = t_\beta + \varepsilon$; then K_β is very small and I claim that K_α does not intersect K_β. In fact, by its definition, the cycle K_α exists before time t_β, for $t < t_\beta$, and I have said that the cycles which cut the small cycle K_β disappear when t falls below t_β. We let t vary from t_β to $t_{\beta+1}$. Between these limits, all the surfaces $W(t)$ are homeomorphic, and since the cycles K_α and K_β do not intersect on one of them, the corresponding cycles K_α and K_β do not intersect on any of them. Since K_α and K_β do not intersect on $W(t_{\beta+1} - \varepsilon)$, we conclude that they also do not intersect on the infinitely close surface $W(t_{\beta+1} + \varepsilon)$, so that the two cycles K_α and K_β do not ever intersect; for $t_{\beta+1} < t < t_{\beta+2}$ all the surfaces $W(t)$ are homeomorphic and, since K_α and K_β do not intersect on one of them, they do not intersect on any of them; and so on.

Then the cycles K_α and K_β do not intersect. Q.E.D.

I add that the cycle K_q does not intersect itself; it does not for $t = t_q + \varepsilon$ because it is then a very small closed curve; then because of the homeomorphism this remains the case for $t_q < t < t_{q+1}$; and hence also for $t = t_{q+\varepsilon}$ because at that time it differs very little from its position at $t = t_{q+1} - \varepsilon$; and so on.

If we let t vary from t_q to 1, the cycle K_q will vary in a continuous fashion; for $t = t_q$ it reduces to a point and for $t > t_q$ to a unique closed curve. It therefore gives rise to a simply connected region that I call A_q.

Two regions A_α and A_β have no common point; and, in fact, if there were, this point would belong to a surface $W(t)$, and on this surface to two cycles K_α and K_β; but we have seen that these two cycles do not intersect.

I again use B_q to denote the partial region generated by K_q when t varies between t_q and $t < 1$ and which, like A_q, is simply connected. We are going to treat K_q, A_q and B_q as cuts; for this purpose we consider two cycles K'_q and K''_q differing infinitely little from K_q; we can assume that these two cycles do not intersect. The very small portion of the surface $W(t)$ between these two cycles will be called $S_q(t)$. The two cycles K'_q and K''_q generate two regions A'_q and A''_q as t varies from t_q to 1 and two regions B'_q and B''_q as t varies from T_q to t.

That being given, we remove the very small regions

$$S_1(t), \quad S_2(t), \quad \ldots, \quad S_p(t)$$

from the closed surface $W(t)$.

After this operation the remaining surface $W - \Sigma S_q$ will no longer be closed, it will be bounded by the $2p$ closed curves K'_q and K''_q.

If we next adjoin the regions B'_q and B''_q to this surface, the result will be a surface

$$W_1(t) = W - \Sigma S_q + \Sigma B'_q + \Sigma B''_q,$$

which will be closed because B'_q admits K'_q as a complete boundary and B''_q admits K''_q.

I claim that the surface $W_q(t)$ obtained in this way is *simply connected* and without singularities. It has no singularities because its different parts do not intersect and no common points other than those of the curves K'_q and K''_q which serve as their boundaries. And, in fact, $W(t)$ can have no point in common with B'_q and B''_q apart from those of K'_q and K''_q; and, since the regions B_α and B_β do not intersect, the regions $B'_q, B''_q, B'_\alpha, B''_\alpha, \ldots$ will not have any other common points.

Concerning the other point, the bounded surface $W - \Sigma S_q$ is homeomorphic to a plane region bounded on the outside by a closed curve and on the inside by $2p-1$ other closed curves (this is none other than the representation of the surface W by a Fuchsian polygon of the third family as in paragraph 3 above) or, what comes to the same thing, a spherical region which results when $2p$ small simply connected regions are removed from a sphere.

On the other hand, the $2p$ regions B' and B'' can be regarded as homeomorphic to $2p$ regions α; we therefore see, noting the way in which they are attached, that the total surface

$$W_1 = W - \Sigma S + \Sigma B' + \Sigma B''$$

is homeomorphic to a whole sphere, i.e., simply connected. Q.E.D.

We now let t vary from 0 to 1 and at the same time imagine that the cycles K'_q and K''_q approach K_q so as to coincide with it when $t = 1$.

I suppose that the successive positions of K'_q and K''_q have no point in common, so that the successive positions of the regions A'_q and A''_q, B'_q or B''_q have not either. Under these conditions, every interior point of V (excepting the points of the region A_q) belongs to exactly one of the surfaces W_1. The points of the bounded surface $W(1)$ belong to $W_1(1)$. For $t = 1$ the regions B'_q and B''_q reduce to A'_q and A''_q, which in turn reduce to A_q because for $t = 1$ the cycles K'_q and K''_q reduce to K_q.

If we then consider a point of A_q, this point will again be found on $W_q(1)$, but this point of A_q will correspond to two points of $W_q(1)$, one being considered as belonging to $B'_q = A'_q$ and the other to $B''_q = A''_q$.

The nested simply connected surfaces $W_1(t)$ generate (*cf.* §1) a simply connected manifold.

We can then say that the p cuts A_q in V render it simply connected. We execute these p cuts, and deform our manifold in such a way as to separate the two sides of these cuts; the new manifold U thus obtained will be simply connected, bounded by a simply connected surface H homeomorphic to a sphere. On this simply connected surface we can distinguish $2p$ simply connected regions which are the two sides of the p cuts; I call these *cicatrices*; they are conjugate in pairs.

One point of V corresponds to each point of U, and similarly one point of U corresponds to each point of V except for the points of the regions a_q, for which there are two points of U situated on conjugate cicatrices.

We consider two manifolds analogous to U; each of these will be simply connected and bounded. It is clear that the two figures formed in this way will be homeomorphic to each other and to the figure consisting of a sphere whose surface carries $2p$ cicatrices in the form of small disjoint circles.

This has the important consequence: all manifolds generated in the manner of V and for which the integer we have called p is the same are homeomorphic.

Suppose that we construct a closed $(2p+1)$-tuply connected surface in ordinary space. This surface divides the space into two regions, interior and exterior. Let R be the interior region. This is a nonclosed three-dimensional manifold susceptible to the same mode of generation as V. Then V is homeomorphic to R if the number p is the same.

I shall call nonclosed manifolds *developable* if they are homeomorphic to a portion of ordinary space; thus *all manifolds generated in the manner of V are developable.*

We can draw another consequence in passing; consider two closed surfaces S and S' in ordinary space, both $(2p+1)$-tuply connected.

Let R be the interior of S, R' the interior of S'. We know that the two surfaces S and S' are homeomorphic; but if we ask if it is the same for the two volumes R and R', we may first be inclined to respond negatively. For the various sheets of the surface S may be entangled with each other in a complicated fashion, forming knots which could not be resolved without leaving the space of three dimensions. Despite this, we are now in a position to conclude that the two volumes R and R' are always homeomorphic, because both can be generated in the manner of V, and two such manifolds are always homeomorphic.

I now come to a question which is important in what follows. I return to the manifold V bounded by the surface $W(1)$, and generated by the surface $W(t)$.

The same manifold can be generated by another surface, which like $W(t)$ reduces to a point when $t = 0$, has a connectivity which increases monotonically and finally reduces to $W(1)$ for $t = 1$. It is evident that V is susceptible to an infinity of such modes of generation. We shall now compare them.

I let $K_1', \dots, K_0'$ denote the p cycles that play the same role in the new generation as $K_1, \dots, K_p$ did in the old.

What is the relation between the cycles K and K'? Can we choose the cycles K' arbitrarily, and what conditions must p cycles on the surface $W(t)$ satisfy in order to be able to play the role of the cycles K'?

1º These cycles must be simple;
2º They must not intersect each other.

But that is not all. The cycle K_q, *relative to the manifold V*, is equivalent to zero, because it forms the boundary of the region A_q which is part of V. We therefore have the equivalences

$$K_1 \equiv K_2 \equiv \cdots \equiv K_q \equiv 0 \pmod V \tag{1}$$

and those which follow from these. I claim that there are no others.

I say that if we have an equivalence of the following form

$$C \equiv 0 \pmod V, \tag{2}$$

where C is a cycle on the boundary surface $W(1)$, which I shall call W for short, then we have, *relative to this boundary surface* W, an equivalence of the form

$$(3) \qquad C = -\alpha_1 + \beta_1 + \alpha_1 - \alpha_2 + \beta_2 + \alpha_2 - \cdots - \alpha_n + \beta_n + \alpha_n \pmod{W},$$

where the α_i are any cycles of W and the β_i are the cycles of that surface such that

$$\beta_i \equiv mK \pmod{W},$$

where m is an integer and K is one of the cycles $K_1, K_2, \ldots, K_q$. And in fact if the equivalence (3) holds we have *a fortiori*

$$C \equiv \Sigma(-\alpha + \beta + \alpha) \pmod{V},$$

whence, because of the equivalences (1), which imply $\beta \equiv 0$

$$C \equiv \Sigma(-\alpha + \alpha) \equiv 0 \pmod{V}.$$

The equivalence (2) is therefore a consequence of the equivalences (1).

To establish the proposition enunciated, I assume that the equivalence (2) holds; it signifies that the cycle C is the boundary of a certain simply connected region D situated in V.

This region cuts the region A_q along a line L_q running from one point of C to another, because the extremities of L_q can only be found on the boundary of A_q, i.e., on W, and since they are also on D they are on the intersection of W and D, i.e., on C. The region D cannot cut A_q except along a number of distinct lines L_q. In all cases the various lines L are disjoint because the various regions A do not intersect and because we can always assume that the regions A and D have no singularity and if necessary deform D a little so that the surfaces D and A do not touch.

Each of the lines L divides the region D into two parts because this region is simply connected. This region D will then be divided into a certain number of partial regions Δ in the following fashion, which I shall illustrate by an example.

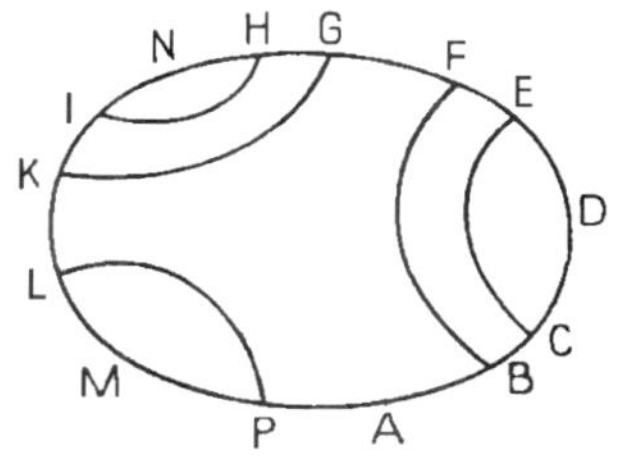

Figure 3

On the figure, the cycle $ABCDEFGHNIKLMPA$ is the cycle C; the lines CE, BF, GK, HI, LP are the lines L.

It is clear that the whole cycle can be replaced by the sum of the following arcs:

$$(4) \quad \Sigma = \begin{cases} (ABCDE + ECBA) + (ABCE + EF + FBA) \\ +(ABF + FGH + HNI + IHGFBA) \\ +(ABFGHI + IK + KGFBA) \\ +(ABFGK + KL + LPA) + (APL + LMP + PA). \end{cases}$$

We see in fact that the last term in each bracket is cancelled by the first term in the following bracket, and by erasing the cancelled terms we recover the whole cycle C.

Now let us take one of these brackets, the third for example; it can be written

$$ABFGH + HNIH + HGFBA.$$

We see that the first and the last term represent the same arc $ABFGH$, described once directly and once in the opposite direction, and that the second term represents the contour of one of the partial areas Δ, namely the region HIN; it is the same for the other brackets in the expression Σ which can be written

$$(5)\quad \begin{cases} (ABC + CDEC - CBA) + (AB + BCEFB + BA) \\ +(ABFGH + HNIH + HGFBA) \\ +(ABFG + GHIKG + GFBA) + (ABFGKLPA) + (AP + PLMP + PA). \end{cases}$$

I now modify the path Σ further. This path is composed of arcs forming part of the primitive contour C and arcs formed by the lines L_q. The latter arcs cancel each other, since we have seen that in the expression (4) the last term of each bracket is cancelled by the first of the next bracket. We can transform these latter arcs; we replace the line L_q by an arc belonging to the cycle K_q and having the same extremities. This is possible because the two extremities of the line L_q are on the cycle K_q, and in addition these new arcs, put in place of the old ones, cancel each other just as the old ones do. And if Σ' denotes the result of this transformation on Σ, the primitive cycle C can then be considered identical with the path Σ' just as with S.

Like Σ, the path Σ' can be put in the form (5); it suffices simply to assume that, e.g., in the second term of (5), CE no longer represents the line L_q between the extremities C and E, but the arc K_q which goes from C to E instead.

The advantage of this transformation is that all the points of the path Σ' are on the bounded surface W, whereas this is not true for all the points of the path Σ.

We now return to the simply connected manifold U defined above.

This manifold is bounded by a simply connected surface which we have called H and which carries $2p$ cicatrices. The parts of H outside the cicatrices then correspond to the surface W and the cicatrices, as we have seen, to the regions A_q.

Any closed curve Q traced on this surface and remaining outside the cicatrices will be equivalent to zero relative to the manifold V, *by virtue of the equivalences* (1). In fact, this curve will envelope a certain number of cicatrices; suppose to fix ideas that it envelopes the two cicatrices A_1 and A_2. Let M be the initial and final point of the closed curve Q; similarly, let M_1 and M_2 be the initial and final points of the two closed curves K_1 and K_2 which bound the two cicatrices A_1 and A_2, respectively.

We join M to M_1 and M_2 by two arcs MM_1 and MM_2; we have

$$Q \equiv MM_1 + K_1 + M_1M + MM_2 + K_2 + M_2M \pmod{W}$$

because the part of the surface H between the curves Q, K_1 and K_2 belongs to that region of H which corresponds to W.

But by virtue of the equivalences (1)

$$K_1 \equiv K_2 \equiv 0 \pmod{V}$$

then

$$Q \equiv MM_1 + M_1M + MM_2 + M_2M \equiv 0 \pmod{V}$$

so that the latter holds, as we have claimed, *by virtue of the equivalences* (1).

Now if we look at the second term in each bracket of the expression (5) for the path S', this term represents a closed curve on W; it likewise represents a closed curve on H. This is not evident and in fact it is no longer true for a closed curve of W which crosses one of the cycles K_q; each point of K_q corresponds to two distinct points on U in such a way that when a continuous path on W meets K_q the corresponding path on U jumps suddenly from one of the two points to the other and becomes discontinuous. But this cannot happen here because the closed curve in question never *crosses* K_q, but goes *around* it.

The second term in each bracket is therefore equivalent to zero *by virtue of the equivalences* (1). It is then the same for the whole bracket, because the first and last terms cancel, and hence the same for the whole path Σ' and consequently for C.

Thus there are no equivalences between the cycles of W other than those which are consequences of the equivalences (1). Q.E.D.

We can then adjoin a third condition to those which we have seen to be necessary in order that p cycles of W can be chosen to play the role of the cycles K'.

The system of equivalences

$$K_1' \equiv K_2' \equiv \cdots \equiv K_p' \equiv 0$$

is no different from the system of equivalences

$$K_1 \equiv K_2 \equiv \cdots \equiv K_p \equiv 0$$

so that each of these systems is a consequence of the other.

Are these three conditions sufficient?

Let $K_1', K_2', \ldots, K_p'$ be p nonintersecting cycles such that we have

$$K_1' \equiv K_2' \equiv \cdots \equiv K_p' \equiv 0 \pmod{V}.$$

The equivalence $K_q' \equiv 0$ tells us that there exists a simply connected region A_q' whose boundary is K_q'. I claim that we can always assume that the various regions A_q' have no common point.

We imagine in fact that $A_1', A_2', \ldots, A_{q-1}'$ do not intersect but that one or more of these $q-1$ regions intersects A_q'. Let E_i be the intersection of the region A_q' and the region A_i'; this intersection has no point on K_q', because K_q' cannot meet A_i' except on W and consequently on K_i', and K_i' does not meet K_q' by hypothesis. This intersection is then entirely in the interior of A_q'; we can assume that the surface A_i' has no singularity and is not tangent to the surface A_q', simply by deforming it slightly; it follows that our intersection is then a curve without double point, it may then be decomposed into a number of disjoint closed curves.

In addition, the intersection of A_i' with A_q' does not meet that of A_k' with A_q' (if $i,\ k < q$) because A_i' does not cut A_k' by hypothesis.

The various intersections E_i and E_k are therefore composed of a certain number of disjoint closed curves. If we envisage two of these curves, either one is exterior to the other, or both are exterior to each other; the words "interior" and "exterior" being understood in relation to the simply connected region A_q'. Among these curves we retain those curves which are not interior to any other; they are then all exterior to each other. Let h_i be one of the closed curves retained, belonging to E_i. This curve h_i will bound a simply connected part of A'_q which I call G_i; similarly it will bound a simply connected part of A'_i which I shall call M_i.

We can trace a closed curve h'_i on A'_q which is very little different from h_i and exterior to it. Then h'_i will bound a simply connected part of the region A'_q that I shall call G'_i; the same h_i will bound a simply connected M'_i which differs infinitely little from M_i but which does not cut the region A'_i.

We then form a region A''_q by removing all the regions G'_i from A'_q and adjoining all the regions M'_i:

$$A''_q = A'_q + \Sigma M'_i - \Sigma G'_i.$$

We see that A''_q, like A'_q, will be a simply connected region bounded by K'_q, because we replace each G'_i by another simply connected region also bounded by h'_i. But the region A''_q does not meet $A'_1, A'_2, \ldots,$ or A'_{q-1}.

We can then assume that the first q regions A'_i do not intersect, and continuing in this way we can assume that none of the p regions A'_i meet, as required.

An analogous argument shows that the cycles K' do not intersect themselves and we can always assume that the regions A'_i are surfaces without double curves.

I now propose to establish that the cycles K' which satisfy the three conditions enunciated can play the role of the cycles K, i.e.,

1° That V can be generated by a surface $W'(t)$ which reduces to a point for $t = 0$ and to W for $t = 1$, and which is $(2q + 1)$-tuply connected for $t_q < t < t_{q+1}$;

2° That for $t_q < t < t_{q+1}$, $W(t)$ meets the regions $A'_1, A'_2 \ldots, A'_q$ along a single closed curve and does not meet the regions $A'_{q+1}, A'_{q+2}, \ldots, A'_p$.

I suppose that this has been proved for a developable manifold V, i.e., one homeomorphic to a portion of ordinary space and bounded by a $(2p-1)$-tuply connected surface, and I propose to prove it also for a developable manifold V bounded by a $(2p+1)$-tuply connected surface W.

We perform the cut A'_p on V and, after separation, obtain a new developable manifold V_1 bounded by a $(2p-1)$-tuply connected surface; this surface W_1 will be composed of two parts, one of which corresponds to the surface W on which the cut K'_p was performed, the other consisting of the two cicatrices corresponding to the two sides of the cut A'_p.

We can particularize this developable manifold V_1 and this surface W_1 in the following fashion. We consider an interior point of V and let δ be the (shortest) distance of this point from the bounded surface W or the cut A'_p. The points such that $\delta > \varepsilon$ (ε small) form the region V_1; the points such that $\delta = \varepsilon$ form the surface W_1; the points such that $\delta < \varepsilon$ form the region $V - V_1$. We easily see that the surface W_1 is $(2p - 1)$-tuply connected, and that it is homeomorphic to the boundary of the region obtained by performing the cut A'_p in V.

The surface W_1 will not cut the region A'_p and it will cut each of the other regions A'_q along a cycle K''_q differing little from K'_q. The cycles K''_q will be simple and disjoint from each other. Moreover, the cycle K''_q will cut a simply connected region A''_q from the simply connected region A'_q. This region A''_q is the portion common to A'_q and V_1; this shows that $K''_q \equiv 0$ relative to V_1. It shows that the cycles K'' satisfy the conditions of the theorem relative to V_1; but this theorem was assumed to hold for V_1.

Then V_1 can be generated by a surface $W'(t)$ which reduces to a point for $t = 0$, reduces to W_1 for $t = u$ (where $t_{p-1} < u < t_p$) and which for $t_q < t < t_{q+1}$ cuts $A''_1, A''_2, \dots, A''_q$ and consequently $A'_1, A'_2, \dots, A'_q$ along a single closed curve and does not cut $A''_{q+1}, A''_{q+2}, \dots, A''_{p-1}$ nor consequently $A'_{q+2}, \dots, A'_{p+1}$ nor A'_p, because this region has no point in common with V_1 and the A'_i have no points in common with V_1 except those of A''_i.

We now envisage the region $V - V_1$ bounded from the outside by the surface W which is $(2p+1)$-tuply connected and which does not cut the region A'_p, which lies entirely in the interior of $V - V_1$. It is clear that we can take a point M on a'_p and construct a surface W_2 entirely inside $V - V_1$ with M as a conical point and no other point in common with A'_p, then generate the region $V - V_1$ by a surface $W'(t)$ which reduces to W_1 for $t = u$, is $(2p-1)$-tuply connected for $u < t < t_p$, does not intersect A'_p and cuts the other A'_q along a single closed curve; which reduces to W_2 for $t = t_p$; which is $(2p+1)$-tuply connected for $t_p < t < 1$ and cuts all the A'_q, including A'_p, along a single closed curve; and which, finally, reduces to W_1 for $t = 1$.

We see that our manifold V can be generated by a surface $W'(t)$ satisfying the enunciation of the theorem; the theorem is therefore demonstrated and the three conditions enunciated are not only necessary, but sufficient.

In addition, in proving this theorem, I have appealed only to the fact that the equivalences $K' \equiv 0$ are a consequence of the equivalences $K \equiv 0$, and I have not appealed to the converse fact that the equivalences $K \equiv 0$ are a consequence of the equivalences $K' \equiv 0$. Then if the cycles K' are disjoint and simple and if the equivalences $K \equiv 0$ entail the equivalences $K' \equiv 0$, then conversely the equivalences $K' \equiv 0$ entail the equivalences $K \equiv 0$. We could also verify this directly.

§6

We now consider a three-dimensional manifold V whose skeleton reduces to a line segment along which the variable t varies from 0 to 1. The system $W(t)$ consists of a unique manifold; this manifold will be a closed orientable surface, which reduces to a point for $t = 0$ and $t = 1$, and whose connectivity increases between $t = 0$ and $t = \frac{1}{2}$, and decreases between $t = \frac{1}{2}$ and $t = 1$. *Our manifold V is then closed.*

We have $2p$ critical values of t, satisfying the inequalities

$$0 < t_1 < t_2 < \cdots < t_p < \frac{1}{2} < \cdots < t'_2 < t'_1 < 1,$$

such that when t passes from the value $t_p - \varepsilon$ to the value $t_1 + \varepsilon$ the surface $W(t)$ increases in connectivity from $2q-1$ to $2q+1$, and when t passes from the value $t'_q - \varepsilon$ to the value $t'_q + \varepsilon$ the surface decreases in connectivity from $2q+1$ to $2q-1$.

The manifold V can be decomposed into two others, V' and V'', the first corresponding to the values of t between 0 and $\frac{1}{2}$ and the second to values of t between $\frac{1}{2}$ and 1. Each of these two partial manifolds satisfy the conditions of the preceding paragraph; they are therefore developable, and it is the manifold V formed from their union that we are now concerned with.

These two manifolds V' and V'' have the surface $W(\frac{1}{2})$, which is $(2p+1)$-tuply connected, as their common boundary. I shall call it W for simplicity.

On this surface I can trace the p cycles

$$K'_1, \quad K'_2, \quad \dots, \quad K'_p$$

defined with respect to the manifold V' just as the cycles $K_1, K_2, \dots, K_p$ were defined with respect to the manifold V in the preceding paragraph.

These p cycles are disjoint and simple. In addition we have the equivalences

(1) $$K'_1 \equiv K'_2 \equiv \dots \equiv K'_p \equiv 0 \pmod{V'}.$$

On the other hand, on that same surface W I can trace the p cycles

$$K''_1, \quad K''_2, \quad \dots, \quad K''_p,$$

defined with respect to V'' as the cycles K_q were defined with respect to the manifold V in the preceding paragraph.

The p cycles K'' are disjoint and simple; and we have the equivalence

(2) $$K''_1 \equiv K''_2 \equiv \dots \equiv K''_p \pmod{V''}.$$

The cycles K' will then be the *principal* cycles of V' and the cycles K'' will be those of V''.

We now consider any cycle C in the interior of V; if M is any point of this cycle we can envisage the corresponding point N of the skeleton. When the point M describes the whole cycle, the point N remains on the line 01 which constitutes the skeleton, and it will execute a series of oscillatory movements on this line, ending in the return to its point of departure.

Let A and B be two extreme positions of the point N in this oscillation. We suppose that the point A lies between t_q and t_{q+1} and that when the point N leaves the value $t_{q+1} - \varepsilon$ it decreases to the value A before returning to the value $t_{q+1} - \varepsilon$. Let H be the arc corresponding to the cycle C; let M be a point of that arc H—it belongs to the surface $W(t)$ where t is between A and $t_{q+1} - \varepsilon$. We know that the surface $W(t)$ remains homeomorphic to itself when t varies from $t_{q+1} - \varepsilon$ to $t_q + \varepsilon$ and consequently, when t varies from $t_{q+1} - \varepsilon$ to A, because$A$ is greater than t_q. Then $W(t)$ is homeomorphic to $W(t_{q+1} - \varepsilon)$. Then let M' be the point of $W(t_{q+1} - \varepsilon)$ which corresponds to M. When the point M describes the arc H, the point M' will describe the arc H' entirely situated on $W(t_{q+1} - \varepsilon)$. I claim that we have the equivalence

$$H \equiv H' \pmod{V}.$$

In fact, consider the different surfaces $W(t)$ where t is an intermediate value between that corresponding to the point M and the value $t_{q+1} - \varepsilon$ corresponding to M'. On each of these different surfaces, which are all homeomorphic to each other, we consider the point which corresponds to M. This point generates a line L whose extremities are M and M', when the point M describes the arc H this line L will generate a simply connected region which will have the two arcs H and H', thus demonstrating the equivalence claimed.

We now consider the two surfaces $W(t_{q+1} - \varepsilon)$ and $W(t_{q+1} + \varepsilon)$; on the first we have the arc H' whose two extremities D and E belong to the arc H and consequently to the cycle C; on the second we have the two points D' and E' infinitely close to D and E and which also belong to C, in such a way that DD' and EE' are two infinitely small arcs of the cycle C. We can trace an arc H'' on the surface $W(t_{q+1} + \varepsilon)$ going from D' to E' and differing infinitely little from H'. The latter point merits attention. The two surfaces $W(t_{q+1} - \varepsilon)$ and $W(t_{q+1} + \varepsilon)$ do not have the same connectivity; it may then happen that, because the two surfaces differ very little from each other, we cannot trace a continuous line on one differing very little from a continuous line traced on the other. If we have a continuous line L on the surface of higher connectivity passing close to the singular point and cutting the cycle which reduces to a point for $t = t_{q+1}$, then we cannot trace a line differing only slightly from L on the other surface. For example, consider the three surfaces which differ very little from each other: the hyperboloid of one sheet, the cone and the hyperboloid of two sheets. The cycle which reduces to a point is the ellipse at the waist of the one-sheeted hyperboloid. A rectilinear generator of the one-sheeted hyperboloid cuts this ellipse, and it will be impossible to trace a similar line on the third surface. But this difficulty will not arise in our situation because it is $W(t_{q+1}+\varepsilon)$ which has the higher connectivity. Our arc H'' will therefore always exist, and we will have

$$H'' \equiv D'D + H + EE' \pmod{V}.$$

We then put

$$C = C_1 + D'D + H + EE'$$

in such a way that our cycle C decomposes into two arcs, the first C_1 and the second $D'D + H + EE'$, both having extremities D' and E'. We have

$$C \equiv C_1 + H'' \pmod{V}.$$

We have thus replaced the cycle C by an equivalent cycle $C_1 + H''$ which enjoys the same properties, but which differs inasmuch as the representative point N, instead of going as far as A in its oscillators, does not go past $t_{q+1} + \varepsilon$.

Suppose first of all that the value $t = \frac{1}{2}$ lies between the extremes A and B of the oscillation of the point N, that the point A is between t_q and t_{q+1}, and that the point B is between t'_h and t'_{h+1}. By the procedure described above, we can replace the cycle C by another, where the point N oscillates between B and the point $t_{q+1}+\varepsilon$, the latter point being between t_{q+1} and t_{q+2} and not between t_q and t_{q+1}, as A is. In other words, we have pulled the point A back between t_{q+1} and t_{q+2}, and we can continue to pull it between t_{q+1} and t_{q+3}, etc., until it is finally between t_p and $\frac{1}{2}$. Operating on B and V'' as we have on A and V', we can pull this point B between $\frac{1}{2}$ and t'_p.

In summary, we have replaced the cycle C by an equivalent cycle C' such that the point N always remains between t_p and t'_p. But when t is between t_p and t'_p the surface $W(t)$ remains homeomorphic to itself and in particular to $W(\frac{1}{2})$ or W. Then let M be any point on the cycle C' belonging to $W(t)$ and M' the corresponding point of W. When the point M describes the cycle C' the point M' will describe a cycle C'' situated on W and we shall have

$$C' \equiv C''$$

by an argument completely similar to that by which we showed that $H \equiv H'$. We then have

$$C \equiv C'' \pmod{V}.$$

If the points A and B are not situated on opposite sides, if for example t remained $< \frac{1}{2}$ over the whole cycle C, we replace the cycle C by the equivalent cycle

$$C + \alpha - \alpha,$$

where the arc α, which is traversed out and back, leads from the final point of the cycle C to any point of V for which $t > \frac{1}{2}$. We are then reduced to the preceding case. Thus we are led to the following general conclusion:

Every cycle of V *is equivalent to a cycle of* W.

Now between the cycles of W we have the equivalences (1) and (2)

$$K_i' \equiv 0 \pmod{V'}, \quad K_i'' \equiv 0 \pmod{V''},$$

so we have *a fortiori*

$$K_i' \equiv K_i'' \equiv 0 \pmod{V}. \tag{3}$$

I now claim that there are no others.

In fact, if we have an equivalence

$$K \equiv 0 \pmod{V},$$

where K is a cycle of W, this means that there is a simply connected region A in V whose boundary is the cycle K. That being given, we distinguish the points in A which belong to V' and form the region A' from those which belong to V'' and form the region A''. If the region A' is not a single piece, it will consist of several separate regions $A_1', A_2', \ldots$, each of which is a single piece. The same holds for A''. If we consider one of these partial regions, A' for example, it may happen that it is not simply connected. Suppose for example that it is triply connected, and consequently bounded from the outside by a closed curve L and from the inside by two closed curves L' and L'' (the words "outside" and "inside" are understood to be in relation to the total region A). We join a point of L to a point of L' by a cut P' and similarly, a point of L to a point of L'' by a cut P''; the two cuts P' and P'' will be the arcs of a curve situated on A_1', and it will render A_1' simply connected. Let B_1' be the simply connected region obtained in this way; and let D_1' be its boundary consisting of the closed curves L, L' and L'' and the two cuts traversed once directly and once in the opposite direction. The region B_1' will be entirely contained in V, and we have

$$D_1' \equiv 0 \pmod{V'}.$$

We operate in the same way on the regions $A_2', \ldots$ and obtain a series of equivalences

$$D_2' \equiv \cdots \equiv 0 \pmod{V'}.$$

We also operate in this way on the regions $A_2', \ldots$ and obtain a series of equivalences

$$D_2' \equiv \cdots \equiv 0 \pmod{V'}.$$

We also operate in this way on the regions $A_1'', A_2'', \ldots$ which together form A'' and this gives equivalences

$$D_1'' \equiv D_2'' \equiv \cdots \equiv 0 \pmod{V''}.$$

Since the region A is the union of the regions $B_1', B_2', \ldots;\ B_1'', B_2'', \ldots;$ the equivalence $K \equiv 0$, where K is the boundary of A, will be a consequence of the equivalences

$$D_1' \equiv D_2' \equiv \cdots \equiv 0 \pmod{V'},$$
$$D_1'' \equiv D_2'' \equiv \cdots \equiv 0 \pmod{V''},$$

where $D_1', D_2', \ldots, D_1'', D_2'', \ldots$ are the boundaries of the partial regions $B_1', B_2', \ldots,$ $B_1'', B_2'', \ldots$.

Our equivalence is therefore a consequence of the various equivalences with respect to V' and V'', respectively. But all equivalences with respect to V' are consequences of the equivalences (1), by the preceding paragraph, and all equivalences with respect to V'' are consequences of the equivalences (2). Thus our equivalence is a consequence of the equivalences (1) and (2) or, in other words, of the equivalences (3). Q.E.D.

Being in possession of all possible equivalences, it is easy to deduce all possible homologies without division; it suffices to allow the terms in these equivalences to commute. Let

$$C_1, \quad C_2, \quad \dots, \quad C_{2p}$$

be the $2p$ fundamental cycles of the surface W.

We have homologies of the form

$$K_i' \sim m_{i,1}' C_1 + m_{i,2}' C_2 + \cdots + m_{i,2p}' C_{2p} \pmod{W}$$

and the form

$$K_i'' \sim m_{i,1}'' C_1 + m_{i,2}'' C_2 + \cdots + m_{i,2p}'' C_{2p} \pmod{W}$$

(where the m' and m'' are integers), so that the equivalences (3) yield the following homologies,

$$\text{(4)} \quad \left\{ \begin{array}{l} m_{i,1}' C_1 + m_{i,2}' C_2 + \cdots + m_{i,2p}' C_{2p} \sim 0 \\ \\ m_{i,1}'' C_1 + m_{i,2}'' C_2 + \cdots + m_{i,2p}'' C_{2p} \sim 0 \end{array} \right. \qquad \pmod{V},$$

and no others.

To discuss these homologies, we form the determinant Δ of the integers m' and m''. Three cases need to be distinguished:

1° We have

$$|\Delta| > 1$$

Then Δ is an integer which is not equal to $0, +1$ or -1; we can then deduce the homologies (4) to be

$$\Delta C_i \sim 0 \quad (i = 1, 2, \dots, 2p)$$

without using division, and with division

$$C_i \sim 0.$$

The Betti number relative to homology with division then equals 1.

But we cannot obtain $C_i \sim 0$ without using division, so that the "torsion coefficients" *are not equal to* 1.

2° We have

$$|\Delta| = 1,$$

and we then deduce, *without division*

$$C_i \sim 0.$$

In this case, *not only the Betti number, but also the torsion coefficients equal* 1.

The Betti number and the torsion coefficients are therefore the same as those for a simply connected manifold. However, this does not mean, as we shall see shortly, that the manifold V is simply connected.

3° We have

$$\Delta = 0.$$

The homologies (4) then yield

$$C_i \sim 0$$

without the use of division.

The Betti number is then greater than 1.

It is equal to 2 if the determinant is zero but all its first order minors are not.

It is equal to k if the determinant of all the determinants of the first $k-2$ orders are zero, but those of the $(k-1)$-th order are not.

We return to the case where Δ is equal to ± 1. In that case we can ask whether the manifold is simply connected, because it has the same Betti number and the same torsion coefficients as a simply connected manifold. It is the principal objective of this work to show that this is not always the case, and it will therefore suffice to give an example.

Suppose that $p = 2$, i.e., that the surface W is 5-tuply connected. I suppose in addition that the cycles K_1' and K_2' are the two fundamental cycles of W, namely

$$K_1' = C_1, \quad K_2' = C_3.$$

I represent the surface W by a Fuchsian polygon R_0 of the third family, bounded by three nonintersecting circles. After this, we then have to trace the two cycles C_1 and C_3 on the surface W so that they do not intersect, cut the surface along these two cycles and map onto a plane.

The cycles K_1'' and K_2'' will then be represented on this plane by a certain number of arcs of the curve going from one point on the perimeter of the Fuchsian polygon R_0 to another.

Here are the conditions to which these arcs of the curve are subject:

1° They must not intersect each other; this is a necessary and sufficient condition for the cycles K_1'' and K_2'' to be simple and disjoint.

2° We then consider the series of arcs whose totality represents the cycle K_1''. These arcs must be followed in a certain order and each of them must go from a certain initial point on the perimeter of R_0 to an final point likewise situated on the perimeter of R_0. We observe that the perimeter of R_0 is composed of four circles conjugate in pairs; each point on one of the circles corresponds to a conjugate point which I call *the* conjugate of the first.

That being given, the final point of each of the arcs which represents K_1'' must be conjugate to the initial point of the arc which follows, and the final point of the final arc must be conjugate to the initial point of the first arc.

Similarly for the arcs which represent K_2''.

Here is the explanation of the figure: the perimeter of R_0 is represented by the four circles $+A, -A, +B, -B$; the circles $+A$ and $-A$ are conjugate and correspond to $K_1' = C_1$; the circles $+B$ and $-B$ are conjugate and correspond to $K_2' = C_3$; the cycles K_1'' and K_2'' are represented by the arcs of curve running between points on the perimeter of R_0.

The arcs which represent K_1'' are shown as unbroken lines; those which represent K_2'' are dotted. The arrows indicate the sense in which they are traversed.

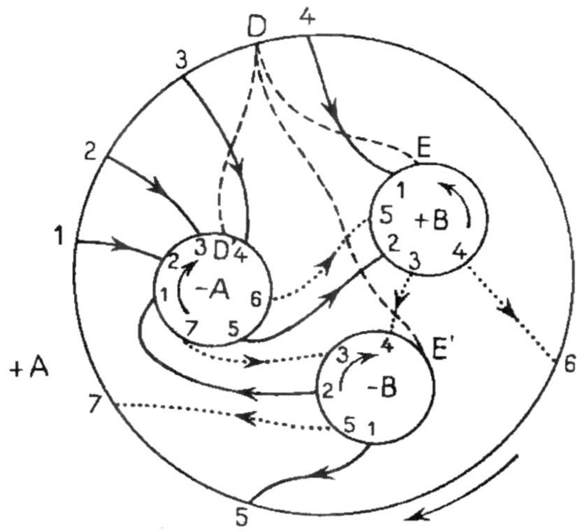

Figure 4

The points where the arcs meet the circles $\pm A$ and $\pm B$ are designated by numbers; these numbers also tell us which points are conjugate; thus the point $+B5$ is conjugate to $-B5$, $+A5$ to $-A5$.

Near each of the four circles $\pm A$, $\pm B$, we have an arrow which expresses the fact that when a point describes the circle in the sense of the arrow the conjugate point describes the conjugate circle in the sense of its arrow.

We verify easily that, if we follow either $+A$ or $-A$ in the sense of the arrow, we encounter in succession

$$1, \quad 2, \quad 3, \quad 4, \quad 6, \quad 5, \quad 7,$$

and when we follow $+B$ or $-B$ in the sense of the arrow, we encounter successively

$$1, \quad 5, \quad 2, \quad 3, \quad 4.$$

The order of conjugation of our points is therefore properly chosen.

The cycle K_1'' then is represented by seven arcs which succeed each other in the following order:

$$+A1 \text{ to } -A2; \quad +A2 \text{ to } -A3;$$
$$+A3 \text{ to } -A4; \quad +A4 \text{ to } +B1; \quad -B1 \text{ to } +A5;$$
$$-A5 \text{ to } +B2; \quad -B2 \text{ to } -A1.$$

The cycle K_2'' is represented by the five arcs

$$+B3 \text{ to } -B4; \quad +B4 \text{ to } +A6; \quad -A6 \text{ to } +B5;$$
$$-B5 \text{ to } +A7; \quad -A7 \text{ to } -B3.$$

It is easy to see that the 12 arcs on the figure do not intersect; our two cycles are therefore disjoint and simple.

We can then construct a manifold V' admitting the cycles $K_1' = C_1$ and $K_2' = C_3$ as principal cycles and a manifold V'' admitting the cycles K_1'' and K_2'' as principal cycles. The union of the two manifolds V' and V'' gives us V.

We can give a better account of it as follows.

We return to our figure and cut the polygon R_0 along the 12 arcs which represent K_1'' and K_2''; we have then decomposed R_0 into 10 partial polygons, which we glue together along the various conjugate arcs of the circles $\pm A$, $\pm B$, for example the $+A3, +A4$ to the arc $-A3, -A4$; in this way we obtain a new polygon which just as truly represents W as does R_0, with the difference that the cuts are made along the cycles K_1'' and K_2'' instead of K_1' and K_2'.

The figure obtained in this way is entirely similar to Figure 1, but the interpretation is different. The circles $+A$ and $-A$ represent K_1'' and not K_1', the circles $+B$ and $-B$ represent K_2''; the arcs drawn with solid lines inside the figure represent K_1' and not K_1''; the dotted arcs represent K_2'.

The points

$$\pm A1, \quad \pm A2, \quad \pm A3, \quad \pm A4, \quad \pm A6, \quad \pm A5, \quad \pm A7$$

represent the respective points

$$\pm A1, \quad \pm A2, \quad \pm A3, \quad \pm A4, \quad \pm B1, \quad \pm A5, \quad \pm B2.$$

The points

$$\pm B1, \quad \pm B5, \quad \pm B2, \quad \pm B3, \quad \pm B4$$

represent respectively

$$\pm A6, \quad \pm B5, \quad \pm A7, \quad \pm B3, \quad \pm B4.$$

We need not worry about the sign $\pm$; the two points $+A1$ and $-A1$ in fact correspond to the same point of the surface W.

The identity of the two figures shows us that the surface W is homeomorphic to itself in such a way that the cycles

$$K_1', \quad K_2', \quad K_1'', \quad K_2''$$

correspond to the cycles

$$K_1'', \quad K_2'', \quad K_1', \quad K_2'.$$

We wish to express K_1'' and K_2'' in terms of the fundamental cycles C_1, C_2, C_3, C_4. To do this, we have to apply the rule at the end of paragraph 4, and to facilitate the application of this rule we have used double dotted lines to trace the three cuts DD', DE, DE' which figure in its enunciation.

The arc $+A1 - A2$ leaves from $+A$, which gives $+C_2$.

The arc $+A2 - A3$ leaves from $+A$, which gives $+C_2$.

The arc $+A3 - A4$ leaves from $+A$, which gives $+C_2$,

and meets DD' and goes to the right, which gives $+C1$.

The arc $+A4 + B1$ leaves from $+A$ which gives C_2,
and crosses to the right of DE which gives $-C3$.

The arc $-B1 + A5$ leaves from $-B$ which gives $+C4$.

The arc $-A5 + B2$ leaves from $-A$ which gives $-C2$,

and crosses DE' from left to right, which gives $-C4 - C3 + C4$.

The arc $-B2 - A1$ leaves from $-B$ which gives $+C4$.

Then

$$K_1'' \equiv 3C_2 + C_1 + C_2 - C_3 + C_4 - C_2 - C_4 - C_3 + 2C_4.$$

We turn to K_2''.

The arc $+B3 - B4$ leaves from $+B$, which gives $-C_4$, and crosses DE' from right to left, which gives $-C_4 + C_3 + C_4$.

The arc $+B4 + A6$ leaves from $+B$ which gives $-C_4$.
The arc $-A6 + B5$ leaves from $-A$ which gives $-C_2$,

and crosses DE' from left to right, which gives $-C_4 - C_3 + C_4$.
The arc $-B5 + A7$ leaves from $-B$ which gives $+C_4$.
The arc $-A7 - B3$ leaves from $-A$ which gives $-C_2$.

Then

$$K_2'' \equiv -2C_4 + C_3 - C_2 - C_4 - C_3 + 2C_4 - C_2.$$

We now have

$$\begin{aligned} K_1' &\equiv C_1, \quad K_2' \equiv C_3, \\ K_1'' &\equiv 3C_2 + C_1 + C_2 - C_3 + C_4 - C_2 - C_4 - C_3 + 2C_4, \\ K_2'' &\equiv -2C_4 + C_3 - C_2 - C_4 - C_3 + 2C_4 - C_2. \end{aligned}$$

In relation to V we have the equivalences

$$(1) \quad \begin{cases} C_1 + C_2 - C_1 - C_2 + C_3 + C_4 - C_3 - C_4 \equiv 0, \\ K_1' \equiv K_2' \equiv 0, \quad K_1'' \equiv 0, \quad K_2'' \equiv 0. \end{cases}$$

The equivalences $K_1' \equiv K_2' \equiv 0$, where $C_1 \equiv C_3 \equiv 0$, permit us to simplify the others; the equivalence (1) reduces to

$$C_2 - C_2 + C_4 - C_4 \equiv 0,$$

i.e., an identity; the equivalence $K_1'' \equiv 0$ reduces to

$$(2) \qquad 4C_2 + C_4 - C_2 + C_4 \equiv 0,$$

and the equivalence $K_2' \equiv 0$ reduces to

$$(3) \qquad -2C_4 - C_2 + C_4 - C_2 \equiv 0.$$

In summary, two distinct cycles, C_2 and C_4, remain and the only equivalences between them are (2) and (3).

If we transform these equivalences into homologies we have

$$3C_2 + 2C_4 \sim 0, \quad -C_4 - 2C_2 \sim 0.$$

The determinant is equal to -1; we are therefore in the case $\Delta = \pm 1$, where the Betti numbers and torsion coefficients are equal to 1. Nevertheless, V is not simply connected since its fundamental group does not reduce to the identity; in other words, the equivalences (2) and (3) do not imply

$$C_2 \equiv 0, \quad C_4 \equiv 0.$$

To show this, we adjoin to (2) and (3) the equivalence

$$(4) \qquad -C_2 + C_4 - C_2 + C_4 \equiv 0,$$

whence

$$C_4 - C_2 + C_4 - C_2 \equiv 0.$$

From (2), (3) and (4) we deduce

$$(5) \quad \begin{cases} -C_2 + C_4 - C_2 + C_4 \equiv 0, \\ 5C_2 \equiv 0, \quad 3C_4 \equiv 0. \end{cases}$$

But the relations (5) are the relations of the structure in which the substitutions C_2 and C_4 generate the *icosahedral group*. We then know that $C_2 \equiv 0$, $C_4 \equiv 0$ cannot be deduced, so *a fortiori* these equivalences cannot be deduced from (2) and (3) alone.

We therefore have two cycles on V which are not equivalent to zero, so V is not simply connected.

One question remains to be dealt with:

Is it possible for the fundamental group of V to reduce to the identity without V being simply connected?[29]

In other words, can we trace the cycles K_1'' and K_2'' in such a way that they are simple and disjoint, such that the equivalences

$$K_1' \equiv K_2' \equiv 0, \quad K_1'' \equiv K_2'' \equiv 0,$$

entail the equivalences

$$C_1 \equiv C_2 \equiv C_3 \equiv C_4 \equiv 0,$$

but that nonetheless the surface W cannot be mapped homeomorphically onto itself in such a way that the cycles C_1, C_2, C_3, C_4 correspond to the cycles C_1', C_2', C_3', C_4'; such that the equivalences

$$K_1' \equiv K_2' \equiv 0$$

entail $C_1' \equiv C_3' \equiv 0$ and conversely; and finally such that the equivalences

$$K_1'' \equiv K_2'' \equiv 0$$

entail $C_2' \equiv C_4' \equiv 0$ and conversely?

However, this question would carry us too far away.

[29]Here at last is a (correct) statement of the Poincaré conjecture. (Translator's note.)

Index

Abel, N.H., viii
abelian functions, 44
Alexander, J.W., xiv
algebraic curve, vii, 6, 53, 135
algebraic surface, 6, 53, 135
  connectivity of, xiii
  cycles on, 143, 151
  first Betti number, xiii, 143
  reduction, 164, 166
  singularities, 164
algebraic topology, xi
analysis situs, 6
  and algebraic curves, 6
  and celestial mechanics, 6
  and differential equations, 6
analytic continuation, 10
  of manifolds, 11

Betti numbers, 15, 75, 113
  and connectivity, viii
  and Euler characteristic, viii
  and Euler polyhedron theorem, 62
  and fundamental equivalences, 41
  and homology groups, xi
  and periods, 18
  derived from table, 98
  equidistant from the ends, xi, 30, 75, 100, 171
  from table invariants, 119
  fundamental theorem, 30, 75, 96, 100, 171
  insufficient for homeomorphism, x, 52, 179
  invariance of, xiv
  of algebraic surfaces, xiii, 53, 143
  of surfaces, 16, 44, 186, 190
  reduced, 81
    invariant under subdivision, 85
  ways of defining, 75
Betti, E., viii, 6, 18
Bianchi, L., xv
boundary, viii, xvi, 8
  and congruence, 78
  and connectivity, viii
  notation, 78
Brouwer, L.E.J., xiv

Cairns, S.S., xiii, xiv
Cayley, A., viii
celestial mechanics, 6
cells, 92, 112, 153
cicatrices, 209
closed chain, 19
conditions of integrability, 17
cone point, 142, 164
congruence, xvi, 78, 79, 81, 82
  and boundary, 103
  and homology, 82, 83, 103, 111
congruent figures, 136, 189
congruent polyhedra, 62, 63
conjugate edges, xvi, 184
conjugate faces, xvi, 31
connected chain, 11
connected network, 11
connectedness, 8
connectivity, 15
  multiple, 53
  of algebraic surfaces, xiii
  of surfaces, viii, 16, 187
  simple, 44
    and fundamental group, 52, 179, 223
cubing, 66
curve, 9
  algebraic, vii, 6, 53, 135
cut, 28
cycle, 32
  equivalence, 189
  invariant, 157
  of q dimensions, 151
  on surface, 187
  on algebraic surface, 143, 151
  one-dimensional, 169
  simple, 190, 196
    equivalent to, 200
    homologous to, 199
  subsisting, 171
  three-dimensional, 155
  two-dimensional, 162

nonsingular, 169
vanishing, 164

Darboux, G., xii
de Jonquieres, E., 62
defining relations, x
of 3-manifold group, 41
of surface group, x
of torus group, x
Dehn, M., xiii
homology spheres, xiv
Descartes, R., vii
differential equations, 6
domain, 8
Dyck, W., viii, 6

edges, 92, 112, 153
elliptic functions, ix, 136
equivalences, 40
differ from homologies, 40
improper, 201
proper, 200
Euler characteristic, vii
and Betti numbers, viii
and curvature, vii
and genus, viii
of Riemann surface, viii
of torus, vii
Euler polyhedron formula, vii
generalized, xi, 62
Euler polyhedron theorem, 6, 34, 61
generalized, 62
second proof, 70
Euler, L., vii

faces, 31, 92, 112, 153
admissible conjugation of, 35
conjugate, 31
free, 31
form boundary, 31
Freedman, M.H., xv
Fuchsian function, ix, 44, 135
Fuchsian group, 32, 36, 44, 135
as fundamental group, 44
defining relations, x
generated by Fuchsian polygon, 135
of first family, 45
of second family, 139
of third family, 45
Fuchsian polygon, 135
edge pairing, 184
for surface, 188
of second family, 139
of third family, 188
function
abelian, 44
elliptic, ix, 136
Fuchsian, ix, 44
hyperelliptic, 135

fundamental contours, 41
fundamental domain, ix, 37
fundamental equivalences, 41
fundamental group, x, 38, 137
abelianization, xii
and simple connectivity, 52, 179, 223
defining relations, xi
equivalences in, 40
generators and relations, 41
of 3-manifold, 41
of surface, 44, 189
defining relation, x, 190
of torus, x
fundamental homologies, 41

Garnier, R., xii
Gauss-Bonnet theorem, vii
genus, viii, 45, 135, 152
geometrisation conjecture
proved by Perelman, xv
geometrization conjecture, xv
geometry
and group theory, 5
of n dimensions, 5
group, 5
continuous, 6
discontinuous, ix, 36
Fuchsian, 32, 44, 135
fundamental, 38
homology, xii, xiii
hypercubic, 43
icosahedral, xiv, 224
of homeomorphisms, 9
Picard, 161, 164

Hamilton, R., xv
Heegaard, P., xii, 75
criticism of Poincaré, 76
diagrams, xiii
example, 76, 123, 147
holoedric, 40, 140, 141
homeomorphism, 9
and Betti numbers, 45
conditions for, 44
implies isomorphic groups, 45
of manifolds, 10
of surfaces, 44, 190
homology, xi, 15, 75, 151
and congruence, 82, 83, 111
and equivalence, 40, 189
and linear independence, 15, 76, 81
group, xii, xiii
notation, 15
relation, 15
sphere, xiii, 134
Dehn, xiv
Poincaré, xiii, 223
with division, 129
without division, 101, 118, 129

homomorphism, 140
hyperbolic plane, ix
hyperboloid, 217
hypercells, 112, 153
hyperelliptic functions, 135
hypergeometry, 5
hypersphere, 55
hypersurface, 5, 9
  in five dimensions, 6

implicit function theorem, 106
integrals, 16
  and Betti numbers, 18
  periods of, 18
intersection
  of manifolds, 23
invariance
  of Betti numbers, xiv
  of dimension, xiv, 9
  of fundamental group, xi
invariant cycle, 157
invariant subgroup, 138
invariants
  of table, 117, 118
    give Betti numbers, 119
inversion, 135
isomorphic groups, 44
isomorphism, 40, 140
  of fundamental groups, 50, 51

Jordan, C., 6

Kirby, R., xiv
Klein, F., ix, 6

Leibniz, G.W., ix
lemniscate, 56
Leray, J., xii
linear independence, 15
  and homology, 76, 81
linear substitution, 46
  elliptic, 47
  hyperbolic, 47
  parabolic, 47

manifold
  boundary of, 8
  bounded, 9
  closed, 9, 31
  connected, 8
  developable, 210
  differentiable, xii
    polyhedral structure, xiv
  dimension of, 8
  finite, 8
  first definition, 7
  four-dimensional, xv
  irregular, 63, 84
  nonorientable, xiii, 19, 54
    definition, 19
  open, xiv
  orientable, 19, 131
    condition for, 33
    examples, 21
    has no torsion, 133
  orientation of, 13
  regular, 63
  second definition, 10
  self-opposite, 23
  simply connected, xii
  singular, 63, 84
  skeleton of, 180
  subdivision of, xii, xiv, 105
  three-dimensional, viii, xi, 31, 179
    polyhedral structure, xiv
  two-dimensional, 44, 53
  unbounded, 9
  via discontinuous group, 36
  with no polyhedral structure, xiv
  with torsion, 120
  without torsion, 120
Maxwell, J.C., viii
meriedric, 40, 140
Möbius band, viii, xiii, 184
Möbius, A.F., viii
  classification of surfaces, viii
Moise, E.E., xiv

Nielsen, J., xiii
nodal points, 90
Noether, E., xii, xiii
non-Euclidean
  geometry, xiii, 135
  line, 135, 201
  metric, xiii
  symmetry, 135, 136
nonorientable chain, 20
null-homologous, 143, 156
null-homotopic, 77, 143, 156

orientation
  opposite, 13

Perelman, G., xv
periods
  of abelian functions, 44, 135
  of elliptic functions, 136
  of integral, 18
Picard, E., xiii, 6
  algebraic surface reduction, 164, 166, 172
  group, 161, 164
  on algebraic surfaces, 53, 151
  on one-dimensional cycles, 169, 171
  Théorie des fonctions algébriques de deux variables, 75, 143
Poincaré, H., vii
  Analysis situs, x
  Compléments, xii

conjecture, xiv, 111, 224
and homology sphere, xiv
incorrect version, xiii, 134
duality theorem, xi, xii, 30, 75
homology sphere, xiii, 134
polyhedra, xi, 31
congruent, 63
conjugate faces of, 31
dual subdivision of, xii
of first kind, 120
of second kind, 120
of third kind, 120
reciprocal, xii
subdivision of, 84
polyhedron, 62
derived, 62
schema of, 78
projective plane, 184
protective circle, 138

Ricci curvature flow, xv
Riemann surface, vii, 44, 151
sheets of, 174
Riemann, G.F.B., vii, 6
connectivity of surfaces, 16

Siebenmann, L., xiv
simple curve
on surface, xiii, 200
simply connected, xii, 44
definition of, 52
skeleton, 180
Smale, S., xiv
star, 34, 65, 80
surface
algebraic, 6, 53, 135
connectivity, 16
cycles on, 187
Fuchsian polygon for, 188
fundamental group, 189
defining relation, 190
nonorientable, 21, 184
of constant curvature, vii
orientable, 184
has odd Betti number, 186
Riemann, 44, 151
simple curves on, xiii, 200

table, 96
invariants, 117
reduced, 97
reduction of, 114
transformations of, 97
yields Betti numbers, 98
tetrahedron, 62
generalized, 65, 80, 107
truncated, 107
rectilinear, 108
Thurston, W., xiii, xv
geometrisation conjecture, xv
torsion, xiii, 120
in algebra, xiii
of manifolds, 131
only if nonorientable, 133
reason for name, xiii, 134
torsion coefficients, xi, xiii, 129, 179
equidistant from the ends, 130
torus, 31, 39, 184
equation of, 13
Euler characteristic, vii
fundamental group, x
with single face, 120
transform, 36, 50

variety *see also* manifold xvi
vertices, 92, 112, 153

Whitehead, J.H.C., xiv